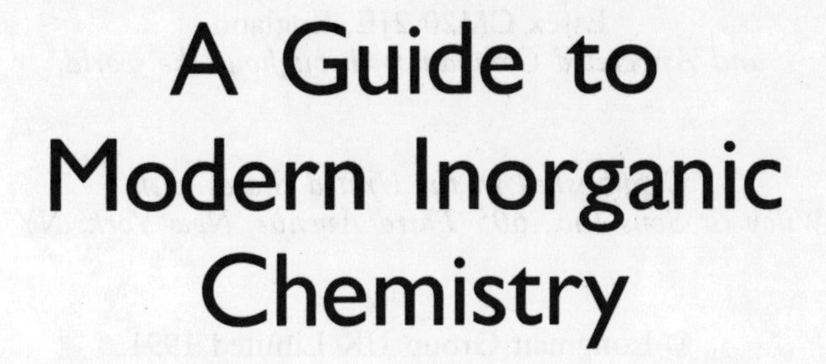

A Guide to Modern Inorganic Chemistry

S. M. OWEN AND A. T. BROOKER

University of Cambridge

Longman Scientific & Technical
Longman Group UK Limited
Longman House, Burnt Mill, Harlow
Essex CM20 2JE, England
and Associated Companies throughout the world.

Copublished in the United States with
John Wiley & Sons Inc., 605 Third Avenue, New York, NY 10158

First published in 1991
Second impression 1992
Third impression 1994

British Library Cataloguing in Publication Data
Owen, S.
A guide to modern inorganic chemistry.
1. Inorganic chemistry
I. Title II. Brooker, A.
546
ISBN 0-582-06439-2

Library of Congress Cataloging-in-Publication Data
Owen, S. (Steven), 1964–
A guide to modern inorganic chemistry / S. Owen and A. Brooker.
p. cm.
Includes index.
ISBN 0-470-21694-8 (USA only)
1. Chemistry, Inorganic. I. Brooker, A. (Alan), 1966–
II. Title.
QD151.2.095 1991
546––dc20 90–43776
CIP

Produced by Longman Singapore Publishers Pte Ltd
Printed in Singapore

Contents

To our parents Michael and Dorothy, Roger and Susan; and Caroline and Peter with much love.

Foreword

I believe that this is an important and pioneering book, on at least three counts.

Firstly, it attempts to give the *essentials* of modern degree-level Inorganic Chemistry in a concise yet readable way; naturally, courses differ but there *are* areas which are covered world-wide. Clearly, much larger, more factual and more comprehensive texts are available. Here is a text to which the student can turn for a general summary, whether due to his or her uncertainty or as a refresher. Although the book is not intended as a 'crammer', it will be of great use for revision and for final clarification.

Secondly, the two authors are unusually well-qualified to write this text. Both are quite recent graduates, and they have just finished Ph.D. research degrees. Thus they would anyway recall very well the areas of Inorganic Chemistry which perplexed *them* during their undergraduate years. However, focussing this recall, both have given numerous tutorials to undergraduates over the three years of their research degrees. They know which topics cause difficulty, and they have had to work on explanations and clarifications. This history comes over in the tone of the book: it is presented in much the style one might use in a tutorial, a bit 'chatty' and informal, yet retaining conciseness.

Thirdly, few books can have been market-researched so well. Of course, the publishers sent drafts to several eminent inorganic chemists for their comments. In addition, however, the authors themselves have sought views from numerous chemists visiting Cambridge (from Europe, North America, Australasia), from several others working and lecturing here and in other U.K. universities, and, most importantly, from a selection of their potential market – current undergraduates.

The three aspects noted above combine to make this book something of a 'first' in its conception, and an extremely useful addition to the chemical literature. I believe that the authors have succeeded admirably in their purpose, and I recommend this book with enthusiasm.

R. Snaith
Lecturer in Inorganic Chemistry
and Fellow of St. John's College
University of Cambridge

Preface

This is a *guide* to inorganic chemistry, not a comprehensive factual account of the subject. Our aim in writing it was to enable the student to understand the fundamental ideas, which are common to courses throughout the world. We believe that a firm grasp of these topics will provide the foundation necessary for understanding and rationalising the wider aspects of the subject developed in individual courses.

We are especially grateful to Dr. Ron Snaith of the Chemistry Department Cambridge who has believed in our project from the outset and given us his full support and advice throughout. Many other lecturers at Cambridge have made invaluable contributions, in particular Drs. E. C. Constable, P. P. Edwards, C. E. Housecroft, M. J. Mays, P. R. Raithby and D. S. Wright. Our thanks go to them, and to the following visitors to the department at Cambridge who showed a keen interest in the book and provided us with helpful comments: Drs. A. M. Brodie (Melbourne University), M. Kubota (Hawaii University), R. Smutzler (Bräunsweig University), D. Osella (Turin University), and D. Barr (Associated Octel Company Ltd.).

S.M.O. and A.T.B.
Cambridge, November 1990

CHAPTER 1

Atomic Structure and Stability

Section A: Atomic Structure

Electronic structure of atoms

Atoms may be viewed as consisting of a positively charged nucleus around which negatively charged electrons orbit. It is the number and the distribution of these electrons which has a fundamental influence on the chemistry of the elements and their compounds. The properties of each electron may be described by of a set of four quantum numbers. These quantum numbers (q.n.) can be regarded as just convenient labels for the electrons.

SYMBOL	NAME	VALUE
n	Principal q.n. or shell number	Integers from 0 to ∞ (in reality, max. $n = 7$)
l	Orbital angular momentum q.n.	Integers from 0 to $(n-1)$
m_l	Magnetic or orientation q.n.	Integers from $-l$ to l
m_s	Spin q.n.	±1/2

The maximum number of electrons in the shell with principal quantum number n is $2n^2$.

The various values of l are assigned letters,

$l = 0$ 1 2 3 4 5, and so on
s p d f g h, and so on

e.g. an electron in a 2s orbital has $n = 2$, $l = 0$, $m_l = 0$ and can have either of two spin values $m_s = \pm 1/2$. An electron in a 3p orbital has $n = 3$, $l = 1$, $m_l = +1, 0, -1$ and $m_s = \pm 1/2$.

The regions of space where there is a high probability of finding an electron are called orbitals. There are four types of orbitals: s, p, d and f orbitals.

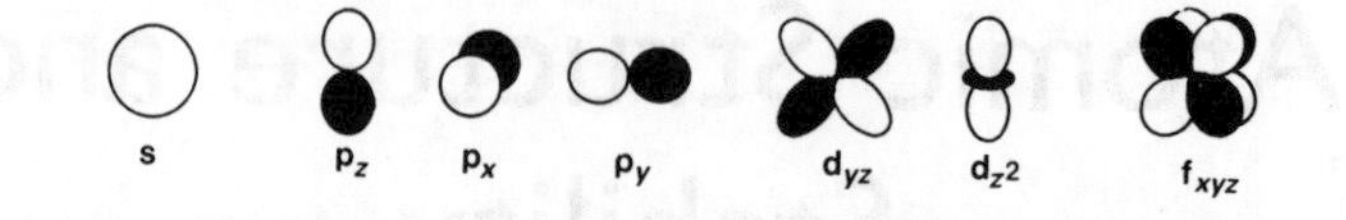

Only two of the five d orbitals, and one of the f orbitals, are shown

Orbitals with differing principal and/or orbital angular momentum quantum numbers have different energies (*except in the hydrogen atom*), i.e. a 1s orbital is lower in energy than a 2s orbital, which, in turn, is lower in energy than a 2p orbital.

THE PAULI EXCLUSION PRINCIPLE No two electrons in an atom can have the same set of four quantum numbers. What this means is that only two electrons can occupy an orbital and these must have opposite spins, i.e. since n, l and m_l are the same for a particular orbital, m_s values must be different.

THE AUFBAU (BUILDING-UP) PRINCIPLE Electrons fill orbitals from the lowest energy orbital upwards. The order for filling orbitals is

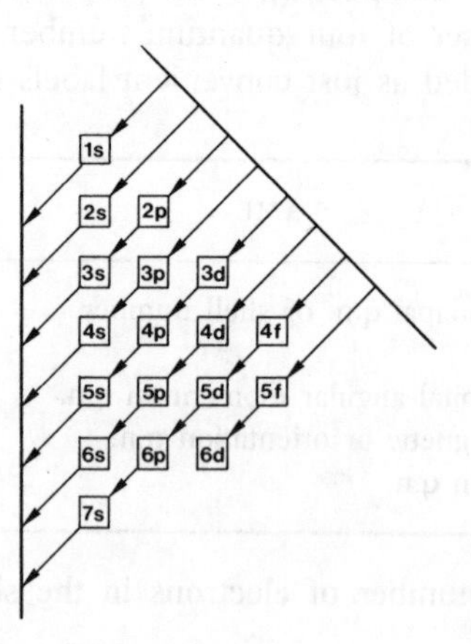

THE RULE OF MAXIMUM MULTIPLICITY (HUND'S FIRST RULE) Electrons fill orbitals of the same energy so as to give the maximum number of

unpaired electrons. This is because electrons repel each other and want to be as far apart as possible. The electrons are arranged to give the maximum number of parallel spins.

EXAMPLE

The electronic configuration of N is $1s^2\ 2s^2\ 2p^3$, with two electrons in the 1s orbital, two in the 2s and one in each of the $2p_x$, $2p_y$, $2p_z$, i.e.

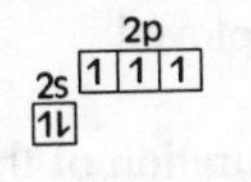

Size of atoms

Atomic size decreases across a period. Consider the period from Li to F; as we move from one element to the next we are adding one proton to the nucleus and one electron to the valence shell (neutrons can be ignored since they have no charge). Electrons are able to *shield* (insulate) each other from the attractive force of the nucleus, i.e. they can be thought of as interposing themselves between each other and the nucleus. Electrons in the same shell (approximately the same distance from the nucleus) shield each other very ineffectively from the attractive effect of the nucleus and so the **effective nuclear charge** ($z_{eff.}$) felt by the outer electrons increases across the period. They are therefore held more tightly and pulled in more closely, decreasing the size of the atom.

SHIELDING This is the ability of an electron to shield other electrons from the attractive effect of the nucleus; electrons do not do this 100 per cent efficiently. The order of shielding ability is s > p > d > f. This is the result of penetration (see below) and the shapes of the orbitals; as we go from s to p to d to f the orbitals become much more directional; the spherically symmetrical s orbital is able to shield in all directions but the angular volume over which a p orbital has any shielding effect is smaller.

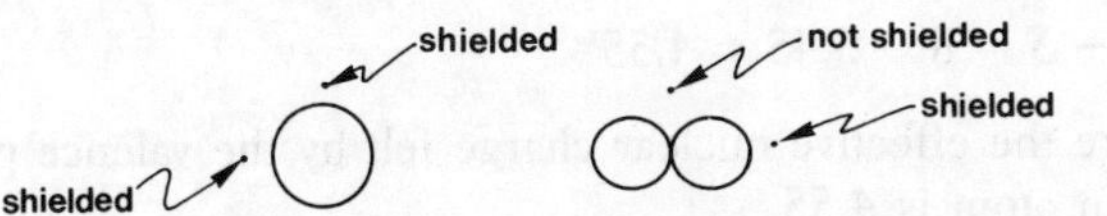

EFFECTIVE NUCLEAR CHARGE If electrons shielded each other completely an outer electron would feel a nuclear charge of +1 unit. This is due to all the other electrons cancelling out the positive charges of all but one of the protons in the nucleus. This does not happen and the electrons experience an effective nuclear charge of greater than +1, how much greater depends on the extent of the shielding. The effective nuclear charge can be estimated using **Slater's rules**.

SLATER'S RULES Slater's rules are a set of rules which can be used for estimating the effective nuclear charge, $z_{eff.}$, experienced by the valence electrons in an atom. We must first work out the shielding constant, S, which tells us how much the valence electrons are shielded by electrons in inner shells, and then subtract this number from the total nuclear charge, z, i.e. $z_{eff.} = z - S$.

To calculate the shielding constant for an electron in an *n*s or *n*p orbital, the following rules should be employed:

1. Write out the electronic configuration of the element in the following order and groupings: (1s) (2s, 2p) (3s, 3p) (3d) (4s, 4p) (4d) (4f) (5s, 5p), etc.
2. Electrons in any group to the right of the (*n*s, *n*p) group we are considering, contribute nothing to the shielding constant.
3. All of the other electrons in the (*n*s, *n*p) group shield the valence electron to an extent of 0.35 each.
4. All electrons in the $(n-1)$ shell shield to an extent of 0.85 each.
5. All electrons in the $(n-2)$ or lower shell shield completely; i.e. their contribution is 1.0 each.

   When the electron being shielded is in an *n*d or *n*f group, rules 2 and 3 are the same but rules 4 and 5 become:
6. All electrons in groups lying to the left of the *n*d or *n*f group contribute 1.0.

EXAMPLE

Consider oxygen and work out z_{eff} on a valence p electron.
The electronic configuration is $1s^2\ 2s^2\ 2p^4$, i.e. $(1s^2)\ (2s\ 2p)^6$.
There are no electrons to the right of the 2p orbital and therefore rule 2 can be ignored. There are five other electrons in the (2s 2p) grouping so they shield the last electron to a total extent of 5(0.35), i.e. 1.75.
The two electrons in the 1s shell shield a total of 2(0.85), i.e. 1.70.

$S = 1.75 + 1.70 = 3.45.$

The total nuclear charge, z, is 8.

$z_{eff.} = z - S = 8 - 3.45 = 4.55$

Therefore the effective nuclear charge felt by the valence p electron in an oxygen atom is 4.55.
Note: this is only an approximate method for determining the effective nuclear charge. For example, rule 2 cannot be strictly true because of the phenomenon of penetration (see below).

Atomic size increases down a group. This is because electrons occupy shells which are further from the nucleus.

Variations of atomic sizes with oxidation state and coordination number

Anions are bigger, and cations smaller, than the parent atom.

ANIONS Introducing more electron density increases the shielding of the outer electrons and the electron–electron repulsions. More electron–electron repulsion for the same nuclear charge causes the electron cloud to expand and therefore anions are bigger than the parent atom.

CATIONS Removing electrons from the atom reduces the shielding felt by the outer electrons, i.e. causes an increase in the effective nuclear charge. The electron–electron repulsion is less, therefore for the same nuclear charge the electrons can be pulled in closer to the nucleus and the electron cloud contracts. Cations are thus smaller than the parent atoms.

OXIDATION STATE This is a purely formal concept which regards *all* compounds (even methane, CH_4!!) as totally ionic and assigns charges to the components accordingly. It provides a guide to the relative charges on atoms in covalent compounds.

The more electronegative atoms are always assigned a negative oxidation state. The values are such as to give them a full outer electronic shell if the compound is ionic. The less electronegative atom is assigned a positive oxidation state of a suitable value so as to give electroneutrality, or the required charge if the species is an ion, e.g. CF_4 where C is less electronegative than F and will have a positive oxidation state. For F to have a full outer shell it must be F^-, therefore if we remove four F^- from CF_4, we are left with C^{4+}, i.e. the oxidation state of C in CF_4 is +4.

The radius decreases with increasing positive oxidation state and increases with increasing negative oxidation state. For an atom in a positive oxidation state the situation is similar to that found for cations but not as marked. This is because oxidation state is a purely formal concept and does not represent the actual charge that exists on the atom/ion. Similarly, size increases with increasing negative oxidation state, as is observed for anions.

COORDINATION NUMBER In a covalent compound the coordination number is the number of atoms to which an atom is directly bonded; in an ionic compound, the coordination number is the number of ions surrounding a given central ion (Chapter 2).

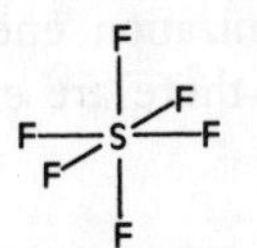

SF_6, coordination number of S is 6

To appreciate how coordination number affects the atomic (covalent) radius it is important to realize how atomic radii are measured. What is actually measured is the inter-nuclear distance (bond length), from which the radius is derived: smaller inter-nuclear distances correspond to smaller atomic radii.

For a given atom the atomic radius increases with increasing coordination number. The more ligands around a central atom the greater the inter-ligand repulsion. The ligands therefore move outwards to reduce this repulsion. The hole in the centre of the ligands is thus bigger for more ligands and this is what we define as the atomic radius of the central atom.

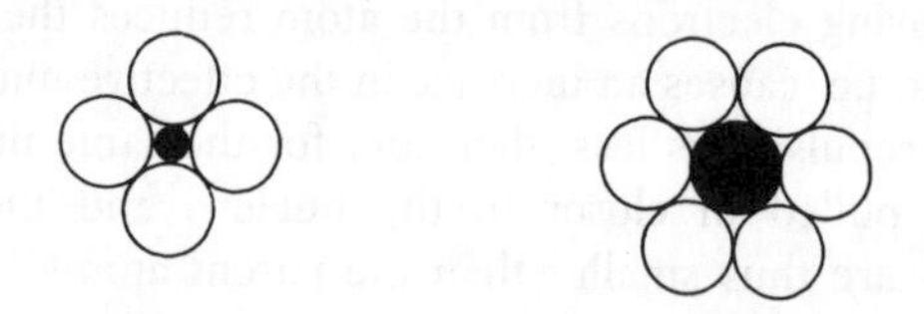

In both cases the central atom is the same

CHARGE AND IONIC POTENTIAL In Group 1 the increasing size of the cation down the group would be expected to lead to Cs^+ having a larger enthalpy of hydration than Li^+, i.e. the enthalpy change for the process,

$$M^+_{(g)} + (aq) \rightarrow M^+_{(aq)}$$

This is because Cs^+ is larger and we would expect to be able to pack more water molecules around it than for Li^+. This is indeed true for the first hydration shell, i.e. there are six water molecules around the larger Cs^+ and only four around the smaller Li^+. However, the overall number of water molecules around the Li^+ is greater than for Cs^+, i.e. there are more coordination shells for the Li^+. This is because the ionic potential (charge/radius ratio) is larger for the Li^+ ion. Therefore water molecules in the outer coordination shells can get closer to the positive charge (at the nucleus) for Li^+ and, therefore, are held more tightly. The result of this is that $Li^+_{(aq)}$ is larger than $Cs^+_{(aq)}$.

IONIZATION ENERGY (IE) This is the energy for the process,

$$M_{(g)} \rightarrow M^+_{(g)} + e^-$$

where $M_{(g)}$ and $M^+_{(g)}$ are in their ground states. This process always requires energy, i.e. it is endothermic.

The general trend is that ionization energy *increases across a period* and *decreases down a group*. However, there are exceptions.

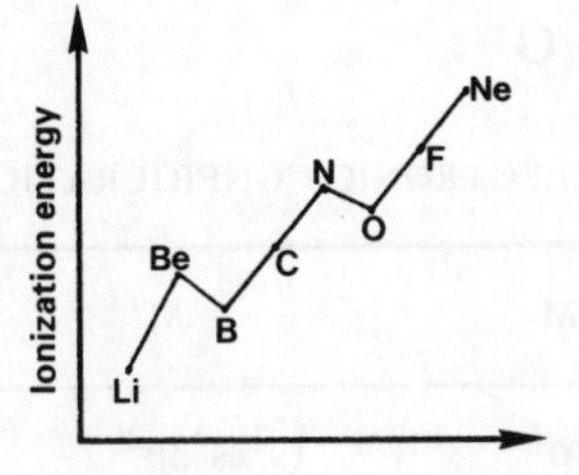

The ionization energy is affected by four factors:

1. Effective nuclear charge.
2. Which orbital an electron occupies, i.e. it is easier to remove an electron from an *n*p than from an *n*s orbital.
3. Electron–electron repulsion.
4. Exchange energy.

EXAMPLE:

Why is Be → Be$^+$ more difficult than B → B$^+$?
The electronic configuration of Be is $1s^2\ 2s^2$ and that of B is $1s^2\ 2s^2\ 2p^1$. Why is it easier to remove a valence electron from a 2p than from a 2s orbital?

This is due to a phenomenon called **penetration**. The diagram gives **radial distribution functions** (RDFs) for some orbitals.

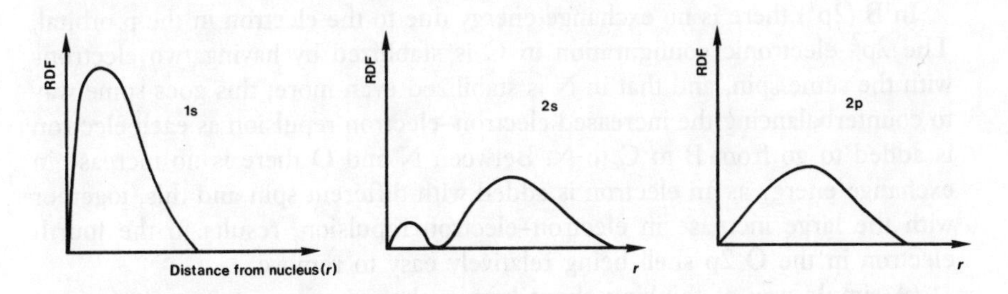

RDFs show the probability of finding an electron at various distances from the nucleus. The 2s electron can penetrate the 1s orbital and has a significant probability of being found close to the nucleus where it experiences a higher nuclear charge and is therefore held more strongly. The 2p orbitals do not have this extra electron density close to the nucleus, i.e. 2p orbitals do not penetrate as well as 2s orbitals, and 2p orbitals are higher in energy. Therefore B→B$^+$ is easier than Be→Be$^+$ because for B we are removing an electron from a higher energy p orbital but for Be we are removing an electron from an s orbital.

$N \rightarrow N^+$ versus $O \rightarrow O^+$

ELECTRONIC CONFIGURATION

ATOM	M	M^+	IE ($kJ\ mol^{-1}$)
N	$1s^22s^22p^3$	$1s^22s^22p^2$	1402
O	$1s^22s^22p^4$	$1s^22s^22p^3$	1314

Although the ionization energy would be expected to be larger for O than for N, because O has a larger nuclear charge, it is actually lower. This is because, for O, the electronic configuration is $2s^2\ 2p^4$, i.e. there are two electrons in the same p orbital which suffer a large amount of inter-electron repulsion (constrained to occupy the same region of space); therefore it is easier to remove one of these electrons than if it had been in an orbital by itself, which is the case for N (**rule of maximum multiplicity**). This effect more than counterbalances the increase in effective nuclear charge from N to O and so the ionization energy is larger for ~~O~~ N.

A complementary explanation of the reversal in the values of the ionization energy between N and O considers **exchange energy**. Exchange energy is a non-classical (derived from quantum mechanics) term by which an electronic configuration is stabilized according to the number of electrons with the same spin – more electrons with the same spin give greater stabilization.

In B ($2p^1$) there is no exchange energy due to the electron in the p orbital. The $2p^2$ electronic configuration in C is stabilized by having two electrons with the same spin, and that in N is stabilized even more; this goes some way to counterbalancing the increased electron–electron repulsion as each electron is added to go from B to C to N. Between N and O there is no increase in exchange energy as an electron is added with different spin and this, together with the large increase in electron–electron repulsion, results in the fourth electron in the O 2p shell being relatively easy to remove.

(A simple way of thinking about how exchange energy comes about is in terms of the Pauli exclusion principle: no two electrons with the same spin can occupy the same region of space, therefore electrons of the same spin tend to keep further apart, so that electron–electron repulsion is less between electrons of the same spin.)

ELECTRON AFFINITY (EA) The first electron affinity is the energy *released* in the process,

$$M_{(g)} + e^- \rightarrow M^-_{(g)}$$

with $M_{(g)}$ and $M^-{}_{(g)}$ in their ground states.

The use of the word 'released' in this definition means that a positive electron affinity corresponds to the addition of an electron to an atom/ion being an exothermic process; a negative electron affinity means that the process is endothermic.

The first electron affinity is usually positive (energy is released), consistent with bringing an electron into a region where it experiences the attraction of the nucleus. The fact that electrons do not shield each other perfectly means that as an electron is brought in close to an atom it experiences a net positive charge.

Second electron affinities ($M^-{}_{(g)} + e^- \rightarrow M^{2-}{}_{(g)}$) are always negative (require energy). This is due to the difficulty in bringing together a negatively charged ion and a negatively charged electron, e.g.

		EA (kJ mol^{-1})	
O →	O^-	+141	(Exothermic)
O^- →	O^{2-}	−780	(Endothermic)

Electron affinities increase across a period in accordance with the increasing effective nuclear charge. Variations in electron affinities down a group are less obvious and there are two factors that govern their value:

1. The size of the attractive force between the nucleus and the incoming electron.
2. The size of electron–electron repulsions.

Factor (1) decreases down a group as the electron is brought into a shell which is further away from the nucleus; thus electron affinity should decrease down a group.

Factor (2) also decreases down a group – as the orbitals get bigger the electrons are further apart and repel each other less. Less inter-electron repulsion should make it easier to bring another electron into the sphere of influence of the nucleus, thus (2) should cause electron affinity to increase in magnitude down a group.

Factors (1) and (2) thus act in opposition to each other. The general trend in electron affinities is to decrease down the group ((1) wins) but this is not a smooth change and, for example for Groups 16 and 17, the electron affinity values peak at S and Cl, respectively, and then decrease after that.

Electronegativity

This is the ability of an atom in a molecule to attract electron density towards itself.

There are various scales of electronegativity, of which the Pauling scale is the most commonly encountered. *Always stick to one scale, never mix scales.* The Pauling scale relates relative values to bond energies using the equation,

$$c(\chi_A - \chi_B)^2 = b(\text{A—B}) - \sqrt{\{b(\text{A—A}) \times b(\text{B—B})\}}$$

where:

$(\chi_A - \chi_B)$ is the electronegativity difference between A and B;
b(A—B), b(A—A) and b(B—B) are the bond enthalpies of A—B, A—A and B—B, respectively;
c is 96.5 kJ mol^{-1}

Fluorine, the most electronegative element, is assigned an electronegativity of 4.0 and other electronegativities are worked out relative to this.

VARIATION OF ELECTRONEGATIVITY Electronegativity *increases across a period* due to the increase in effective nuclear charge.

For *main group elements*, electronegativity *decreases down a group* due to the increasing size of the atom. Electrons in the bond are therefore further from the attraction of the nucleus.

For *transition metals*, electronegativity based on the Pauling scale *increases down a group*. This is the result of the very poor shielding ability of the d and f electrons and hence a large increase in effective nuclear charge down the group. (Note that the sizes of the second and third row transition elements tend to be very similar, again due to poor shielding by f electrons – see 'the lanthanide contraction,' in Chapter 6.)

ELECTRONEGATIVITY OF GROUPS In a molecular environment, the electronegativity of an atom depends on what other atoms are attached to it, e.g. the electronegativity of the C in the —CF_3 group is greater than that of the C in —CH_3. This is because F is more electronegative than H, making the C in CF_3 more positively charged so that it has a greater tendency to attract electron density towards itself.

ELECTRONEGATIVITY AND HYBRIDIZATION Hybridization (see Chapter 2) can affect electronegativity, i.e. an sp hybrid C atom in C_2H_2 is more electronegative than an sp^3 hybrid C in C_2H_6. More s character in the hybrid orbital means that the electrons in the C—H bond are held closer to the C (the electron density in an s orbital is closer to the nucleus than that in a p orbital) and there is thus more negative charge on the C in C_2H_2. This results in C_2H_2 being a stronger acid than C_2H_6 because the greater electronegativity of the sp C makes its hydrogens more $\delta+$ and therefore more readily lost as protons, H^+ (C_2H_2 forms salts such as Ag_2C_2 but C_2H_6 does not).

PROBLEMS OF ELECTRONEGATIVITY Electronegativity depends on the other groups attached (see above) and the oxidation state of an atom. For a higher positive oxidation state the 'atom' attracts electrons more because it is more positively charged; therefore its electronegativity is larger. Consider the table below.

COMPOUND	OXIDATION STATE OF S	HYBRIDIZATION	PAULING ELECTRONEGATIVITY OF S
H_2S	−2	sp^3	2.20
SO_2	+4	sp^2	3.44
SCl_4	+4	sp^3d	3.16

Section B: Stability

When talking about stability we must always try and answer two questions:

1. Kinetic or thermodynamic stability?
2. Stable with respect to what?

For example, C_2H_6 is kinetically stable with respect to reaction with O_2 (the reaction requires, e.g. a spark to initiate it) but thermodynamically unstable with respect to the products of this reaction ($2CO_2 + 3H_2O$). However, C_2H_6 is thermodynamically stable with respect to its constituent elements ($C_{(s)} + 3H_{2(g)}$).

What is stability?

Consider the reaction

$A + B \rightarrow CD \rightarrow E + F$

If we look at how the energy of the system varies as the reaction proceeds from reactants to products we get a **reaction profile** as follows:

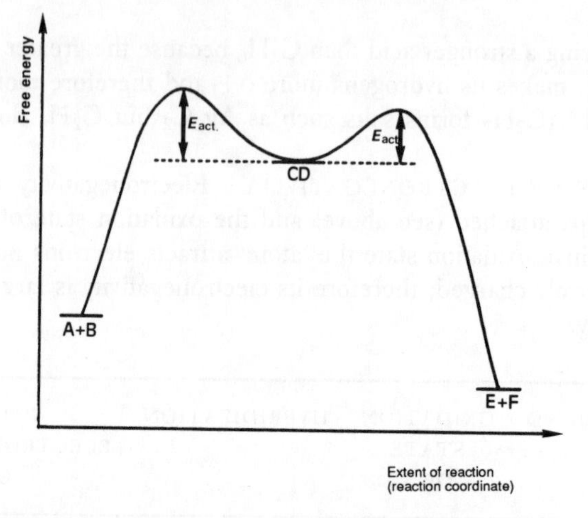

A reaction profile consists of a series of hills and valleys. Stable species sit in the valleys, i.e. stable compounds are represented by local energy minima along a reaction profile.

CD is an intermediate and sits in a shallow local minimum. CD is thermodynamically unstable with respect to decomposition to A+B or E+F but its kinetic stability depends on the height of the activation energy (E_{act}) barriers to the forward and back reactions. If the valley occupied by CD is very shallow then there is a high probability that CD will quickly pick up sufficient energy, e.g. by collision, to overcome the activation energy almost immediately it is formed, i.e. CD will be a very short-lived (transient) species. The lifetime of CD increases as the height of the activation energy barrier increases since then there is a lower probability of the CD unit having sufficient energy to proceed to products or re-form reactants.

What factors affect thermodynamic and kinetic stability?

THERMODYNAMICS For a reaction to be thermodynamically feasible the Gibbs free energy change, ΔG, must be negative. ΔG is the free energy difference between the reactants and products (see diagram opposite).

ΔG is related to the equilibrium constant, K, of a reaction by the equation,

$$\Delta G = -RT \ln K$$

and so ΔG gives us information about the position of equilibrium. In the equilibrium,

$$\text{Reactants} \rightleftharpoons \text{Products}$$

as ΔG becomes more negative the equilibrium lies further over towards the products.

ΔG can be divided into two components,

$$\Delta G = \Delta H - T\Delta S$$

where ΔH is the enthalpy change of a reaction and ΔS is the entropy change of a reaction.

ΔH is what we are used to calling the energy change of a reaction and terms like 'lattice energy' are more correctly called lattice enthalpies. Thus ΔH depends on bond strengths, ionization energies, etc. (see Chapter 2).

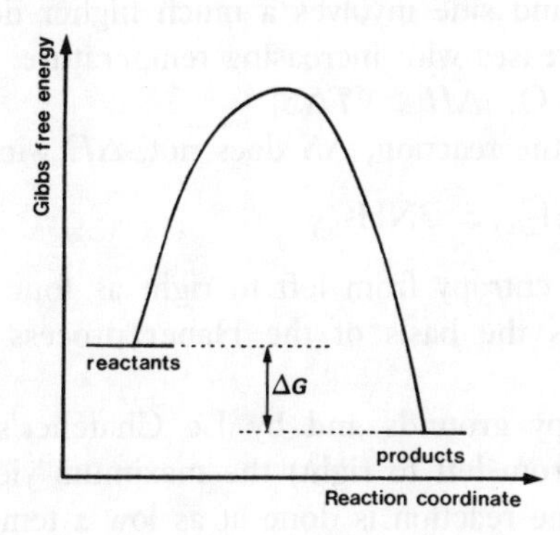

S, the entropy, in simple terms, is a measure of the disorder/randomness of a system. S is essentially zero for solids, for liquids it is a little larger and for gases it is much larger. In solids the atoms/molecules are very constrained, being able to undergo just vibrational motion about a fixed point; in a liquid there is more freedom of movement and in a gas the atoms/molecules are completely free to move in all directions – a gas is extremely disordered. A reaction involving release of a gas is very favourable on entropy grounds since ΔS will be large and positive, e.g.

$$CaCO_{3(s)} \rightarrow CaO_{(s)} + CO_{2(g)}$$

There are three ways in which a negative ΔG can be obtained:

1. $\Delta H < 0, \Delta S > 0$

   i.e. both enthalpy and entropy favour the reaction, e.g. thermal decomposition of $CaCO_3$. As the temperature is raised we will expect the reaction to be more favourable on entropy grounds ($T\Delta S$ larger). However, the reaction is exothermic from left to right, therefore by Le Chatelier's principle ('if a system in equilibrium is subjected to a change then the equilibrium will shift so as to counteract this change as far as is possible'), we expect the equilibrium to move more to the left as the temperature is raised. But enthalpy changes are generally much larger than entropy changes and therefore they are the dominant factor.

2. $\Delta H > 0$, $\Delta S > 0$ but $|T\Delta S| > \Delta H$

   i.e. although ΔS favours the reaction ΔH does not, but, at a particular temperature, ΔS wins. For this system the reaction becomes more favourable as the temperature is raised since $T\Delta S$ is positive and it is also favoured by Le Chatelier. An example of a reaction in which entropy is dominant is dissolving an ionic solid such as KCl in water

   $$KCl_{(s)} + (aq) \rightleftharpoons K^+{}_{(aq)} + Cl^-{}_{(aq)} \qquad \Delta H = +17\ \text{kJ mol}^{-1}$$

   The right-hand side involves a much higher degree of disorder. The solubilty increases with increasing temperature.

3. $\Delta H < 0$, $\Delta S < 0$, $|\Delta H| > |T\Delta S|$

   ΔH favours the reaction, ΔS does not, ΔH wins, e.g.

   $$N_{2(g)} + 3H_{2(g)} \rightleftharpoons 2NH_{3(g)}$$

   (decrease in entropy from left to right as four gas molecules become two). This is the basis of the Haber process for the production of ammonia.

   On entropy grounds and by Le Chatelier's principle (reaction is exothermic from left to right) the maximum yield of ammonia will be obtained if the reaction is done at as low a temperature as possible.

ΔG tells us whether a particular reaction is thermodynamically feasible but it does not tell us whether the reaction will actually occur. Thus, a reaction with a very large negative value for ΔG may not occur if the activation energy is too large. Activation energy is a kinetic phenomenon.

KINETICS The larger the activation energy barrier, the less chance there is of reaction occurring at a particular temperature. The Arrhenius equation relates the rate coefficient (generally, the larger the rate coefficient the faster the reaction) for a reaction to the activation energy and the temperature,

$$k = Ae^{-E_{act.}/RT}$$

where A is essentially constant for a particular reaction.

Thus, as the temperature is increased the rate of reaction increases. On qualitative grounds, this is due to more reactant molecules having energy greater than the activation energy and hence being able to surmount the hill (see diagram opposite).

The rate of reaction can also be increased by the presence of a catalyst, which works by providing a lower activation energy pathway for a reaction.

Here we have considered thermodynamics and kinetics separately.

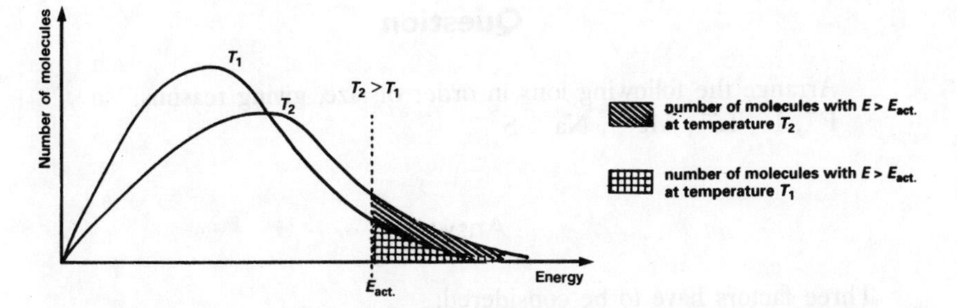

However, the stability of compounds and whether reactions occur depends on an interplay of the two factors. We saw above that to get the maximum yield from the Haber process the reaction should be done at low temperature. However, kinetics tells us that the rate of formation of product increases as the temperature is raised. In practice a compromise is reached – the reaction is run at a moderate temperature using a catalyst.

Summary

1. Electrons are assigned four quantum numbers. No two electrons in an atom can have the same set of quantum numbers.
2. Electrons in an atom occupy orbitals from the lowest energy up. When two or more orbitals have the same energy they are occupied so as to give the maximum number of like spins.
3. Atomic size decreases across a row and increases down a group. Size decreases with increasing positive oxidation state and increases with increasing negative oxidation state, i.e. M^+ is smaller than M which is smaller than M^-.
4. Ionization energy increases across a period (albeit irregularly) and decreases down a group.
5. Electron affinities increase across a period. Two opposing factors, effective nuclear charge and inter-electron repulsion, affect the electron affinity down a group.
6. Pauling electronegativity values increase across a period and decrease down a group (except for transition metals). Electronegativities are dependent upon which groups are attached to the atom being considered and on the atom's oxidation state.
7. There are two types of stability: kinetic and thermodynamic. The stability of a compound depends on both kinetic and thermodynamic factors.
8. For a reaction to be thermodynamically favourable ΔG must be negative. The rate of a chemical reaction is dependent on the activation energy and the temperature.
9. $\Delta G = \Delta H - T\Delta S$ $\quad k = Ae^{-E\text{act.}/RT}$

Question

Arrange the following ions in order of size, giving reasons:
F^-, I^-, Li^+, Mg^{2+}, Na^+, S^{2-}

Answer

Three factors have to be considered:

(a) There is a decrease in atomic radius from left to right across a period, due to an increase in effective nuclear charge.
(b) There is an increase in size going down a group due to higher principal quantum number shells being occupied. This effect is generally larger than the variation in size across a period.
(c) Other things being equal, negative ions are larger than positive ions – more electrons, larger inter-electronic repulsion, the charge cloud expands.

Li^+ and Mg^{2+} are going to have the smallest ionic radii. Li^+ is in the second period and has a positive charge. There is an increase in atomic radius in going from Li to Mg (Mg is in the third period) but the higher charge on the Mg^{2+} ion counteracts this increase in size, making the ionic radii of Li^+ and Mg^{2+} very similar (the radius of Mg^{2+} is actually 0.04 Å smaller).

F^-, like Li^+, is in the second period but the negative charge more than counteracts the decrease in size from left to right, so that F^- is significantly larger than Li^+.

There is an increase in size going from the second to the third period so that the atomic radius of Na is much larger than that of F. However, the positive charge on the Na^+ ion and the negative charge on F^- cause the charge clouds to contract and expand respectively, making the ionic radii very similar (the radius of Na^+ is actually 0.03 Å smaller than that of F^-).

S^{2-}, in the third period with a 2− charge is significantly larger than F^-.

I^- is the largest ion, having a 1− charge and being in the fifth period. The order of ionic radii is, thus,

$$Li^+ \approx Mg^{2+} < F^- \approx Na^+ < S^{2-} < I^-$$

CHAPTER 2
Covalent and Ionic Bonding

In this chapter we will discuss covalent and ionic bonding. We shall see below, however, that *ionic and covalent bonding can just be regarded as extreme cases of the same thing.*

Covalent Bonding

Simple molecular orbital (MO) theory

A covalent bond involves the overlap of two **atomic orbitals** (AOs) on different atoms to give bonding and antibonding **molecular orbitals** (MOs).

1. Bonding molecular orbitals
   (a) Electrons in a bonding MO lie in the region of space between the nuclei, though not necessarily on the inter-nuclear axis (e.g. π bonds); they attract both nuclei and therefore hold them together.

   A.O. + A.O. → M.O.

   (b) A bonding MO is *lower* in energy than either of its constituent atomic orbitals.
2. Antibonding molecular orbitals
   (a) In an antibonding MO there is less electron density between the

nuclei than for the two individual AOs. This results in nucleus–nucleus repulsions being greater than nucleus–electron attractions in the region between the atoms.

(b) An antibonding MO is *higher* in energy than either of its constituent atomic orbitals.

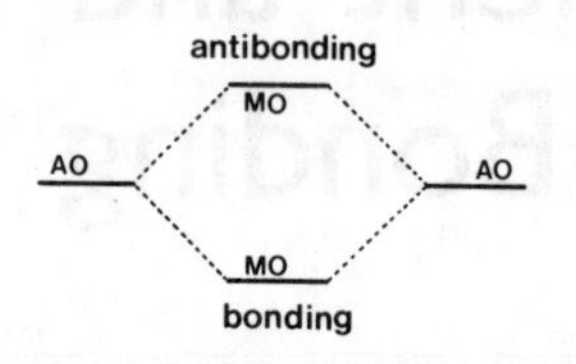

Molecular orbital diagrams

Rule 1 Antibonding molecular orbitals are slightly more antibonding than bonding molecular orbitals are bonding. A simple way of thinking about this is that formation of a bonding MO brings the atom centres closer together and so increases the inter-nuclear repulsion (i.e. the bonding MO is slightly destabilized).

Rule 2 Molecular orbitals are usually constructed as a *L*inear *C*ombination of *A*tomic *O*rbitals (LCAO).

Rule 3 *We must always end up with the same number of MOs as AOs we started with.*

THE LCAO APPROACH A molecular orbital can be described as,

$$\Psi_{MO} = c_A\phi_A + c_B\phi_B + c_C\phi_C + \ldots$$

ϕ_i is an AO on atom i, e.g. a 1s or a 2p orbital and c_i is the fractional contribution of ϕ_i to the MO. Note that $\Sigma c_i^2 = 1$ must be obeyed (the molecular orbital is said to be 'normalized') and therefore, if we have an MO made up of two identical AOs (they contribute equally to the MO) the values of the coefficients will be $1/\sqrt{2}$ and not 1/2, e.g. for H_2,

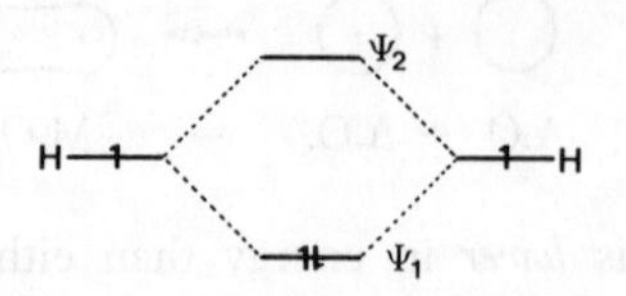

In MOs one and two $|c_A| = |c_B| = 1/\sqrt{2}$, since the AOs are identical and equal in energy and therefore contribute equally to the MOs.

TYPES OF MOs A σ MO has the same symmetry as an s AO, relative to a plane taken perpendicular to the inter-nuclear axis,

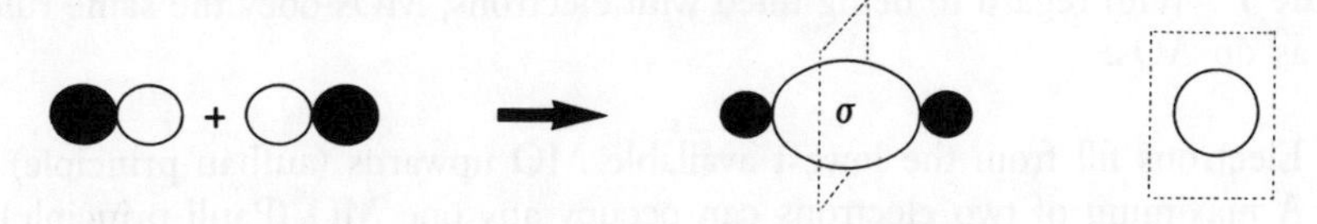

s, p, d and f orbitals can all interact to give σ bonds.

A π MO has the same symmetry as a p AO, again relative to a plane taken perpendicular to the inter-nuclear axis,

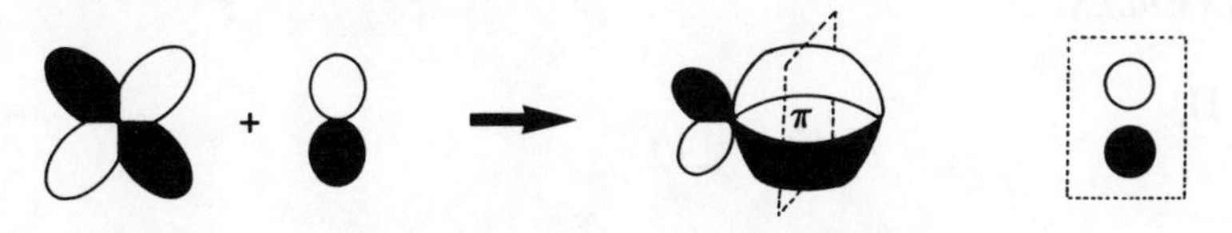

p, d and f orbitals can form π bonds, s orbitals cannot.

A δ MO has the same symmetry as a d AO orbital relative to a plane taken perpendicular to the inter-nuclear axis and it would involve, e.g. the face-on overlap of two d orbitals. Only d and f orbitals can form δ bonds.

Interactions such as

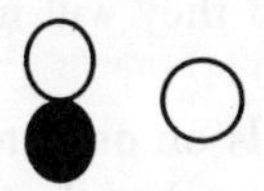

are not allowed as the orbitals are of different symmetries and there is no net bonding or antibonding effect (this does *not* give a non-bonding orbital).

Rule 4 The degree of interaction between orbitals depends on:

1. How close they are in energy – all other things being equal, the closer two orbitals are in energy, the stronger the interaction between them.
2. How diffuse (concentrated) the orbitals are, the more diffuse two orbitals are, generally the less well they overlap.

Rule 5 All other things being equal, σ interactions are stronger than π interactions. This is because a σ interaction results from a head-on overlap of AOs, whereas π is a side-on overlap.

Rule 6

$$\text{Bond order} = \frac{\text{number of electrons in bonding MOs} - \text{number of electrons in antibonding MOs}}{2}$$

The bond order provides a guide to the relative strength of covalent bonds. This is most useful in homonuclear diatomics but is less useful in heteronuclear systems (see below) where it must be used with caution because of ionic contributions to the bonding.

Rule 7 With regard to being filled with electrons, MOs obey the same rules as do AOs:

1. Electrons fill from the lowest available MO upwards (aufbau principle).
2. A maximum of two electrons can occupy any one MO (Pauli principle).
3. If two or more MOs are of the same energy they will be occupied to give the maximum number of parallel (like) spins (Hund's first rule).

EXAMPLES

H_2

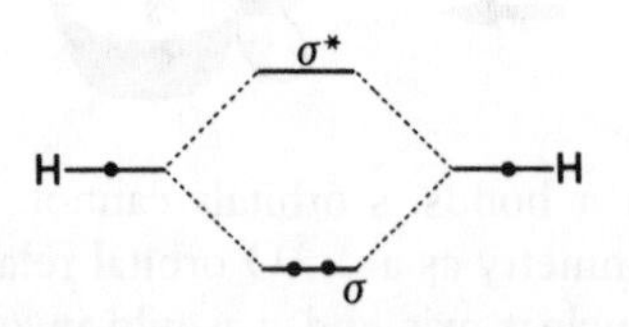

1. The atomic orbitals have the same energy and are identical.
2. Both atomic orbitals are 1s so they will interact to give a σ bonding MO and a σ* antibonding MO.
3. Interactions between 1s orbitals on one atom and a 2s orbital on the other H atom are possible in theory but the energy difference between these two orbitals is so great that the interaction energy is minimal (Rule 4).
4. There are two electrons, one from each H atom, which can be paired up in the lower MO.

Bond order = 1/2 × (2) = 1, i.e. a single bond.

He_2

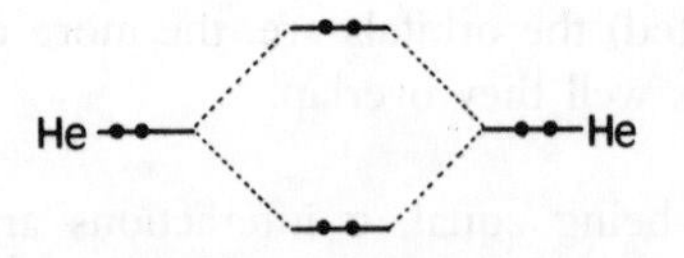

The molecular orbital diagram is constructed in the same way as for H_2. Here, however, there are four electrons, so that two go into the bonding and two into the antibonding molecular orbital.

Overall bond order = 1/2 × (2−2) = 0

Therefore, He_2 does not exist.

For He_2^+ there are three electrons available for bonding and therefore two go into the bonding and one into the antibonding MO. This gives,

Bond order = 1/2 × (2−1) = 1/2

so that He_2^+ can, and does, exist as a discrete molecular ion.

O_2

This is a more difficult example. Here we consider the interactions between the individual atomic orbitals in turn. For O_2 the following interactions are possible:

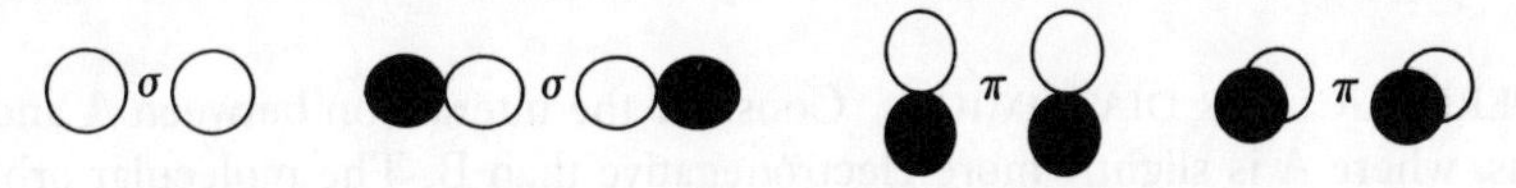

These interactions can be shown on an energy level diagram:

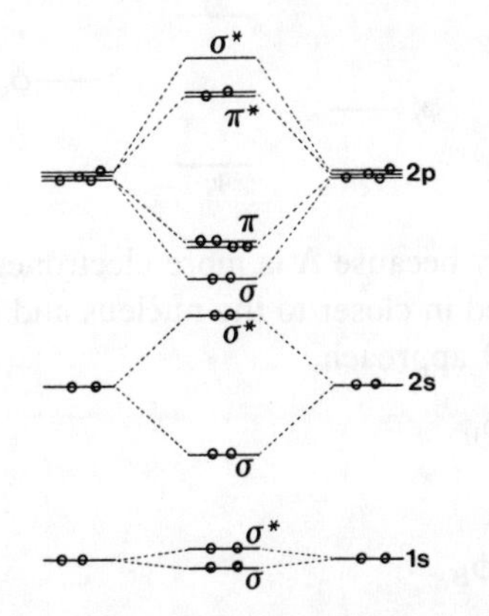

The interaction between the 1s orbitals is weak because these are core orbitals and are held in very closely to the nuclei. The overlap of these two atomic orbitals is thus very slight at the inter-nuclear distance of O_2,

O—O

Applying the rule of maximum multiplicity and the aufbau principle, the electrons occupy the MOs as shown.

Bond order = 1/2 × (10 − 6) = 2, i.e. a double bond

BOND LENGTHS AND BOND ORDERS Consider the series O_2^+, O_2, O_2^-; the O—O bond lengths are 1.12, 1.21 and 1.26 Å respectively.

The MO diagram for O_2^- is the same as that for O_2 except that one more electron is added to the lowest available orbital, i.e. a π* antibonding orbital, in accordance with the aufbau principle. Putting electron density into antibonding orbitals reduces the bonding between two atoms and therefore the bond is weaker and longer. The bond order for O_2^- is 3/2.

O_2^+ in its ground state is derived from O_2 by the removal of an electron from the highest occupied MO in O_2, i.e. O_2^+ has one less electron in the π^* antibonding set of orbitals. Removal of electrons from an antibonding orbital increases the degree of bonding between two atoms and therefore the bond in O_2^+ is stronger and shorter than in O_2; the bond order in O_2^+ is 5/2.

Although the O—O bond is stronger in O_2^+ than in O_2, O_2^+ is *not* more stable than O_2 since energy has to be supplied to O_2 to remove an electron; this energy is more than that released by the formation of a stronger bond.

HETERONUCLEAR DIATOMICS Consider the interaction between A and B atoms, where A is slightly more electronegative than B. The molecular orbital diagram is

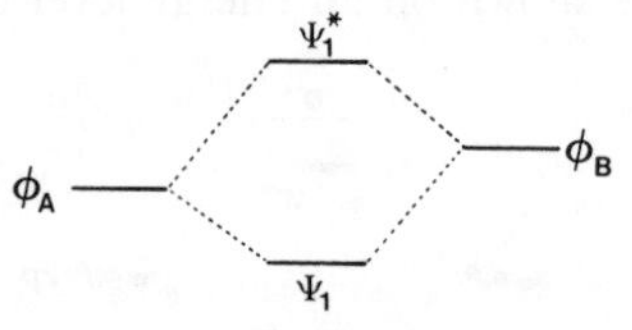

ϕ_A is lower in energy because A is more electronegative than B and therefore the orbitals are pulled in closer to the nucleus and thereby lowered in energy.

Using the LCAO approach,

$$\Psi_1 = c_A\phi_A + c_B\phi_B$$

and

$$\Psi_1^* = c_C\phi_A - c_D\phi_B$$

The bonding MO resembles more closely the A AO, to which it is closer in energy, i.e. $c_A > c_B$. In contrast, Ψ_1^* has more B AO character, therefore $c_D > c_C$.

If there are two electrons in Ψ_1 and none in Ψ^*_1 there will be more electron density residing on atom A than on atom B because Ψ_1 has greater ϕ_A character. This will result in A having a small negative charge and B a positive one, which is consistent with the electronegativity differences between A and B.

If A is *much* more electronegative than B we have the following:

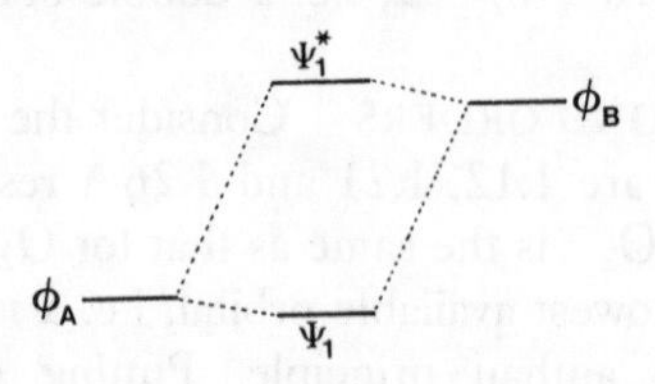

Ψ_1 is virtually identical to the A AO, i.e. $c_A >> c_B$ and $\Psi_1 \approx \phi_A$
Ψ_1^* is virtually identical to the B AO, i.e. $c_D >> c_C$ and $\Psi_{1*} \approx \phi_B$

Now the two electrons essentially occupy an atomic orbital on A, giving A^-B^+, i.e. an ionic bond. It is not actually possible to have a 100 per cent ionic compound because c_B and c_C, as defined above, never fall exactly to zero. It is also not possible to form a 100 per cent covalent bond, e.g. even in H_2 there is some ionic character. This is because the electrons in the bond do not always lie totally symmetrically between the H atoms and at any one time they may both lie closer to one nucleus than to the other, resulting in a temporary polarization (this is how van der Waals' forces come about). In other words, the resonance form (page 34) H^+H^- contributes to the structure (albeit to a very slight extent).

There is thus a gradual change between covalent and ionic bonding depending on the electronegativity difference between the atoms. Most compounds come somewhere in between the two extremes.

Ionic Bonding

Atoms with widely different electronegativities form predominantly ionic compounds, e.g. NaCl (Pauling electronegativities, Na 1.0 and Cl 3.5). These compounds consist of positively and negatively charged ions packed together in a lattice array. The lattice is held together by non-directional electrostatic forces between the oppositely charged ions. There are various ways of packing ions:

1. Packing should maximize the number of contacts between oppositely charged ions.
2. Packing should keep ions of the same charge as far apart as possible.

The structure adopted depends strongly upon the sizes of the ions and the charges on the ions.

Examples of two ways of packing ions in an ionic lattice are:

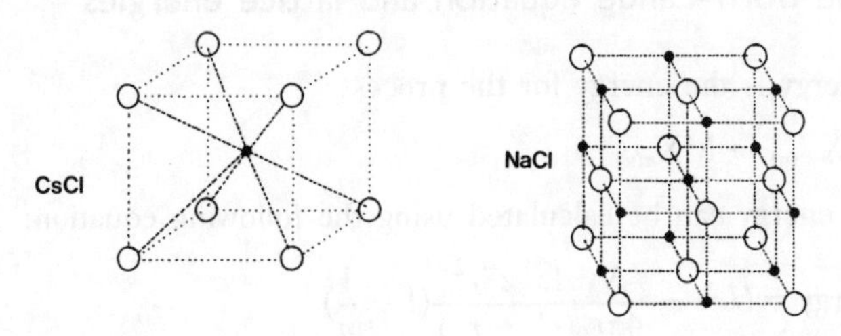

Radius ratio rules

These provide a means of predicting the coordination number of ions in a predominantly ionic compound. The **coordination number** is the number of ions of type B that surround an ion of type A, or vice versa, in an ionic compound A_xB_y.

Radius ratio = r^+/r^-

where r^+ is the ionic radius of the positively charged ion and r^- the ionic radius of the negatively charged ion.

The radius ratio provides a guide to how many cations it is *geometrically* possible to pack around an anion, and vice versa, considering ions as hard spheres (billiard balls).

Maximum coordination numbers are obtained for r^+/r^- approaching 1 (see table).

COORD. NO.	GEOMETRY	RADIUS RATIO (r^+/r^-)
4	Tetrahedral	below 0.414
6	Octahedral	0.414 – 0.732
8	Cubic	0.732 – 1.0
12	Dodecahedral	above 1.0

This approach assumes a purely ionic hard sphere model and deviations occur when there is significant covalent character in the bonding, e.g. r^+/r^- for ZnS = 0.52, hence a coordination number of 6 would be predicted but a coordination number of 4 is found. This is because, in ZnS, the electronegativity difference between Zn and S is not large and therefore there is significant covalent character in the bonding.

The Born–Landé equation and lattice energies

The lattice energy is the energy for the process,

$M^{n+}{}_{(g)} + nX^-{}_{(g)} \rightarrow MX_{n(s)}$

The lattice energy can be calculated using the following equation:

$$\text{Lattice energy} = U = -\frac{N_A A z^+ z^- e^2}{4\pi\varepsilon_0(r^+ + r^-)}(1 - \frac{1}{n})$$

where N_A is Avogadro's number, A is the Madelung constant, z^+ and z^- are the charges on the positive and negative ions respectively, e.g. 1 for Na^+ and

2 for O^{2-}, ε_0 is the permittivity of a vacuum – a constant = 8.85×10^{-12} F m^{-1}, e is the electronic charge, n is the Born exponent, r^+ is the ionic radius of the cation and r^- is the ionic radius of the anion while $r^+ + r^-$ is the inter-ionic distance.

This is the Born-Landé equation and assumes a purely ionic model for the bonding.

Lattice energies are thus proportional to charge and inversely proportional to inter-ionic distances,

$$\textbf{Lattice energy} \propto \frac{z^+z^-}{(r^+ + r^-)}$$

Inter-ionic distances can be obtained by X-ray crystallography (Chapter 10).

The Madelung constant, A, is a number (dimensionless) describing the geometry of the lattice. It depends only on the geometrical distribution of the positive and negative ions and not in any way on the exact nature of the ions (e.g. whether they are K^+ or Na^+).

The Born exponent, n, is a correctional term allowing for the fact that ions are not hard spheres and some repulsive interaction between electron clouds on adjacent ions can occur, i.e. the electron cloud on a positive ion repels the electron cloud on an adjacent negative ion, thus reducing the attractive force between them. It takes a value between 1 and 10. It can be estimated from compressibility experiments, i.e. the fractional change in volume of a crystal per unit change in pressure.

EXAMPLE

Calculate the lattice energy for NaCl using the Born–Landé equation and the data given:

$n = 9.1$, $A = 1.748$, $r^+ + r^- = 2.82\,\text{Å} = 2.82 \times 10^{-10}$ m, where $r^+ + r^-$ = inter-ionic distance of NaCl.

$$U = -\frac{N_A A z^+ z^- e^2}{4\pi\varepsilon_0(r^+ + r^-)}\left(1 - \frac{1}{n}\right)$$

For Na^+Cl^- $z^+ = z^- = 1$

$$\text{Therefore } U = -1.748 \times 6.022 \times 10^{23} \times 1 \times 1 \times (1.6021 \times 10^{-19})^2 \times (1 - 1/9.1)\ 4\pi \times 8.854 \times 10^{-12} \times 298 \times 10^{-12}$$
$$= -765\ \text{kJ mol}^{-1}$$

A more accurate estimation of the lattice energy can be obtained from a more sophisticated version of the Born–Landé equation. This version uses a more precise expression for the repulsion energy (i.e. not just the Born exponent) and includes corrections for the heat capacity of the crystal, van der Waals' and zero point energies.

Because the Born–Landé equation considers a purely electrostatic model, comparison of values obtained from this equation and 'experimental' values for the lattice energy provide a guide to how ionic a compound is. It is not possible to design an experiment to measure lattice energies directly and so 'experimental' values are calculated from Born–Haber cycles.

Born–Haber cycles

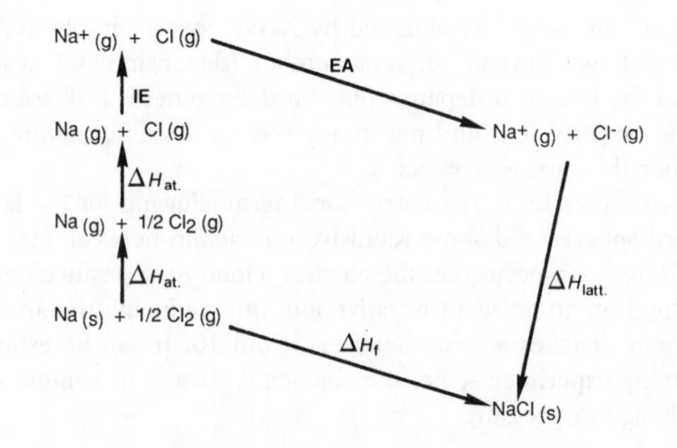

If all other values are known then $\Delta H_{\text{latt.}}$ (lattice energy) can be worked out from Hess' Law, which states that the enthalpy change for the process reactants → products is independent of the pathway taken between the initial and final states.

Good agreement between Born–Haber and Born–Landé values for the lattice energy suggests mostly ionic character; poor agreement suggests polarization of ions and significant covalent character in the bonding. When there is poor agreement between Born–Haber and Born–Landé lattice energies this does not mean that the 'experimental', Born–Haber, value is the correct value as, when significant covalent character is present in a compound, the interpretation of lattice energy becomes more complex and a much less useful concept.

A comparison of the lattice energy values for NaCl, calculated above from the Born–Landé equation ($-765\,\text{kJ mol}^{-1}$) and the Born–Haber cycle ($-776\,\text{kJ mol}^{-1}$), implies that the bonding in NaCl is mostly ionic.

Degree of ionicity and polarizing power/polarizability (Fajan's rules)

Compounds having the most ionic character are formed by relatively large cations with small positive charges in conjunction with anions with small negative charges.

The degree of ionicity depends on:

1. the polarizing power of the cation; and
2. the polarizability of the anion.

If a metal ion is small and highly charged (high charge/radius ratio, i.e. high ionic potential) then it will be very polarizing and will pull electrons from the anion towards itself to give partial covalent bonding. Thus Be^{2+} is more polarizing than Mg^{2+} (Mg^{2+} is larger) leading to Be compounds being mostly covalent, and those of Mg being mostly ionic.

The larger and more negatively charged an anion, the more polarizable it will be, i.e. less well able to hold on to its electrons when next to a cation. Thus I^- is much more polarizable than F^- as the outer electrons in I^- are much further from the nucleus and are therefore less strongly held. Similarly S^{2-} is more polarizable than Cl^- as S^{2-} has a larger negative charge and hence will let go of its electrons more easily.

These two factors together lead to the degree of ionicity going in the order:

$CsF > CsI > LiI$

Physical properties of ionic and covalent compounds

Compounds are usually classified as ionic or covalent depending on their macroscopic physical properties. Thus, ionic compounds have high melting points, are hard and are generally insoluble in organic solvents; covalent substances have low melting points, are soft and are generally soluble in organic solvents.

These physical properties are not a consequence of the *type* of bonding, but depend rather on the *extent* of bonding. 'Traditional' ionic compounds consist of infinite three-dimensional lattices of positive and negative ions held together by strong electrostatic forces. It is the infinite nature of the bonding that leads to the observed physical properties, e.g. in order to melt an ionic solid a large amount of energy has to be supplied to break apart all the ions in the lattice (high melting point). 'Traditional' covalent compounds consist of discrete molecules with the interactions between these molecular units relatively weak (van der Waals', dipole–dipole, hydrogen bonding). When this type of covalent solid is melted, energy has to be supplied to break only the weak intermolecular interactions, and not the covalent bonds within the molecules (low melting point).

Diamond is an example of a different type of covalent substance, in which the bonding involves covalent bonds between all adjacent C atoms to give an infinite three-dimensional structure. Although the bonding is covalent, the extent of bonding is the same as in a traditional ionic salt and diamond exhibits many of the physical properties normally associated with ionic substances, e.g. high melting point, hardness.

Just as it is possible to have infinite covalent substances it is also possible to have molecular ionic ones, e.g. Li_4Me_4 (see Chapter 5). Here the bonding is best considered as involving predominantly electrostatic interactions between Li^+ and Me^- ions within the Li_4Me_4 molecular unit. The large size of Me^- compared to Li^+ inhibits strong interaction between ions in adjacent 'molecules', i.e. the Li^+ ions in one molecular unit are insulated from the attraction of Me^- ions in adjacent units and so the formation of an infinite three-dimensional lattice is prevented. Thus, we have ionic 'molecules' linked by weak intermolecular forces, which results in Li_4Me_4 exhibiting properties more usually associated with covalent substances, e.g. low melting point and solubility in organic solvents.

Hard/soft acid/base (HSAB) theory

A Lewis *ac*id *ac*cepts an electron pair; a Lewis base donates an electron pair. Acids and bases can be categorized as hard or soft and HSAB theory says that *hard acids prefer to bind to hard bases; soft acids prefer to bind to soft bases*.

Factors which affect the hardness/softness of acids and bases are:

1. Size of an ion.
2. Charge/oxidation state.
3. Electronic structure.
4. Nature of attached groups already present.

Hard acids are small and highly charged; if transition metals they tend to have high oxidation states and a small number of d electrons.

Soft acids are large and either neutral or with small charges. If they are transition metals they tend to have many d electrons.

Hard bases are highly electronegative, have low polarizability and high resistance to oxidation.

Soft bases have low electronegativity, high polarizability and are readily oxidized.

Hard and soft acids and bases are depicted in the table. There are also borderline cases, e.g. Br^- and pyridine are borderline bases; Fe^{2+} and GaH_3 are borderline acids.

HARD AND SOFT ACIDS AND BASES

HARD		SOFT	
ACIDS	BASES	ACIDS	BASES
H^+	OH^-	Cu^+	CO
BF_3	H_2O	BH_3	R_3P
Be^{2+}	NR_3	Pt^{2+}	I^-
Mn^{2+}	F^-	Pt^{4+}	
Fe^{3+}			

THE THEORETICAL BASIS OF HSAB Most hard acids and bases are those that might be expected to form ionic bonds. Most soft acids and bases are those that might be expected to form covalent bonds. The stability of hard acid–hard base complexes results from a highly ionic interaction. With soft acid–soft base complexes the stability arises from a predominantly covalent interaction. It is generally considered that π bonding between transition metals with more than six d electrons and ligands contributes greatly to soft acid–soft base interactions.

Hard acids and soft bases, or vice versa, do not form strong complexes because their preferred modes of interaction are incompatible, e.g. CO forms a stable complex with BH_3 but not with BF_3; BH_3 is a soft acid and CO is a soft base but BF_3 is a hard acid. This principle is applicable throughout the whole of inorganic chemistry in the prediction of the stability of complexes.

Valence shell electron pair repulsion theory (VSEPR)

VSEPR theory is a means of predicting the structures of *main group* covalent compounds. It will not predict the structures of transition metal compounds (see Chapter 6).

1. Valence shell electron pairs (lone pairs and bonding pairs) are arranged to minimize repulsions between themselves.
2. The basic geometry is controlled by the number of σ bonding electron pairs, π bonds merely support the σ framework. That is, the two electron pairs which make up the double bond occupy the same region of space and do not act as separate units.

3. The ideal structures which allow *n* valence shell electron pairs to be as far apart as possible are:

NO. OF VSEPs	IDEAL STRUCTURE
2	linear
3	trigonal planar
4	tetrahedral
5	trigonal bipyramidal
6	octahedral

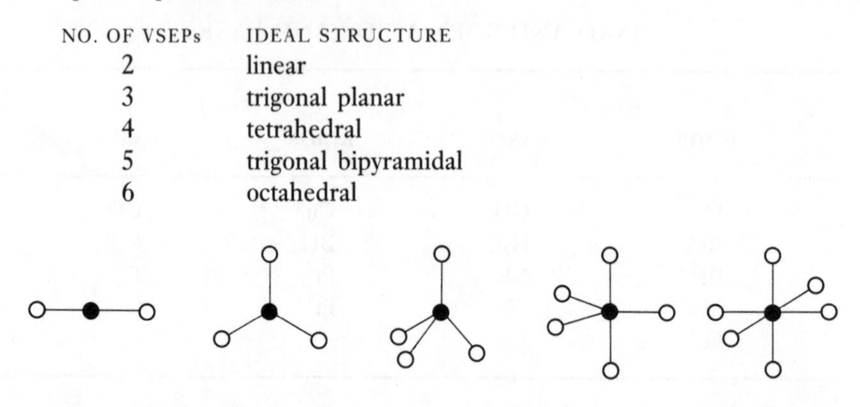

4. The order of repulsive power is lp–lp > lp–db > lp–sb > db–db > db–sb > sb–sb, where lp = lone pair, db = double bond and sb = single bond. This is because electrons in a lone pair are held in more closely to the central atom than are those in a bonding pair.
   A double bond contains twice as many electrons as a single bond.
5. In a trigonal bipyramidal molecule lone pairs go in the equatorial plane. In a trigonal bipyramid all the vertices are not equivalent so that the equatorial ligands must always be different from the axial ones (look at the shape). The appropriate hybridization scheme for a trigonal bipyramid is sp^3d (one-fifth s character) and this can be divided up approximately into sp^2 (one-third s character) trigonal planar in the equatorial plane and pd hybrids perpendicular to the plane. The lone pairs prefer to be in orbitals with more s character as these are closer to the nucleus and lower in energy (an s orbital is lower in energy than a p orbital in the same shell).

EXAMPLES OF HOW TO PREDICT THE SHAPES OF MOLECULES

NH_3

Draw the structure satisfying the valency on the *outer* atoms; do not worry about the central atom.

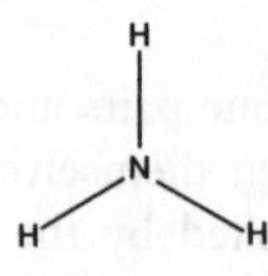

Count electrons in the valence shell of the central atom. N has five electrons already present and three from a half share in three covalent bonds (one from each).

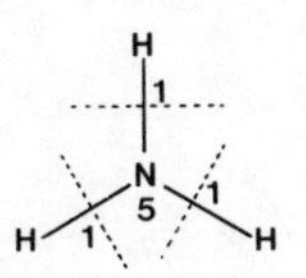

eight electrons = four electron pairs

Categorize as single (S) and double (D) bonds and work out how many lone pairs (L).

S D L
3 0 1 = 4

There are thus four units which have to be kept as far apart in space as possible in order to minimize inter-electron repulsions. The structure adopted is thus based on a tetrahedron, which appears pyramidal, because we cannot 'see' lone pairs.

It will not be a regular tetrahedron as lone pairs are held in closer to the N and therefore will repel the σ bonds more than they will repel each other. The H—N—H bond angles, at 107.8°, are thus less than the ideal tetrahedral value of 109.5°.

ClF_3

The valency of the outer chlorine atoms is satisfied by the following basic structure,

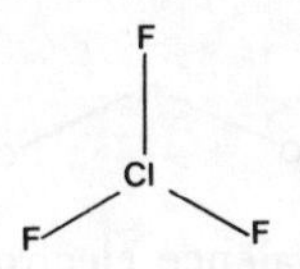

Dividing up the electrons gives 10 valence electrons, i.e. five pairs,

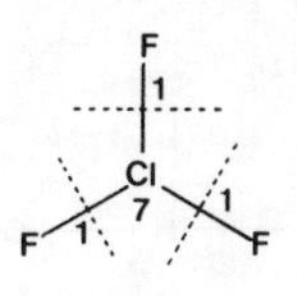

S D L
3 0 2 = 5

Again we have five units to arrange in space and the structure will be based on a trigonal bipyramid. Lone pairs go in the equatorial plane, therefore the structure is actually 'T-shaped'.

Repulsion between a lone pair and a bonding pair is greater than between two bonding pairs (Rule 2). This results in the axial Fs being bent back away from the lone pairs to give a distortion from the perfect T-shape. The bond angles will therefore be:

F_{ax}—Cl—F_{ax} < 180° F_{ax}—Cl—F_{eq} < 90°

The bond angles of NO_2, NO_2^+ and NO_2^-

NO_2 has the following basic structure:

O═N═O

This gives nine valence electrons, i.e. 4.5 pairs.

S D L
0 2 0.5 = 2.5

Half a lone pair acts like a lone pair but it does not repel as much. Therefore there are three units to arrange in space, hence the molecule is based on a trigonal planar structure.

Because half a lone pair (one electron) repels less than a double bond (four electrons) the O—N—O bond angle will be greater than the idealized 120°.

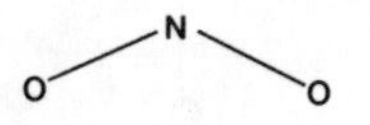

For NO_2^+ there are eight valence electrons.

S D L
0 2 0 = 2

Hence NO_2^+ is linear.

O—N—O

For NO_2^- there are 10 valence electrons.

S D L
0 2 1 = 3

Similarly to NO_2, the structure is based on a trigonal planar geometry.

However this time there is a full lone pair on the N and in accordance with

Rule 4, lone pairs repel more strongly than do double-bonded pairs. Hence the O—N—O bond angle is smaller than the idealized 120° expected for this geometry.

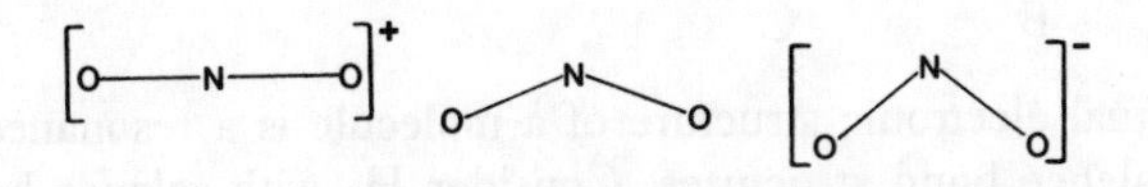

Valence bond theory

This is a 'traditional' two centre–two electron (2c–2e) *localized* bond approach to the bonding in molecules. In a valence bond treatment covalent bonds can be formed in two ways:

1. By interaction between two orbitals, each containing one electron.

2. By one atom donating a lone pair of electrons into a vacant orbital on another atom, i.e. a dative covalent bond.

Once a dative covalent bond has been formed it is indistinguishable from a 'normal' covalent bond, e.g. all the N—H bond lengths in $[NH_4]^+$ are equal.

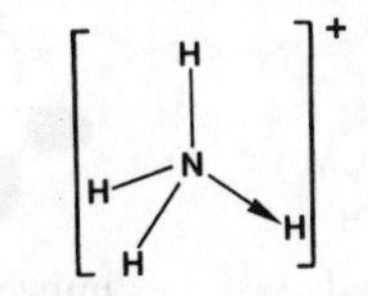

A VALENCE BOND APPROACH TO O_2 The oxygen atom has the outer shell electronic configuration $2s^2\ 2p^2\ 2p^1\ 2p^1$. Two covalent bonds can be formed using these unpaired electrons. The bond formed by p orbitals overlapping head-on is going to be stronger than that formed by their overlapping sideways on, and thus, valence bond theory can account for the overall strength of the O=O double bond not being twice that of the O—O single bond. However, since valence bond theory considers all the electrons in O_2 to be paired, either in atomic orbitals or in bonds, it cannot readily provide an explanation of the observed paramagnetism of O_2 (paramagnetism being due to the presence of unpaired electrons; see Chapter 6). The paramagnetism can be explained by an MO approach to the bonding (see page 21).

RESONANCE It is always possible to draw more than one plausible **valence bond structure** (canonical form) for a molecule, e.g. for H_2 we can have;

H—H	H^+H^-	H^-H^+
A	B	C

The actual electronic structure of a molecule is a resonance hybrid of all possible valence bond structures. Consider H_2 with valence bond structures A, B and C. Form A is a more realistic representation of the bonding in a hydrogen molecule than forms B and C and as such will have a greater contribution to the overall electronic structure of the molecule. Thus, the overall resonance hybrid is:

$$aA + bB + cC$$

where $a >> b,c$ (a, b and c are the coefficients representing how much a particular form contributes to the overall bonding).

It is important to realize that H_2 does not exist in the form H—H one moment and then in the form H^+ H^- the next but is an inseparable mixture of all forms. A useful analogy is that a mule is a resonance hybrid of a horse and a donkey, i.e.

Mule = *a*Horse + *b*Donkey

but it is not a horse one day and a donkey the next, it is always just a mule.

HYBRIDIZATION This is the mixing of orbitals on an atom to generate new orbitals which are able to participate in better covalent bonding by pointing more directly at orbitals on the bonded atoms.

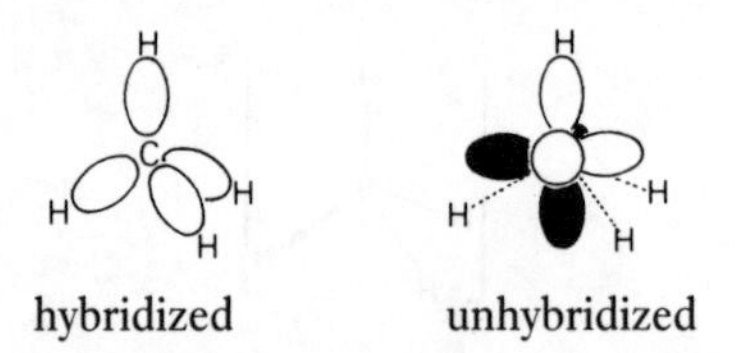

The rules for constructing hybrid orbitals are:

1. You must always end up with the same number of hybrid orbitals as atomic orbitals you started with.
2. The energy of a hybrid orbital is the weighted average of the energies of the constituent atomic orbitals.

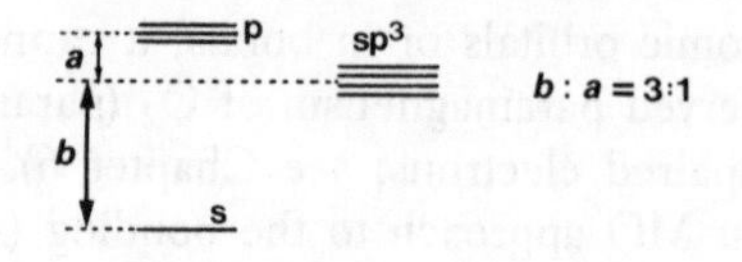

Hybridization can affect electronegativity (see page 10).

Valence bond theory, hybridization and bonding in simple molecules

Here we will consider how the bonding in some molecules of formula EX_n can be described using a localized 2c–2e valence bond approach.

EXAMPLES

CH_4

This is tetrahedral by VSEPR rules,

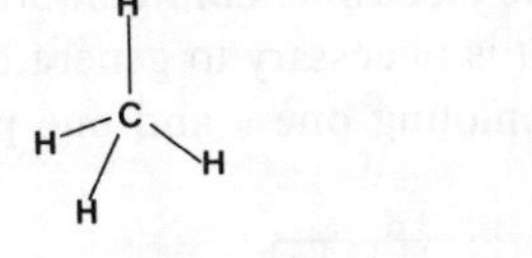

C has the ground state electronic configuration $2s^2\ 2p^2$, and therefore only has two unpaired electrons in its outer shell. Since we generally regard bond formation as involving pairing of electrons from interacting orbitals, some electronic promotion must occur to generate four unpaired electrons so that four bonds can be formed. Promotion is followed by hybridization to give four equivalent sp^3 orbitals.

The promotion step involves putting an electron into a higher energy orbital, which requires energy. This energy is more than compensated for by the extra energy released when C forms four bonds instead of two (as it would if it had the electronic configuration $2s^2\ 2p^2$). The hybridization involves one s and three p orbitals on C mixing to give four sp^3 hybrid orbitals pointing towards the vertices of a tetrahedron.

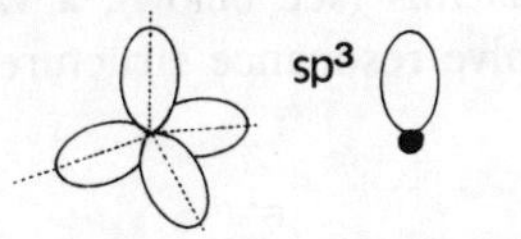

As the energy of the hybrid orbitals is the weighted mean of the energies of the orbitals from which they are derived, this hybridization process involves no overall change in the energy of the electrons (see above).

The hybrid orbitals point towards the hydrogen atoms and covalent bonds are formed by pairing unpaired electrons in the hybrids with those in the 1s orbitals on the hydrogens. We thus have four 2c–2e bonds between the carbons and the hydrogens.

SF_6

By VSEPR this is octahedral,

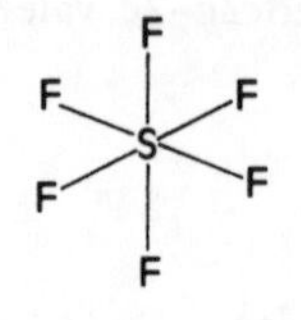

The ground state electronic configuration of S is $3s^2\ 3p^2\ 3p^1\ 3p^1$. In order to form six bonds it is necessary to generate six unpaired electrons, which can be achieved by promoting one s and one p electron to the d set of orbitals.

Hybridization involves combination of the s, three p and two d orbitals to generate six equivalent sp^3d^2 hybrid orbitals, which point towards the vertices of an octahedron. The d orbitals used are the $d_{x^2-y^2}$ and d_{z^2}, which point in the correct direction.

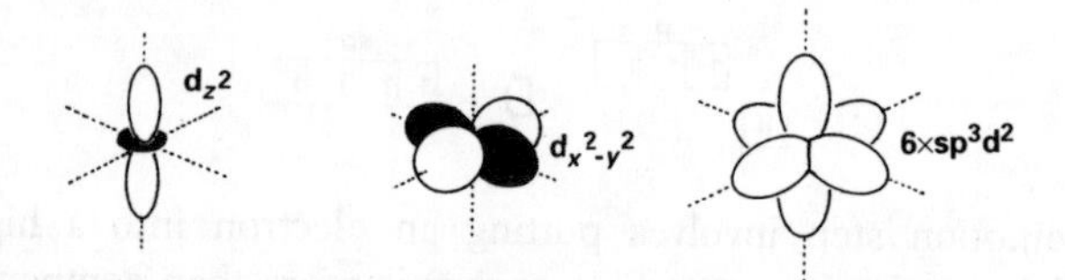

The unpaired electrons in the hybrid orbitals on S are then paired up with the unpaired electrons in the 2p orbitals on the F ligands to generate six 2c–2e bonds.

There is some debate as to the degree of participation of d orbitals in the bonding in systems such as this (see below); a valence bond treatment not using d orbitals would involve resonance structures such as

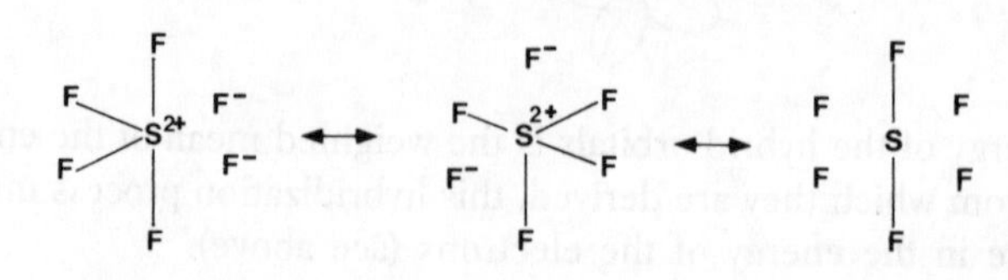

which indicates an S—F bond order of less than one, which some people believe to be the case.

PF_5
VSEPR predicts a trigonal bipyramidal structure,

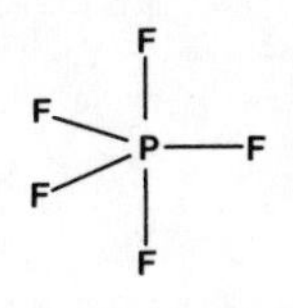

To form five bonds promotion and hybridization must take place,

The vertices in a trigonal bipyramidal structure are not all equivalent (this is not geometrically possible); the equatorial vertices are all equivalent but different from the axial ones (see n.m.r. of PF_5, page 307). Similarly, it is not possible to generate five equivalent hybrid orbitals which point towards the vertices of a trigonal bipyramid. The sp^3d hybrid orbitals can be divided into two sets: sp^2 hybrids in the equatorial plane and pd hybrids (d_{z^2}) in the axial direction.

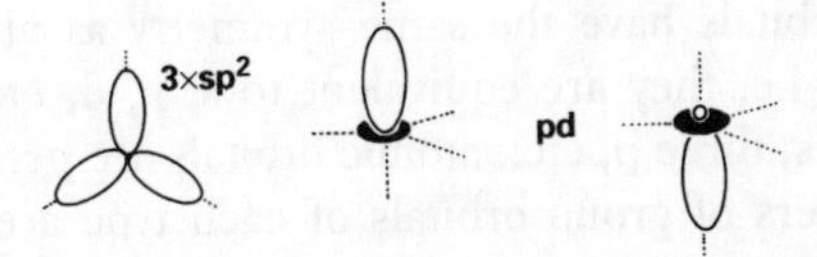

The use of hybrid orbitals with more d character for bonding would be expected to give rise to longer bond lengths because d orbitals are more diffuse than s and p orbitals and extend out further from the central atom. This is experimentally observed; the axial P—F bond lengths are longer than the equatorial ones.

A bonding scheme not involving d orbitals requires resonance structures of the form,

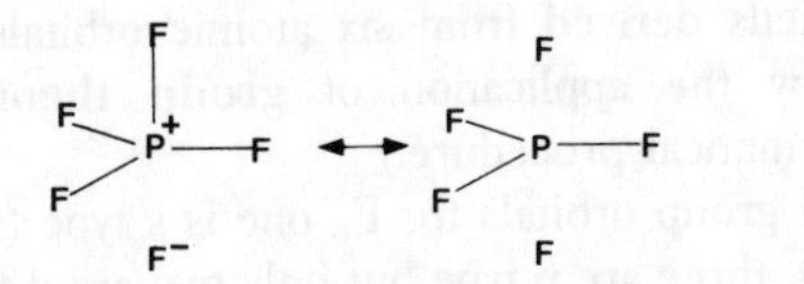

where just a p orbital is used for bonding in the axial direction. This suggests weaker bonding in the axial direction.

Advanced MO diagrams

Here we consider molecular orbital diagrams for complexes of general formula EX_n, with E as the central atom, surrounded by nX atoms. The simplest way to construct an MO diagram is to *consider the* n *Y atoms as a single unit interacting with the X atom.*

EXAMPLE

SF_6

If just σ bonding is considered then each fluorine atom has a σ symmetry orbital containing one electron and pointing towards the centre of the octahedron.

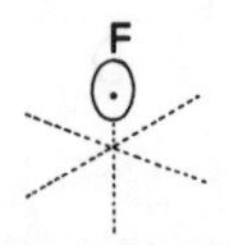

Just as s, p, d, etc., orbitals can be constructed for an atom, *group, or symmetry, orbitals can be constructed for a ligand unit.* In order to construct symmetry orbitals the following rules should be obeyed:

1. We must always end up with the same number of group orbitals as atomic orbitals we started with, e.g. six F atomic orbitals give six group orbitals.
2. These group orbitals have the same symmetry as atomic orbitals on the central atom, i.e. they are equivalent to s, p, d, etc., atomic orbitals.
3. Just as only one s, three p, etc., atomic orbitals are permitted in atoms, only certain numbers of group orbitals of each type are permitted.

Applying these rules, the following symmetry orbitals can be constructed for an octahedral F_6 unit in SF_6,

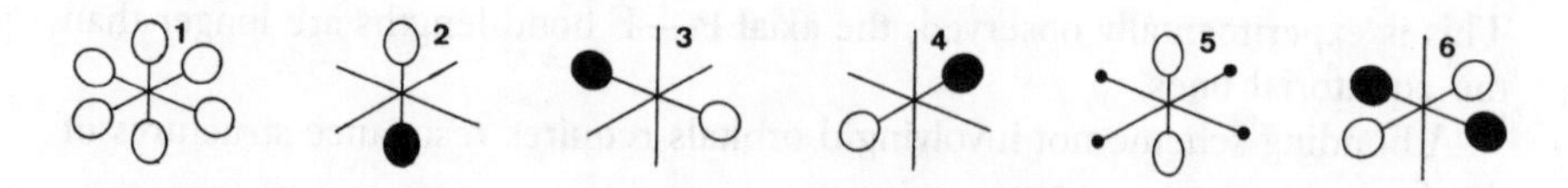

i.e. six group orbitals derived from six atomic orbitals. (These orbitals are actually derived by the application of **group theory**, which involves a complicated mathematical procedure.)

Thus, of the six group orbitals for F_6 one is s type (the same symmetry as an s atomic orbital), three are p type but only two are d type (corresponding to $d_{x^2-y^2}$ and d_{z^2} atomic orbitals). This is because the other three d atomic orbitals (d_{xy}, d_{xz}, d_{yz}) have lobes lying between the axes and therefore group orbitals corresponding to these cannot be constructed from σ orbitals lying along the axes.

These six orbitals can be regarded as degenerate and of the same energy as the atomic orbitals from which they were derived. This is because the distance between the F atoms is too great for interaction to occur between them (remember that the F atoms are interacting with the S and not with each other). However, the six group orbitals are lower in energy than the sulphur 3s orbital because F is more electronegative (its atomic orbitals are pulled in more strongly and lowered in energy).

These group orbitals interact with the valence atomic orbitals on the central atom, obeying the following set of rules:

1. Only orbitals of the same symmetry can interact:

   1 with s
   2, **3**, **4** with the p_z, p_y, p_x
   5 and **6** with d_{z^2} and $d_{x^2-y^2}$

2. Interaction energy is inversely proportional to the energy difference between the two interacting orbitals (Rule 4, page 19), i.e. the strongest interaction is between orbital **1** and the s atomic orbital on sulphur because they are closest in energy. This gives one bonding and one antibonding molecular orbital.

The three sulphur p orbitals can interact with **2**, **3** and **4** respectively to give three bonding and three antibonding molecular orbitals, e.g.

We now come to a fork in the road; there are two ways of proceeding from here. Some people prefer the use of d orbitals in molecular orbital diagrams and others do not.

MO DIAGRAMS WITHOUT d ORBITALS If the d orbitals on the S are not involved in bonding then group orbitals **5** and **6** have no sulphur atomic orbitals of the correct symmetry with which to interact. Therefore they remain non-bonding, i.e. with the same energy as the six group orbitals. Using this scheme, the molecular orbital diagram (**I**) can be drawn:

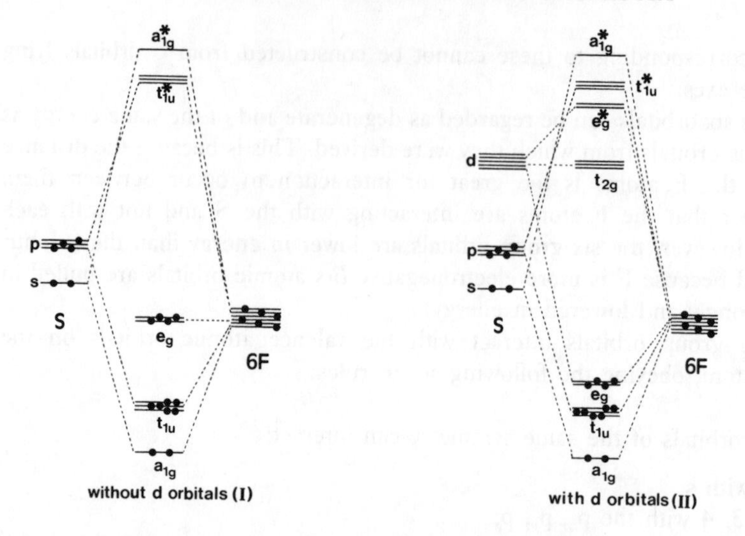

(See below for explanation of the labels on the orbitals.)

For SF_6 there are six electrons on S and one from each F, i.e. 12 electrons available. Molecular orbitals are filled with electrons from the lowest energy orbital up, therefore these 12 electrons occupy the six lowest molecular orbitals.

The overall bond order = 1/2 (number of bonding electrons − number of antibonding electrons).

= 1/2 (8 − 0) = 4

Therefore the bond order per S—F bond = 4/6, i.e. 2/3.

Note: non-bonding electrons contribute nothing to the bond order.

MO DIAGRAMS WITH d ORBITALS Of the five 3d orbitals on the S, only two are of the correct symmetry to interact with the ligand group orbitals; the d_{xy}, d_{xz} and d_{yz} will always be non-bonding for σ interactions.

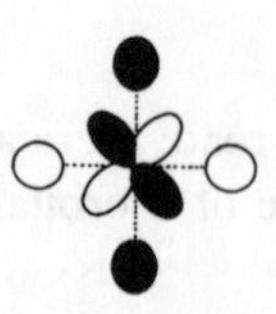

The $d_{x^2-y^2}$ and d_{z^2} can interact with the two d type group orbitals to produce two bonding and two antibonding molecular orbitals. The interaction between the d type orbitals and the d atomic orbitals is the smallest because they are furthest apart in energy – MO diagram (II) above. The set of three orbitals labelled t_{2g} are the d_{xy}, d_{xz} and d_{yz} orbitals, which remain non-bonding. The overall bond order = 1/2 (12−0) = 6, i.e. the bond order for each S—F bond is 1.

There is differing opinion as to whether d orbitals are involved in the bonding in systems such as these. Some people believe that the d orbitals are too high in energy to interact with orbitals on the ligands. However, if the central atom has a high positive charge (oxidation state), as does the S in SF_6, then the d orbitals are pulled in closer to the nucleus and lowered in energy, thereby becoming compatible for bonding.

Note that an MO treatment considers the bonding as totally delocalized. If the diagrams of the interactions between group and atomic orbitals are examined it is seen that the *interactions cannot be assigned to any one S—F bond but apply to the molecule as a whole.*

Note that a good way to work out the symmetry orbitals is to observe the shape of the molecule and then work out the hybridization at the central atom. Group orbitals are generated to match the atomic orbitals involved in the hybridization, e.g. if the central atom would be sp^3d^2 hybridized for a particular geometry then we generate one s, three p and two d type symmetry orbitals.

SYMMETRY LABELS The labels on the molecular orbital diagrams are derived from group theory and are used to describe the symmetry and degeneracy of the molecular orbitals:

- Singly degenerate orbitals are labelled *a* or *b*.
- Doubly degenerate orbitals are labelled *e*.
- Triply degenerate orbitals are labelled *t*.

In molecules with a centre of symmetry the orbitals can be given labels to describe the **parity**. Consider an s orbital; the sign (phase) of the orbital at a point (x,y,z) is the same as that at the point $(-x,-y,-z)$, the point obtained by inverting (x,y,z) through the centre of symmetry. An s orbital is thus given the label *g* (gerade, in German this means even).

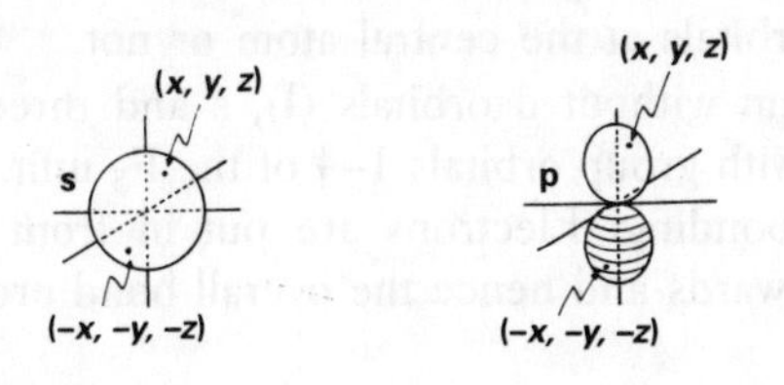

For a p orbital the sign at (x,y,z) is opposite to that at $(-x,-y,-z)$ and the orbital is labelled *u* (ungerade, odd).

AO	PARITY
s	g
p	u
d	g
f	u

The same procedure can be used to assign parity labels to molecular orbitals.

The 1 and 2 subscripts in the molecular orbital labels refer to symmetry with respect to mirror planes/rotation axes in the molecule and can be worked out from group theory.

EXAMPLES

PF_5

For this trigonal bipyramidal molecule the hybridization at the central atom is sp^3d, where the d orbital used is the d_{z^2}. Therefore we must generate one s, three p and one d type symmetry orbitals for the F_5 unit, i.e.

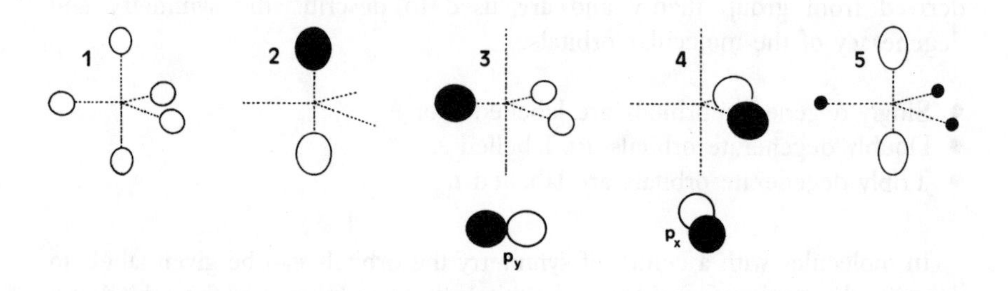

Group orbitals 3 and 4 look a bit strange, but it can be seen that they really do have the same symmetry as p_x and p_y orbitals and so interact with them.

Similarly to the SF_6 example we can consider two cases, whether the bonding involves d orbitals at the central atom or not.

In the MO diagram without d orbitals (I), s and three p orbitals on the phosphorus interact with group orbitals 1–4 of the F_5 unit. Group orbital 5 in this scheme is non-bonding. Electrons are put in from the lowest energy molecular orbitals upwards and hence the overall bond order is 4, or 4/5 per P—F bond.

In the MO diagram with d orbitals (II), the d_{z^2} orbital on the P interacts with symmetry orbital 5 of the F_5 unit. The overall bond order is hence 5, or 1 per P—F bond.

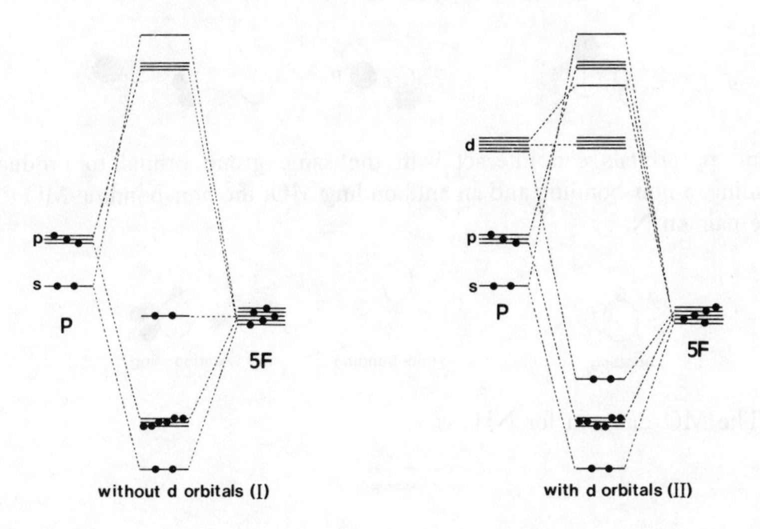

CH_4

A tetrahedron can be derived from a cube by taking alternate vertices. The hybridization at the carbon atom appropriate for a tetrahedral structure is sp^3, therefore we must generate one s type and three p type group orbitals for the H_4 unit.

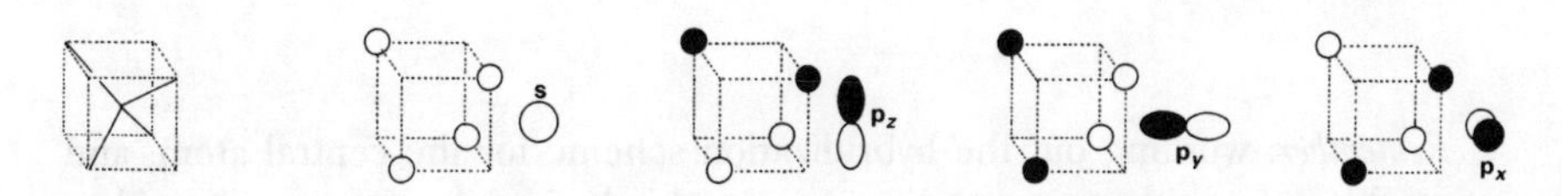

Thus, following the same procedure as above, we can generate an MO diagram.

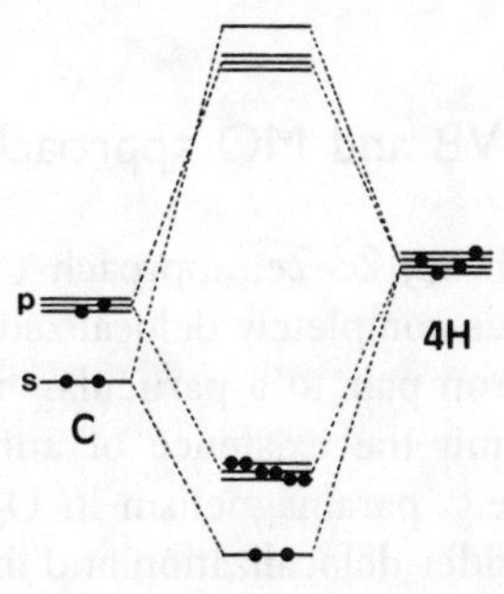

NH_3

The N here is sp^3 hybridized so that we might think that we ought to generate one s and three p type group orbitals for the H_3 unit. However, since we have only three atomic orbitals from the hydrogens we can only generate three group orbitals (Rule 1): one s type and two p type orbitals. These orbitals are:

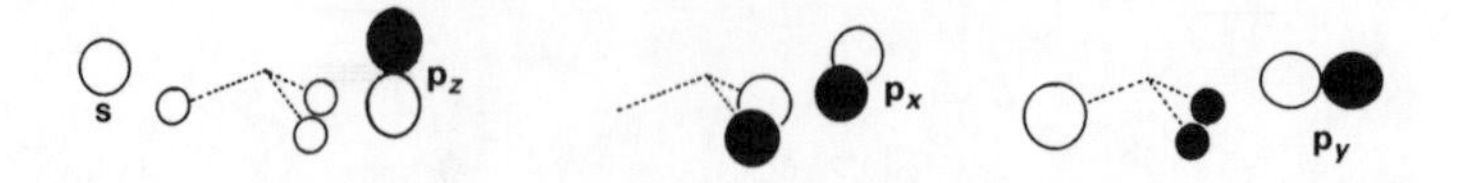

s and p_z orbitals can interact with the same group orbital to produce a bonding, a non-bonding and an antibonding MO; the non-bonding MO is the lone pair on N.

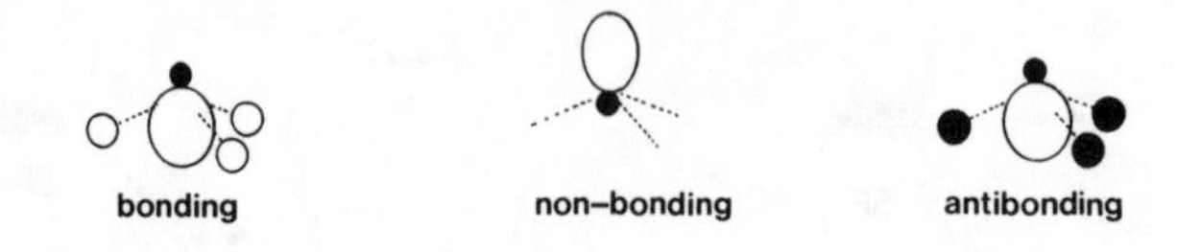

The MO diagram for NH_3 is

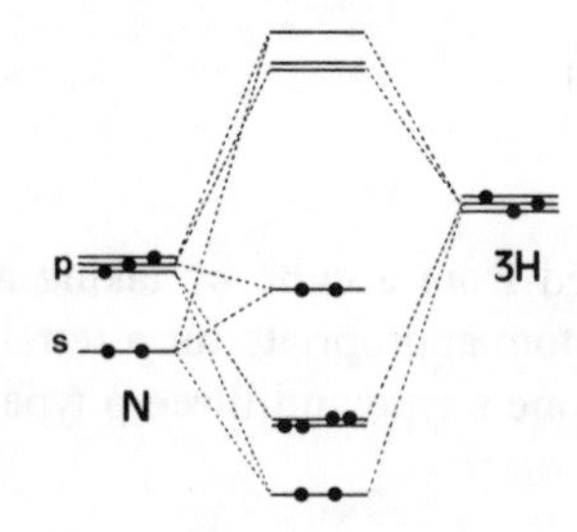

Remember: working out the hybridization scheme for the central atom, and using this as a guide to generating group orbitals, is only a convenient aid to memory and it is not intended to imply anything about the bonding in the molecule.

Comparison of VB and MO approaches to bonding

VB theory provides a localized, 2c–2e, approach to bonding. In MO theory the bonding is considered as completely delocalized, so that it is not possible to assign a particular electron pair to a particular bond.

VB theory does not admit the existence of antibonding orbitals and, as such, cannot account for, e.g. paramagnetism in O_2.

VB theory does not consider delocalization and in order to explain, e.g. the equal C—C bond lengths in benzene, it is necessary to resort to resonance.

Photoelectron spectroscopy studies (PES; see page 315) on CH_4 show two peaks, corresponding to the removal of electrons from two different energy levels. VB theory would predict four equivalent C—H bonds and therefore just one peak in the spectrum. MO theory, however, predicts the existence of two sets of molecular orbitals (a_1 and t_1) and, therefore, two peaks.

These two approaches represent *chemists'* way of looking at the bonding in the molecule; whether a VB or an MO approach is used does not change the *actual* bonding in the molecule.

Summary

1. Covalent bonds involve the overlap of atomic orbitals (AOs) to give molecular orbitals (MOs).
2. MOs are usually described by the LCAO approach.
3. All other things being equal, the strongest interaction occurs between AOs of the same energy.
4. The same rules (Pauli, maximum multiplicity, aufbau) apply to filling AOs and MOs with electrons.
5. Bond order = 1/2(total number of bonding electrons − total number of antibonding electrons).
6. There is a gradual change between covalent and ionic bonding in accordance with electronegativity differences between the bonded atoms.
7. Ionic compounds exist when there is a large electronegativity difference between the atoms. They exist as a 3-D array in which packing maximizes contacts between oppositely charged ions and minimizes interactions between like charges.
8. Coordination numbers in ionic compounds depend largely on radius ratio rules.
9. Born–Landé lattice energy equation:

$$U = -\frac{N_A A z^+ z^- e^2}{4\pi\varepsilon_0(r^+ + r^-)}(1 - \frac{1}{n})$$

10. Comparison of lattice energies from both Born–Landé equations and Born–Haber cycles gives an estimate of the degree of ionicity in the bonding.
11. VSEPR predicts shapes of main group compounds – valence shell electron pairs arrange themselves so as to minimize inter-electron repulsions.
12. In advanced MO diagrams we consider the surrounding atoms to act as a single unit which generates symmetry orbitals which can interact with the central atom.

Questions

1 The lattice energy of CsF_2 has been estimated as $2250 \pm 150\,\text{kJ mol}^{-1}$. To what extent do this estimate and the data below enable you to predict whether CsF_2 can be prepared? Would any additional data have been useful in enabling you to make a more confident prediction?

	kJ mol^{-1}
$\Delta H_{\text{sub.}}$ Cs(s)	76
First IE of Cs(g)	376
Second IE of Cs(g)	2422
$\Delta H_{\text{at.}}$ F(g)	79
EA of F(g)	339
$\Delta H_{\text{latt.}}$ energy of CsF(s)	720

Answer

To work out the heat of formation of CsF_2 from the thermodynamic data given we have to construct a Born–Haber cycle, i.e. (remember that EA is defined as the energy released and therefore the process $F_{(g)} + e \rightarrow F^-_{(g)}$ is exothermic).

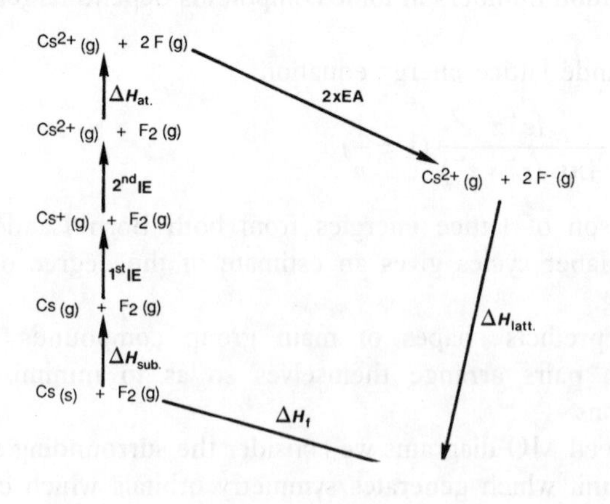

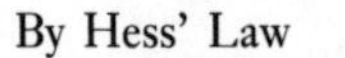
By Hess' Law

$$\Delta H_f = +76 + 376 + 2422 + 79 - 2 \times 339 - 2250\ (\pm 150)$$
$$= +25 \pm 150\ \text{kJ mol}^{-1}$$

$\Delta H_{formation}$ for CsF_2 is not unreasonably endothermic but a true understanding of whether it can be formed can only be obtained by consideration of ΔG for the process. For a reaction to be thermodynamically feasible ΔG must be negative.

$\Delta G = \Delta H - T\Delta S$ and thus we also require entropy data for the reaction, to predict whether the reaction is feasible.

The process involves a solid and a gas forming a solid, therefore there is going to be a large decrease in entropy, i.e. ΔS is negative, making the ΔG value more positive and, hence, the formation process less favourable.

We are not given any entropy data and so must continue with the analysis of the ΔH values and see whether any other factors may be important.

Consider the decomposition process,

$$CsF_2(s) \rightarrow CsF(s) + 1/2F_2(g)$$

It is possible to work out the enthalpy change for this process but we first have to work out the heat of formation of CsF using another Born–Haber cycle, i.e.

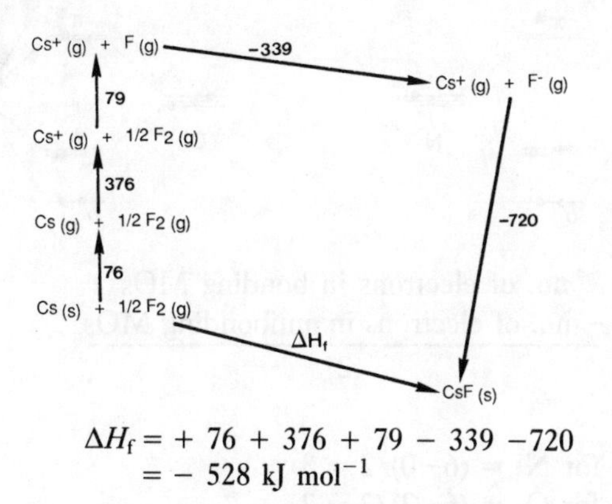

$$\Delta H_f = + 76 + 376 + 79 - 339 - 720$$
$$= - 528 \text{ kJ mol}^{-1}$$

Now we can work out the enthalpy change for the decomposition, i.e.

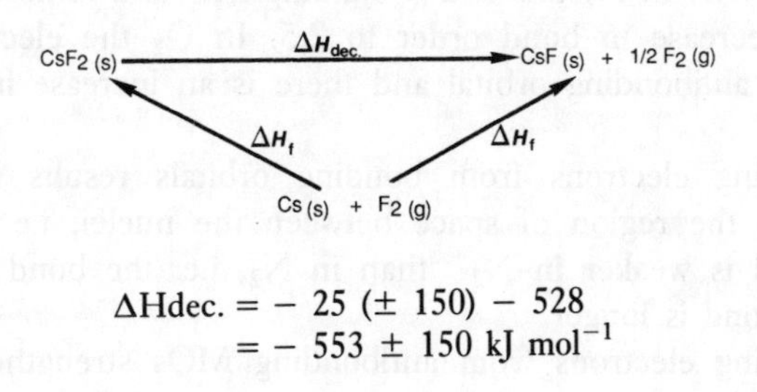

$$\Delta H\text{dec.} = - 25 (\pm 150) - 528$$
$$= - 553 \pm 150 \text{ kJ mol}^{-1}$$

The decomposition is thus extremely favourable on enthalpy grounds and also on entropy grounds (solid → solid + gas) and ΔG will therefore be large and negative. $\Delta G = -RT\ln K$, where K is the equilibrium constant, thus the equilibrium for the decomposition is going to lie almost totally in favour of CsF.

There could possibly be a kinetic block to the decomposition, i.e. a high activation energy, but there isn't and CsF_2 has never been prepared.

2 What are the bond orders in N_2 and O_2? What is the effect on bond length, bond energy and magnetic properties of removing an electron from N_2 and O_2 to give N_2^+ and O_2^+?

Answer

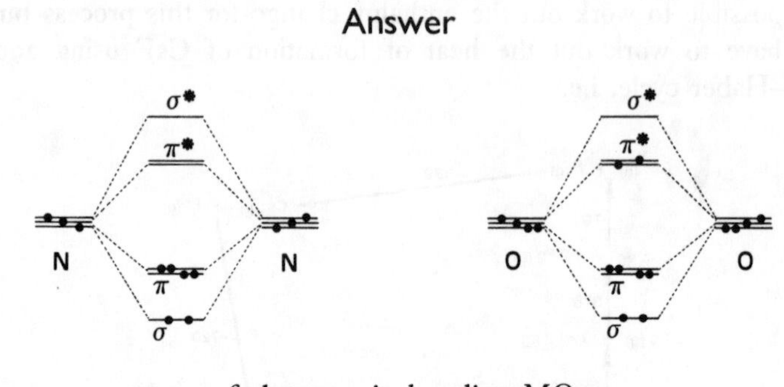

$$\text{Bond order} = \frac{\text{no. of electrons in bonding MOs} - \text{no. of electrons in antibonding MOs}}{2}$$

Thus,

bond order for $N_2 = (6-0)/2 = 3$
bond order for $O_2 = (6-2)/2 = 2$

To generate the positive ions an electron is removed from the highest occupied MO. In N_2 this is a π bonding MO and removal of an electron gives a decrease in bond order to 2.5. In O_2 the electron is removed from a π antibonding orbital and there is an increase in bond order to 2.5.

Removing electrons from bonding orbitals results in less electron density in the region of space between the nuclei, i.e a weaker bond. The bond is weaker in N_2^+ than in N_2, i.e. the bond energy is lower and the bond is longer.

Removing electrons from antibonding MOs strengthens a bond and the bond in O_2^+ is stronger than in O_2. The O—O bond energy is larger and the O—O bond length shorter in O_2^+ than in O_2.

N_2 with all its electrons paired in molecular orbitals is diamagnetic. Removal of an electron generates a half-filled orbital and N_2^+ is paramagnetic.

O_2 has two unpaired electrons and is paramagnetic. Removal of one unpaired electron will reduce the magnetic moment of the O_2 – simply, it is made less magnetic.

3 How would you expect the X—O bond lengths to vary along the series, $[SiO_4]^{4-}$, $[PO_4]^{3-}$, $[SO_4]^{2-}$, $[ClO_4]^-$?

Answer

The best way to approach this is by drawing out valence bond structures, i.e.

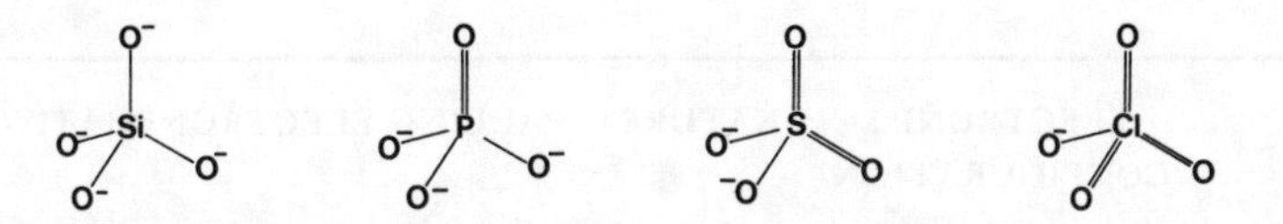

Thus, in each case, by a combination of covalent bonds and negative charges we have satisfied the valency requirements of the oxygens (given them full outer shells). In $[SiO_4]^{4-}$ this has been done without recourse to double bonds and with all single bonds we have an average bond order of 1 and the longest X—O bonds. In $[PO_4]^{3-}$ the average X—O bond order is 1 1/4 – four single bonds and a double bond shared between four bonds, therefore the bond length is shorter than in $[SiO_4]^{2-}$. The average X—O bond order in $[SO_4]^{2-}$ is 1 1/2 – four single bonds and two double bonds shared between four bonds, and the bond is shorter than for $[PO_4]^{3-}$. $[ClO_4]^-$ has the largest bond order, 1 3/4, and the shortest X—O bond.

CHAPTER 3
Main Group

Down the Groups

Group 1

	ELECTRONIC CONFIGURATION	NATURE	PAULING ELECTRONEGATIVITY
Li	[He] $2s^1$	metal	0.98
Na	[Ne] $3s^1$	metal	0.93
K	[Ar] $4s^1$	metal	0.82
Rb	[Kr] $5s^1$	metal	0.82
Cs	[Xe] $6s^1$	metal	0.79

The elements are all metals. They are characterized by weak metallic bonding with a very open (body centred cubic) structure and hence have low melting points and are soft. The weak metallic bonding is due to the fact that only one valence electron is available for bonding (ns^1). Thus, if metallic bonding is simply considered as positive ions sitting in a 'sea' of electrons, then this bonding is going to be weakest when the ions only have a single positive charge and each atom only contributes one electron to the sea (cf. lattice energies, where the strength of ionic bonding is proportional to the charges on the ions).

There is a regular decrease in melting point down the group. This is attributable to three factors:

1. The repulsion between electrons on adjacent ions increases down the

group, because there are more of them.
2. There is an increase in the size of the atoms down the group so that the electrons involved in metallic bonding are further from the positively charged nucleus, therefore attractive forces are lower.
3. The regularity of the decrease is due to a regular increase in the size of the atoms and the fact that all the elements have the same structure. If they had different structures, the atoms (ions) would be arranged differently in the metallic lattice, which would affect the overall bonding.

The ionization energies are low because one electron in the outer shell leads to the lowest effective nuclear charge in a period (as we go to the right we add electrons to the same shell; these shield each other ineffectively and there is an increase in effective nuclear charge from left to right across a period). This results in the Group 1 elements being very electropositive. The ionization energy decreases down the group because the atoms are bigger, therefore the outer electron is further from the nucleus and easier to remove.

The alkali metals react with virtually all other elements to form a variety of predominantly ionic compounds. This is because $\Delta H_{at.}$ (cf. melting points) and the ionization energy, both endothermic quantities, are small so that lattice energy more than compensates for these terms in a Born–Haber cycle.

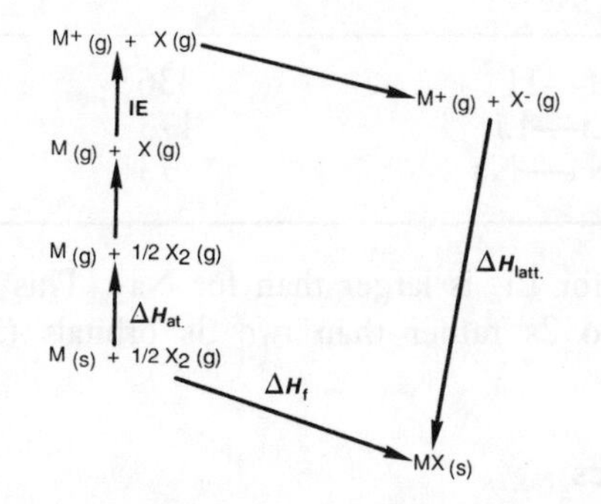

GENERAL BONDING The alkali metals exist almost exclusively in the +1 oxidation state (there are some compounds which contain the M^- ions – see below).

$$M \rightarrow M^+ + e^-$$

is a low energy process but

$$M \rightarrow M^{2+} + 2e^-$$

requires a great deal of energy because the second electron is being removed from an inner quantum shell. This large input of energy would not be compensated for by $\Delta H_{latt.}$ or by $\Delta H_{hyd.}$ of the 2+ ion.

There is only a small amount of covalent character in the bonding with the predominancy of ionicity being due to the relative ease of formation of the +1 ion and the weakness of any covalent bonds (lower $z_{eff.}$ at the left-hand side leads to orbitals being more diffuse, giving poorer overlap and weaker covalent bonds).

The degree of covalency is largest for lithium and this covalent character decreases down the group. This is because Li^+ is the smallest and most polarizing cation and so will have the greatest tendency to attract electron density back from the negative ions, i.e. introducing covalency into the bonding. As we go down the group, the size increases and hence the polarizing power of the positive ion decreases. Also, the strength of covalent bonds decreases down the group as the valence orbitals on the metal become more diffuse. This can be seen in two examples.

Vapours above liquids

$3M_{(l)} \rightarrow M_{(g)} + M_{2(g)}$

where $M_{2(g)}$ is a covalently bonded diatomic molecule. Li_2 is the most stable alkali metal diatomic molecule, which can be seen from the bond energies.

	BOND ENERGY ($kJ\ mol^{-1}$)
H—H	436
Li—Li	173
Na—Na	73

The bond energy for Li_2 is larger than for Na_2. This is because overlap is larger between two 2s rather than two 3s orbitals (3s orbitals are more diffuse).

Organometallics

Consider MR compounds, where R = alkyl, aryl, etc. As the group is descended the degree of M^+R^- character increases.

Organolithiums are generally considered to show some covalency, i.e. they are sublimable, distillable and generally quite soluble in organic solvents. They also tend to associate, i.e. $n\text{RLi} \rightarrow (\text{RLi})_n$ as in $(CH_3)_4Li_4$ (see Chapter 5).

Organorubidiums have high melting points and are effectively insoluble in organic solvents, i.e. they possess the general characteristics of ionic compounds.

IONIC BONDING Radius ratio rules provide a guide to the lattice structures that are adopted, e.g. Li^+ is predicted to be four coordinate in most compounds. However, there are exceptions, e.g. LiCl, LiBr and LiI all have the NaCl structure (six coordinate Li^+). Deviations arise because of covalent character in the bonding and the uncertainty in the values of ionic radii used in the radius ratio calculations (page 24).

They are all colourless salts, except when the anion is coloured.

Solubility:

$$MX_{(s)} + nH_2O \rightarrow M^+{}_{(aq)} + X^-{}_{(aq)}$$

This process involves the breaking up of the lattice (endothermic, $\Delta H_{latt.}$) and the formation of hydrated ions (exothermic, $\Delta H_{hyd.}$), i.e. two opposing energy terms.

$$\Delta H_{soln.} = \Delta H_{latt.} + \Delta H_{hyd.}$$

$\Delta H_{soln.}$ usually has a small positive value, i.e. $\Delta H_{latt.}$ and $\Delta H_{hyd.}$ almost cancel each other out. Therefore, dissolving these salts is an endothermic process. But it is the change in Gibbs free energy, ΔG, which dictates whether a process is thermodynamically feasible (page 6).

$$\Delta G = \Delta H - T\Delta S$$

Therefore, the favourability of the process is governed by the increase in entropy upon breaking up the lattice to give free ions in solution. At higher temperatures the $T\Delta S$ term becomes larger and hence ΔG becomes more negative, consequently the solubility increases on increasing the temperature. $\Delta H_{hyd.}$ decreases down the group (see page 6) and this change is larger than the change in lattice energies and so some Rb and Cs salts are not very soluble because $\Delta H_{hyd.}$ is low but the lattice energies are still quite high.

SOLVATION OF IONS The hydrated radius decreases down the group (Chapter 1).

We do not get salt hydrolysis, i.e. in the reaction,

$$[M(H_2O)_n]^+ + H_2O \rightleftharpoons M(H_2O)_{n-1}(OH) + H_3O^+$$

the equilibrium lies largely to the left because the charge:radius ratio of the M^+ ion is not large enough to cause sufficient polarization of the water. This results in solutions of the alkali metal ions being neutral (as long as the counterion is neutral).

FORMATION OF ANIONS Anions, M^-, have been prepared for all the alkali metals except Li. Anion formation is possible because of the exothermic first electron affinity of the alkali metals, i.e. the process,

$$M_{(g)} + e^- \rightarrow M^-{}_{(g)}$$

is thermodynamically favourable. However, the ions produced are thermodynamically and kinetically unstable with respect to formation of oxides and can only be stabilized by the presence of special ligands (cryptands), which surround the metal and inhibit the attack of oxygen.

Summary of Group I metals

1. They are low melting point metals, the melting point decreases uniformly down the group.
2. They are highly electropositive metals and their compounds are predominantly ionic.
3. They exist almost exclusively in the +1 oxidation state and the degree of covalency in their compounds is largest for lithium.
4. The salts generally dissolve in high dielectric constant solvents (water). The degree of solubility increases with increasing temperature.
5. The ions in aqueous solution are neutral provided the anion is neutral.
6. The negative ions, M^-, can be formed.

Group 2

	ELECTRONIC CONFIGURATION	NATURE	PAULING ELECTRONEGATIVITY
Be	[He] $2s^2$	metal	1.57
Mg	[Ne] $3s^2$	metal	1.31
Ca	[Ar] $4s^2$	metal	1.00
Sr	[Kr] $5s^2$	metal	0.95
Ba	[Xe] $6s^2$	metal	0.89

MELTING POINTS The melting points are higher for Group 2 than for Group 1 because, for Group 2, there are two electrons participating in metallic bonding.

There is a decrease in melting point down the group, but it is not regular because the elements have different structures.

IONIZATION ENERGIES Effective nuclear charge increases across a period, therefore the atomic radii of the Group 2 elements are smaller than for their Group 1 counterparts and hence the first ionization energy is greater. The second ionization energy for Group 2 is low enough, because we are still removing an electron from the outer shell, so that the energy needed for formation of the +2 ion can be more than compensated for by the larger lattice energy associated with compounds containing the smaller and more highly charged +2 ions.

DIAGONAL RELATIONSHIPS There exists a 'diagonal relationship' between magnesium and lithium. This refers to the stability of predominantly ionic compounds, e.g. their carbonates, hydroxides and nitrates all readily decompose to the oxides (Na and K nitrates decompose to nitrites). The driving force for the decomposition reaction is the large lattice energies obtained when the small Li^+/Mg^{2+} cations are in combination with the small O^{2-} anion.

The sizes of the Li^+ and Mg^{2+} ions are very similar (size increases down the Periodic Table but decreases from left to right and with increasing positive charge – the increase and the decrease cancel each other out), thus their coordination numbers in ionic lattices are similar (radius ratio rules) and they tend to form stable compounds with the same anions.

Lithium and magnesium show similarities in the degree of ionicity/covalency in their compounds. The polarizing power of the cations leads to significantly more covalent character in, e.g. organometallic compounds, than for later elements in the groups.

BONDING PATTERNS Beryllium is very different from the other Group 2 metals (more so than lithium in Group 1). The compounds of beryllium are nearly always four coordinate covalent species, e.g. $BeCl_2$ and $BeMe_2$ have the same basic polymeric structure,

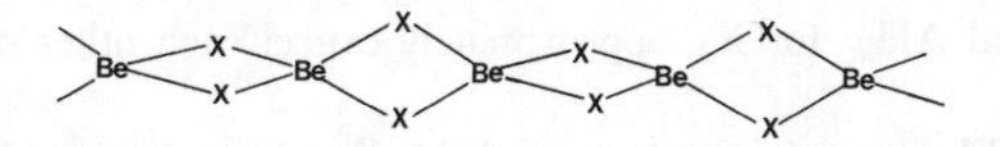

This is because Be^{2+} would be extremely polarizing and polarizes an ionic bond so much as to make it predominantly covalent. However, if the effective size of Be^{2+} is increased it is possible to get ionic salts, e.g. $[Be(H_2O)_4]^{2+}[SO_4]^{2-}$, in which the bonding between Be^{2+} and H_2O has a high degree of covalent character (dative covalent bond involving donation of a lone pair from the water), but that between the 'big beryllium cation', $[Be(H_2O)_4]^{2+}$ and the $SO_4{}^{2-}$ anion is predominantly ionic. Part of the stability of this compound comes from the energy released in formation of bonds between the Be^{2+} and the water, compensating in part for the energy required to remove two electrons from Be.

Magnesium is fairly similar to lithium, in that the ions have fairly similar ionic radii, and the degree of ionicity/covalency exhibited in its compounds is, as for lithium, quite dependent upon the groups to which it is attached, e.g. MgF_2 is very ionic but Mg alkyls are appreciably covalent in character.

Ionization energy and polarizing ability of the ion (charge:radius ratio) decrease down the group and therefore compounds of Ca, Sr and Ba are predominantly ionic in character.

OXIDATION STATES OTHER THAN +2 M^{3+} does not exist for Group 2 as an electron would have to be removed from an inner shell – the large amount of extra energy required to do this would not be counterbalanced by the larger lattice and hydration energies that would be obtained.

Formation of M^+ would involve $ns^2 \rightarrow ns^1$. This obviously requires less energy than $M \rightarrow M^{2+}$ but no Group 2 compound containing an M^+ ion has been isolated. This is because the difference between the lattice energies for MX_2 and MX is greater than the second ionization energy, therefore the increased $\Delta H_{latt.}$ more than offsets the increased ionization energy.

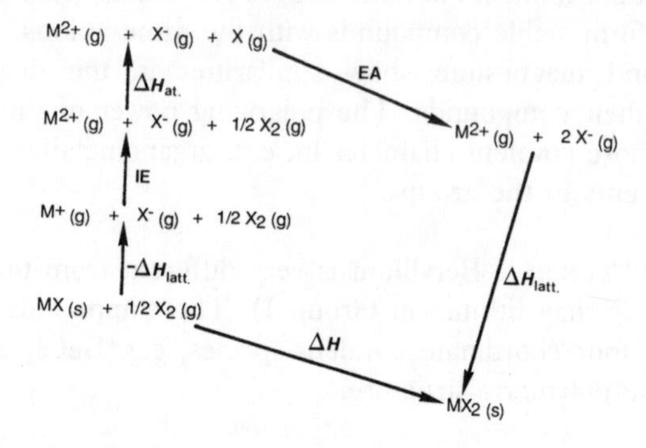

EA and $\Delta H_{at.}$ for X_2, approximately cancel each other out

Thus, if M^+ could occur it would probably disproportionate, i.e.

$$2M^+ \rightarrow M + M^{2+}$$

This is another reason for the 'stability of the full shell electronic configuration'.

$Be \rightarrow Be^{2+}$ would require the most energy in Group 2, and if an M^+ ion is going to occur it is most likely for beryllium. Indeed,

$$BeF_{2(g)} \xrightarrow{\text{high temp.}} BeF_{(g)} + 1/2F_{2(g)}$$

Thus, BeF does exist, albeit as a high temperature species.

HYDRATION The hydration energy is exceptionally large for Be^{2+} and the disparity between Be and the other elements in Group 2 is more than was seen for Group 1. This is because Be^{2+} is much smaller than the other Group 2 2+ ions, and indeed much smaller than Li^+.

For Be^{2+} we get:

- crystalline hydrated salts, e.g. $[Be(H_2O)_4]^{2+}[SO_4]^{2-}$
- predominant salt hydrolysis resulting in the formation of acidic solutions (remember, Group 1 salts are neutral), i.e.

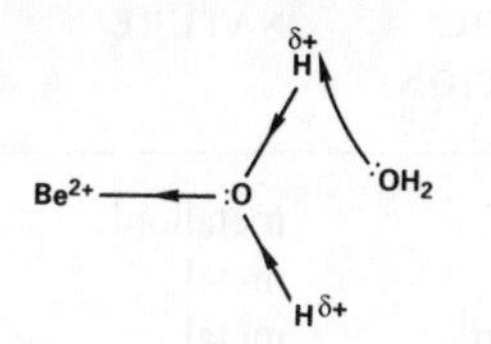

$$[Be(H_2O)_4]^{2+} + H_2O \rightarrow [Be(H_2O)_3(OH)]^+ + H_3O^+$$

For other Group 2 2+ ions, increasing the charge to radius ratio leads to the degree of hydration decreasing down the group (more hydration shells for Be – see page 6), fewer crystalline hydrated salts and no salt hydrolysis, i.e. Mg^{2+} salts are neutral, provided that the anion is neutral.

OTHER COMPLEXES Group 2 compounds can act as Lewis acids in forming complexes with electron pair donating species, e.g. $[Mg(EDTA)]^{2-}$. The EDTA formula is

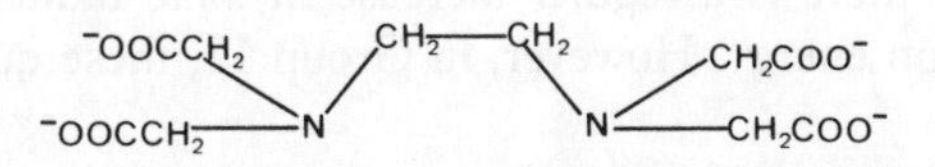

Group 2 compounds cannot act as Lewis bases as they have no lone pairs.

The tendency to form complexes is higher than for Group 1 because of the higher charge on the M^{2+} ion resulting in stronger attractive forces and, therefore, more energy released upon formation of the complex.

Summary of Group 2

1. Melting points for Group 2 elements are higher than for their Group 1 counterparts.
2. Except for Be the Group 2 elements form predominantly ionic compounds.
3. The first ionization energy is higher, but the second ionization energy is lower than for the Group 1 counterpart. The result of this is that Group 2 species exist exclusively in the +2 oxidation state.
4. Be compounds are rather covalent, much more so than Li ones.

5. Be^{2+} ions in aqueous solution are acidic but Mg^{2+} ions give neutral solutions.

Group 13

	ELECTRONIC CONFIGURATION	NATURE	PAULING ELECTRONEGATIVITY
B	[He] $2s^2\ 2p^1$	metalloid	2.04
Al	[Ne] $3s^2\ 3p^1$	metal	1.61
Ga	[Ar] $3d^{10}\ 4s^2\ 4p^1$	metal	1.81
In	[Kr] $4d^{10}\ 5s^2\ 5p^1$	metal	1.78
Tl	[Xe] $4f^{14}\ 5d^{10}\ 6s^2\ 6p^1$	metal	1.62

Boron has a very stable, inert, macromolecular structure. The bonding is somewhere between metallic (free electrons) and covalent (localized electrons), i.e. boron is a semiconductor; all the other elements are metallic. The lack of metallic structure for B reflects its reluctance to form positive ions (high ionization energies).

ATOMIC RADII AND IONIZATION ENERGIES In Groups 1 and 2, as we descend the group, there is a regular increase in ionic radius and a regular decrease in ionization energy. However, in Group 13, these quantities do not vary so regularly.

	ATOMIC RADIUS (pm)	FIRST IONIZATION ENERGY ($kJ\ mol^{-1}$)
B	80	801
Al	125	578
Ga	125	579
In	150	559
Tl	155	589

From the table we can see that there is no general regular increase in atomic size. This is due to interposition of d block elements between Al and Ga and f block elements between In and Tl. Both d and f electrons are much poorer shielders of the outer electrons than are s and p electrons, and the presence of the transition metal series between Al and Ga means that the effective nuclear charge ($z_{eff.}$) felt by the outer electrons in Ga is much larger than would be expected from periodicity considerations. The expected increase in atomic size on going form Period 3 to Period 4 is thus effectively counterbalanced by this large increase in $z_{eff.}$ and the result is that Al and Ga have very similar atomic radii. The same explanation can be used to rationalize the similar sizes of In and Tl, these elements being bridged by the lanthanide, f block, elements.

BONDING PATTERNS As with Groups 1 and 2, boron is very different from the other Group 13 elements.

If we compare the series,

$$Li \rightarrow Be \rightarrow B$$

Li shows some covalent character in its bonding, Be shows a predominance of covalency and B is exclusively covalent.

The B^{3+} cation is unknown as the very high ionization energy required to attain it would not be compensated for by $\Delta H_{latt.}$. Even $[B(H_2O)_4]^{3+}$ is unknown {cf. $[Be(H_2O)_4]^{2+}$ will exist in dilute acidic solutions}. In mononuclear compounds of boron the oxidation state is usually +3. This is because the promotional energy to go from the ground state electronic configuration, $1s^2\ 2s^2\ 2p^1$, to the first excited state, $1s^2\ 2s^1\ 2p^1\ 2p^1$, is low enough to be compensated for by the extra two covalent bond energies released. These bonds are short and strong because they involve the overlap of boron 2p orbitals:

$$2 \text{ bond energies} > \text{promotional energy} \rightarrow 3+ \text{ oxidation state}$$

Aluminium is an ionic/covalent borderline case. 'Bare' Al^{3+} is unknown and anhydrous aluminium compounds tend to be quite covalent because of the highly polarizing nature (high charge:radius ratio) of the Al^{3+} ion. Hydrated ions, which have smaller charge densities, i.e. a 'big Al^{3+} ion', do form ionic salts, e.g. $[Al(H_2O)_6]_2(SO_4)_3$.

Although for boron the +3 oxidation state is exhibited almost exclusively, as we descend Group 13 the relative stability of the +1 oxidation state increases.

$$Al + 3HCl \rightarrow AlCl_3 + 3/2H_2$$

However, the reaction of thallium with HCl gives the stable ionic compound, TlCl. So, for Tl, the +1 oxidation state is favoured over the +3 oxidation state and Tl^{3+} is a strong oxidizing agent (wants to be reduced to Tl^+). This is known as the **inert pair effect** because it can be considered that the pair of valence s electrons are not involved in bonding.

THE INERT PAIR EFFECT Down a group the separation between the valence s and p electrons increases due to the s electrons being able to penetrate nearer the nucleus, and hence feel the greater effect of the increasing nuclear charge. The promotional energy to go from the $ns^2\ np^1 \rightarrow ns^1\ np^1\ np^1$ thus increases down the group. Bond energies decrease down the group as the atomic size increases and more diffuse orbitals are used for bonding. The result of these two factors is that, at the bottom of Group 13, the energy required for promotion of an electron from an s to a p orbital is often greater than the energy that would be released by the formation of two extra bonds. However, when the bond energy becomes stronger (larger than the promotion energy term), then we see the emergence of the +3 oxidation state, e.g.

$$Tl + 3HF \rightarrow TlF_3 + 3/2H_2$$

with TlF_3 existing as a stable compound. This is due to the Tl—F bond being strong (stronger than Tl—Cl bond) and twice the bond energy being greater than the promotional energy.

Note: TlI_3 is in fact $Tl^+I_3^-$, in accordance with the expected weak TlI bond.

Group 13 +2 ions do not exist in salts for the same reason as there are no known +1 ions in Group 2 salts, i.e. they would disproportionate to give the +3 and +1 oxidation states which would overall be more energetically favourable.

AQUEOUS CHEMISTRY OF Al → Tl Although the bare M^{3+} ions are rare, $[M(H_2O)_6]^{3+}$ ions are common, i.e. there are a lot of hydrated salts of the general formula $MX_3.6H_2O$. The +3 ions show extensive solution hydrolysis {as for $[Be(H_2O)_4]^{2+}$, above}, i.e.

$$[Al(H_2O)_6]^{3+} + H_2O \rightarrow [Al(H_2O)_5(OH)]^{2+} + H_3O^+$$

Thus, solutions of Al^{3+} salts are acidic.

HALIDES, MX_3 Except for Tl in TlI_3, which is $Tl^+I_3^-$, they can all have M in the +3 oxidation state. There is a subtle balance between covalent and ionic bonding for these compounds. For example, AlF_3 and $AlCl_3$ are predominantly ionic solids whereas the analogous bromide and iodide exist as covalent dimers of the form Al_2X_6 in the solid state. This is because Br^- and I^- are large, giving rise to lower $\Delta H_{latt.}$ for ionic compounds. Also Br^- and I^- are easily polarized, especially by the small highly charged Al^{3+} ion, therefore introducing a large degree of covalency into the bonding. If we proceed to Ga, only the fluoride, GaF_3, is ionic and the chloride is predominantly covalent. This is because Ga^{3+} is larger than Al^{3+} and hence will have lower $\Delta H_{latt.}$ if ionic, and therefore only the fluoride, which is the smallest anion, is able to bring out the ionicity of Ga. (This lattice energy effect dominates the effect of the larger Ga^{3+} ion being less polarizing than the smaller Al^{3+} ion, which would be expected to lead to Ga compounds having a smaller degree of covalency in their bonding.)

Another example of the importance of lattice energy is the change in bonding of $AlCl_3$ throughout its different phases. In the solid state it exists as ionic $AlCl_3$; in the liquid phase it exists as the covalent dimeric Al_2Cl_6; and in the gas phase there is an equilibrium between the dimer and the monomer.

$$AlCl_{3(s)} \rightarrow Al_2Cl_{6(l)} \rightarrow AlCl_{3(g)} + Al_2Cl_{6(g)}$$

EXAMPLE

Why do we get Al_2Cl_6 but BCl_3?

$AlCl_3$ and BCl_3 if covalent would both be electron deficient (i.e. six electrons in the outer shell of the B). $AlCl_3$ can make up for this deficiency by formation of a dimer.

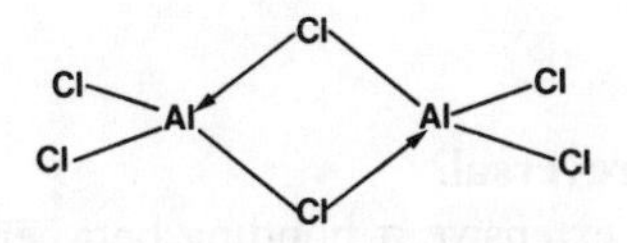

BCl_3, which might also be expected to do this, actually exists as the electron deficient monomer. This is due to two factors:

1. There is fairly strong π bonding ($2p_\pi$–$3p_\pi$) between the vacant B 2p orbital and the filled Cl 3p orbital in the planar configuration, which would be lost upon dimerization – four coordinate B does not have any orbitals available to participate in π bonding.
2. The small size of the B means that a large amount of inter-ligand steric repulsion would result on dimerization, i.e. four large Cl atoms around a small B.

These two factors more than outweigh the advantages of B gaining a full outer shell (i.e. forming four σ bonds instead of three σ bonds + some π).

For Al, the weaker π bonding with Cl ($3p_\pi$–$3p_\pi$) and the larger size (better able to accommodate more groups) mean that dimerization is energetically favourable.

π-BONDING AND LEWIS ACIDITY OF BORON TRIHALIDES (BX_3) The ground state valence electronic configuration of boron is $2s^2\ 2p^1$. Promotion of an electron and hybridization gives three unpaired electrons in three sp^2 hybrid orbitals and a vacant p orbital perpendicular to the sp^2 plane. Boron can pair up these three electrons in three covalent bonds but this still leaves it two short of the 'magical octet'. Thus, BX_3 compounds, with just six electrons in the valence shell of the boron, are termed 'electron deficient'.

The electron deficiency and the vacant p orbital should lead to BX_3 compounds being strong Lewis acids, i.e. they should readily react with electron pair donors to give compounds such as $F_3B \leftarrow NH_3$. Compounds such as this are indeed found but BX_3 tends not to be as strong a Lewis acid as would be expected and the order of Lewis acidity with change of halogen atom is $BBr_3 > BCl_3 > BF_3$, and not the other way around as would be expected from electronegativity considerations (BF_3 should have the most $\delta+$ B). The boron in BF_3 would be expected to be more positive and therefore to have a greater desire to attract a lone pair of electrons than that in BCl_3. However, BF_3 reacts with water to give hydrates such as $BF_3.H_2O$, but the other boron halides are hydrolysed to $B(OH)_3$ (boric acid), i.e. the B—F bonds are quite strong and H_2O is only acting as a donor ligand whereas for the other BX_3 compounds the H_2O cleaves the B—X bond, i.e.

$$BCl_3 \rightarrow H_2O.BCl_3 \rightarrow (OH)BCl_2 + HCl \rightarrow B(OH)_3 + 2HCl$$

EXAMPLE

Why do we get this reversal?

This is because there is extensive π bonding between the vacant p orbital on the boron atom and a lone pair of electrons on the halogen, i.e.

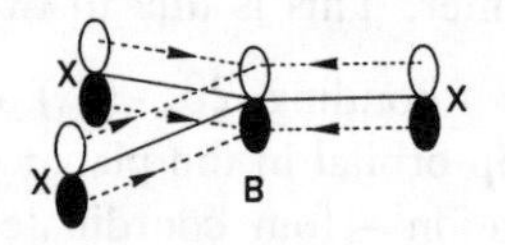

(The π bonding is maximized for a planar structure because the orbitals are better set up for side-on interaction.)

The order of π bond strength (Chapter 4) is

2p–2p > 2p–3p > 2p–4p . . .

This is because:

1. Two 2p orbitals are closer in energy than a 2p and a 3p, etc.; overlap decreasing with increasing energy difference between the π bonding orbitals.
2. 2p Orbitals are quite contracted, therefore good overlap is obtained between them. (The electron density can be thought of as being more 'concentrated' in the lower energy AOs.) As we go to 3p and 4p orbitals the diffuseness increases (less concentrated) and the overlap decreases.
3. π Bonding involves a side-on overlap of orbitals which depends critically on how far apart these orbitals are. As we go from 2p to 3p to 4p, the

size of the atom increases and the distance between p orbitals on adjacent atoms increases, therefore the degree of interaction decreases.

Thus, π bonding is strongest for the B—F bond.

Formation of an adduct with a Lewis base gives a tetrahedral complex (VSEPR). This has no vacant orbital on the boron, therefore no π bonding exists in this molecule. The reluctance of BX_3 compounds to act as Lewis acids arises from the energy required to break the π bonding. The stronger π bonding in BF_3 means that BF_3 is the weakest Lewis acid.

OXIDES B_2O_3 is a macromolecular solid whereas all other Group 13 oxides are ionic, containing discrete M^{3+} ions. The formation of an ionic structure is not a viable option for boron because the large amount of energy needed to form B^{3+} would not be compensated for by $\Delta H_{latt.}$ and the B^{3+} ion would be too polarizing.

Summary of Group 13

1. Boron is metalloid but the other Group 13 elements are metallic.
2. Al and Ga are similar in size and so are In and Tl. However, the latter two are very different in size from the former two.
3. There are no compounds that contain discrete B^{3+} ions.
4. Al compounds are generally ionic/covalent borderline cases, e.g. $AlCl_3$ is predominantly ionic in the solid state but in the liquid and gas phases it is predominantly covalent.
5. The stability of the +1 oxidation state with respect to the +3 oxidation state increases down the group, this is the inert pair effect.
6. Al salts dissolve in water to give acidic solutions but In and Tl salts give neutral solutions.
7. The Lewis acidity of the boron halides shows the order: $BBr_3 > BCl_3 > BF_3$.

Group 14

	ELECTRONIC CONFIGURATION	NATURE	PAULING ELECTRONEGATIVITY
C	[He] $2s^2\ 2p^2$	non-metal	2.55
Si	[Ne] $3s^2\ 3p^2$	metalloid/ semiconductor	1.90

contd

	ELECTRONIC CONFIGURATION	NATURE	PAULING ELECTRONEGATIVITY
Ge	[Ar] $3d^{10}$ $4s^2$ $4p^2$	metalloid/ semiconductor	2.01
Sn	[Kr] $4d^{10}$ $5s^2$ $5p^2$	metallic, although it does have a non-metallic allotrope	1.96
Pb	[Xe] $4f^{14}$ $5d^{10}$ $6s^2$ $6p^2$	metallic	1.87

ALLOTROPY OF CARBON There are two common allotropes of carbon: diamond and graphite,

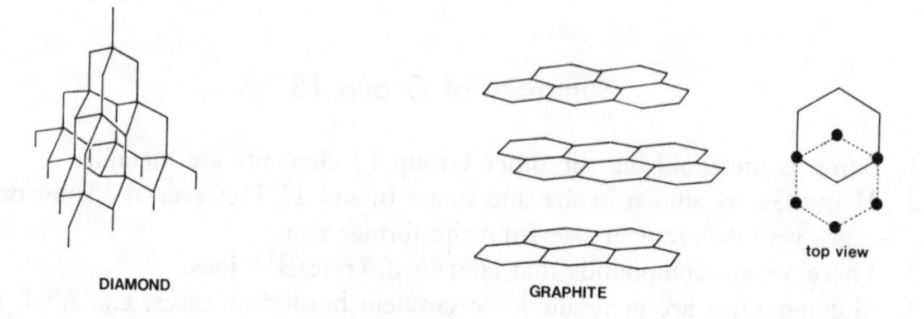

The bonding in diamond involves sp^3 hybridized carbon and all valence electrons tied up in 2c–2e bonds.

In graphite, the carbon atoms are sp^2 hybridized and there is an unpaired electron on each carbon atom residing in a p orbital perpendicular to the sp^2 plane. These p orbitals can overlap between adjacent carbon atoms to give a closely spaced set of molecular orbitals delocalized over each C_n plane, i.e. a half-filled valence band (see page 282) is formed. The electrons can move within this band, under the influence of an applied potential difference, meaning that graphite can conduct electricity along the C_n planes. There is no delocalization of electrons between the layers, therefore graphite is an insulator perpendicular to the layers.

Silicon lies in the middle of Period 3 and is a semiconductor; the elements to the left of it are all metallic and those to the right are non-metallic insulators (see Chapter 9).

On the left-hand side of the period the effective nuclear charge is lower, resulting in:

• the ionization energies being lower, so that positive ions and a sea of

electrons can be formed;
- the orbitals being more diffuse (less concentrated), so that element–element overlap is smaller, hence valence and conduction bands merge to give partially filled bands.

On the right-hand side of the period the effective nuclear charge is higher and:

- the ionization energy is too high to allow the formation of positive ions;
- the orbitals are much more contracted (concentrated), therefore there is strong element–element bonding and the energy gap between valence and conduction bands is large, i.e. the substances are insulators (see Chapter 9).

ATOMIC RADII AND IONIZATION ENERGIES As for Group 13, the d and f blocks make their presence felt by causing the atomic radii and ionization energy for Si and Ge to be quite similar, and likewise for Sn, and Pb.

	ATOMIC RADIUS (pm)	FIRST IONIZATION ENERGY ($kJ\ mol^{-1}$)
C	77	1086
Si	117	786
Ge	122	761
Sn	140	709
Pb	154	715

The result of this is that Si and Ge have similar chemistries; the same also applies for Sn and Pb. Thus, for example, Si and Ge hydrides are thermodynamically stable at room temperature whereas the Sn and Pb analogues decompose under the same conditions. Differences are also seen in the catenating abilities of Si/Ge and Sn/Pb (see below). The differences in the radii are larger than for Group 13 and the chemistries are not as similar as is found in Group 13.

OXIDATION STATES Group 14 elements have the general valence electronic configuration $ns^2\ np^1\ np^1$. This can result in the formation of +2 or +4 ions or of two covalent bonds.

The promotion of an electron produces the excited configuration $ns^1\ np^1\ np^1\ np^1$. There are four unpaired electrons and four covalent bonds can be formed, i.e. the +4/−4 oxidation states. The promotion process requires energy, therefore four bonds will be formed if the energy released in the formation of two extra bonds is greater than the promotion energy; this is always the case for covalent compounds.

Most compounds of Group 14 have a high degree of covalent character, the degree of covalency/ionicity in the bonding depends on several factors:

1. The strength of covalent bonds – this decreases down the group as the atoms get bigger and the orbitals become more diffuse.
2. The ionization energy for formation of a cation – this decreases down the group as the outer electrons get further from the nucleus.
3. Polarizing power of the cation – this decreases down the group as the charge to radius ratio for cations of a particular charge decreases.

Formation of a discrete M^{4+} cation is impossible because of the high ionization energy required not being compensated for by $\Delta H_{\text{latt.}}$ or $\Delta H_{\text{hyd.}}$ and because the positive ions involved would be far too polarizing (electrons are pulled away from the anion and as they are shared between the anion and the cation there is covalency in the bonding). However, Pb does form a number of compounds in the +4 oxidation state, in which the bonding is regarded as predominantly ionic, e.g. PbF_4. This is because, being at the bottom of Group 14, the ionization energies for Pb are the lowest. Pb^{n+} ions are not so small as to be very polarizing and Pb is only able to form relatively weak covalent bonds. (More energy is released in formation of the 'ionic bonds' than would be released by the formation of four covalent bonds, cf. CF_4, SiF_4, and GeF_4 which are covalent molecular gases – ionization energies are higher, ions would be too polarizing, and more concentrated orbitals lead to stronger covalent bonds.) We must stress that even compounds such as PbF_4 do not contain a Pb^{4+} ion and the bonding has some covalent character. It is thought that the charge on Pb in PbF_4 is probably about +2.5.

Oxidation states are just formalisms and do not represent the actual charges on atoms in a compound; real charges depend upon the ionization energies and electronegativities of bonded atoms.

M^{2+} ions can be obtained for Sn and Pb with fairly electronegative groups, e.g. $SnCl_2$ and $PbCl_2$.

As we go down the group the +2 oxidation state becomes more stable than the +4: this is the inert pair effect again.

C Si Ge	Sn	Pb
+4 oxidation state and predominantly covalent	+4, +2 oxidation states ionic/covalent	predominantly +2 predominantly ionic

CATENATION This is the ability of an element to bond to itself. There is a whole world of C—C bonds but there are no naturally occurring Si—Si and Ge—Ge bonds except in the elements.

Catenating ability: C >> Si ≈ Ge > Sn ≈ Pb

e.g. E_nH_{2n+2} for C, *n*: 1–100
Si, *n*: 1–10
Ge, *n*: 1–9
Sn, *n*: 1–2
Pb, *n*: 1

Why is this?

1. The E—E bond energies decrease as you descend the group. This is a consequence of the less good overlap between larger orbitals and the longer E—E bonds.

BOND ENERGIES	(kJ mol^{-1})
C—C	350
Si—Si	200
Ge—Ge	160
Sn—Sn	151

2. The stability to oxidation/hydrolysis shows a general decrease down the group (the trend is reversed between Si and Ge). The explanation of this phenomenon requires a consideration of both thermodynamic and kinetic factors.
   (a) Thermodynamics: thermodynamically the reaction shown below is favourable for all Group 14 elements.

$$E_nH_{2n+2} + H_2O \rightarrow EO_2 + H_2O$$

   If we compare ΔG values for C and Si we find that the reaction for Si is slightly more favourable due to the greater strength of the Si—O bond (see below), however, this difference is small. After Si the E—O bond strength decreases down the group.
   (b) Kinetics: the hydrolysis reaction involves attack of a water molecule at the Group 14 atom; two factors make this less favourable for C:
      (i) C is slightly more electronegative than H, therefore in a hydrocarbon C will be δ− with respect to H and less likely to be attacked by a nucleophilic H_2O molecule. All the other Group 14 elements are less electronegative than H and so the polarization of the bond to H is $E^{\delta+}$—$H^{\delta-}$.
      (ii) Attack by H_2O requires the presence of a fairly low energy, vacant, acceptor orbital on E. In alkanes there are no such vacant orbitals and carbon is unable to increase its coordination

number beyond four. However, for Si, Ge, Sn and Pb, there are low energy vacant d orbitals in the valence shell which can accept a lone pair of electrons from the water molecule. Hence, although the reaction of alkanes with water or air is thermodynamically favourable, there is a **kinetic block** prohibiting this reaction. This is not the case for the compounds of the other Group 14 elements and they are prone to this type of attack.

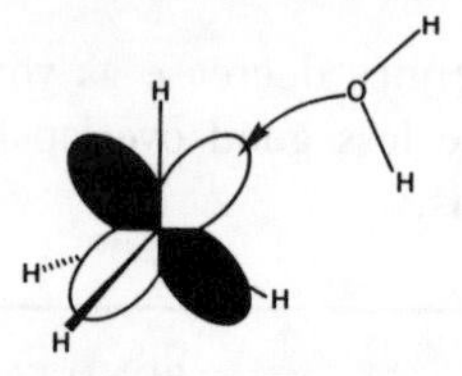

Ge is more electronegative than Si (making the Ge—H bond less polar and therefore less susceptible to attack – kinetic) and the Ge—O bond is weaker than the Si—O bond (making formation of oxides less favourable for Ge – thermodynamic). These factors combine to make germanes less susceptible to oxidation/hydrolysis than silanes.

3. The stability of multiple bonds, i.e. π bonding. Only carbon exhibits extensive π bonding, e.g. C=C in C_nH_{2n}, C=O in CO_2. Among the other Group 14 elements there are few analogues of these compounds. The other elements prefer to form single bonds which, in some cases, can be strengthened by partial π bonding. The reason for π bonding being uncommon for all but carbon is that the π bond strength follows the order:

   $$2p\text{–}2p > 2p\text{–}3p > 3p\text{–}3p > 3p\text{–}4p > \ldots$$

   This order is observed because π bonding involves a side-on overlap of orbitals and depends critically on how far apart the orbitals are. As we go down a group the atoms are larger, the bonds are longer, the p orbitals are therefore further apart and the π bonding is weaker (Chapter 4).

SOME BOND ENERGIES ($kJ\,mol^{-1}$)

C—C	C—O	C=O
300	340	800
Si—Si	Si—O	Si=O
200	370	640

That is, considering bond energies,
C=O > 2C—O but Si=O < 2 Si—O

This is exemplified in the chemistry of C and Si, with CO_2 existing as a discrete gaseous molecule containing C=O double bonds but SiO_2 being a macromolecular solid containing Si—O single bonds.

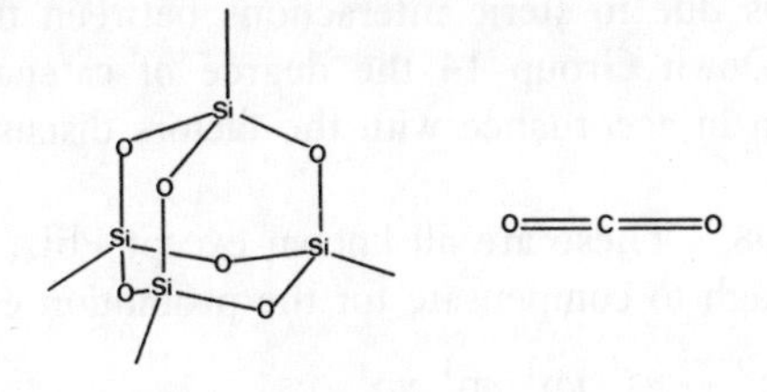

OXIDES CO_2 is a molecular gas but B_2O_3 is a macromolecular solid. This is the result of better π bonding between C and O than between B and O (the orbitals are closer in energy for C and O, the C—O bond is shorter than the B—O bond, the p orbitals on C are more concentrated, i.e. less diffuse), favouring the formation of

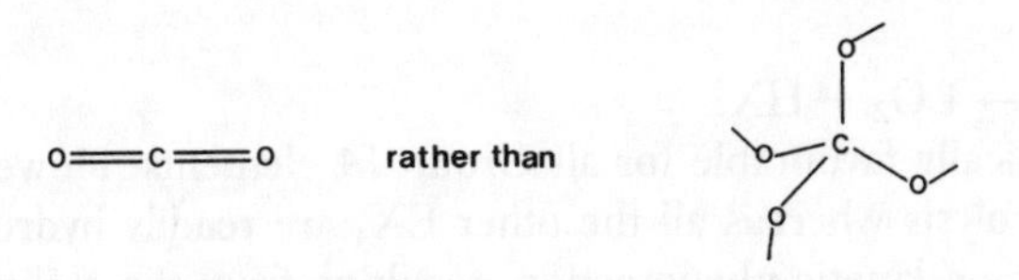

CO_2 is a gas but SiO_2 is a macromolecular solid. The existence of SiO_2 as a discrete gaseous molecule would require $2p_\pi$–$3p_\pi$ π bonding, which is weaker than $2p_\pi$–$2p_\pi$ π bonding (in CO_2), therefore SiO_2 prefers to exist in a structure in which Si and O are linked by single bonds with some multiple bond (d_π–p_π) character (see bond energies above).

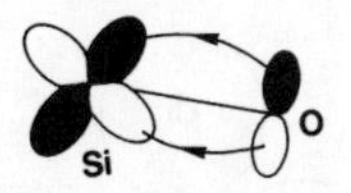

The degree of ionicity of the oxides increases down the group, in accordance with factors discussed earlier, so that PbO is a predominantly ionic solid containing Pb^{2+} ions.

HYDRIDES Catenation is extensive for carbon but far less common for the others (see page 67). The thermal stability decreases down the group due to the strengths of the E—E and E—H bonds, e.g. CH_4 decomposes at 1200 °C, GeH_4 at 280 °C and PbH_4 at 0 °C. The chemical stability with respect to hydrolysis was discussed earlier on page 67.

HALIDES For carbon the catenation varies as follows:

C_nF_{2n+2} n: 1–100 (PTFE – polytetrafluoroethylene)
C_nCl_{2n+2} n: 1–10
C_nBr_{2n+2} n: 1–3

This difference is due to steric interactions between the increasingly bulky halogen atoms. Down Group 14 the degree of catenation decreases for a particular halogen in accordance with the factors discussed above.

MX_4 COMPOUNDS These are all known except PbI_4, the Pb—I bond not being strong enough to compensate for the promotion energy in the process,

$$ns^2\ np^1\ np^1\ np^0 \rightarrow ns^1\ np^1\ np^1\ np^1$$

Except for SnF_4 and PbF_4, which are ionic, all the others are volatile covalent compounds (the orbitals are concentrated enough to form strong covalent bonds).

Hydrolysis

The reaction;

$$EX_4 + H_2O \rightarrow EO_2 + HX$$

is thermodynamically favourable for all Group 14 elements. However, CX_4 is resistant to hydrolysis whereas all the other EX_4 are readily hydrolysed. This must therefore be a kinetic phenomenon, resulting from the ability of Si, Ge, Sn and Pb to extend their coordination number beyond four by using vacant d orbitals. This gives water a fairly low energy pathway for attacking the central Group 14 element. This process is not available for C because of the lack of vacant low energy orbitals, i.e. it cannot extend its coordination number beyond four. There must also be a steric term due to surrounding the small C atom by four halogen atoms, i.e. making it very difficult to squeeze in another group to attack the C.

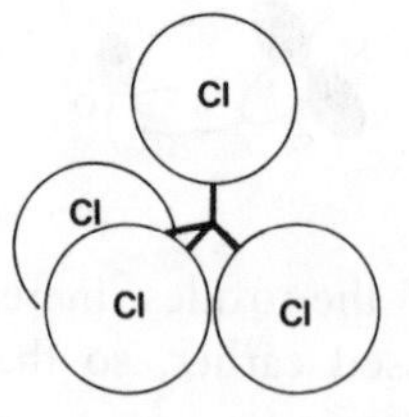

Complex formation

CX_4 cannot act as a Lewis acid as carbon differs from the other elements in having all its valence orbitals tied up in bonding and in its inability to increase its coordination number beyond four. The other EX_4 compounds have valence d orbitals which are capable of accepting electrons, hence they can act as Lewis acids, e.g.

$SiF_4 + F^- \rightarrow SiF_5^-$

Si, Ge, Sn and Pb can show coordination numbers up to six, e.g. in $[SiF_6]^{2-}$. However, $[SiCl_6]^{2-}$ cannot be formed, although the analogous compound in the next group, $[PCl_6]^-$ exists. Thus, although Si is larger than P, the greater negative charge on the chlorines in $[SiCl_6]^{2-}$ means that the inter-ligand repulsion is considerably greater than in $[PCl_6]^-$. In addition to this, it is going to be difficult to bring together negatively charged $[SiCl_5]^-$ and Cl^- ions. $[SiF_6]^{2-}$ can be formed since, although it is still difficult to bring together the negatively charged species, the inter-ligand repulsion is less with the smaller fluorines and also the extra Si—F bond energy is high enough to counterbalance the above endothermic terms.

Summary of Group 14

1. C is non-metallic, Si and Ge are metalloid, Sn and Pb are metallic.
2. Si and Ge are similarly sized as are Sn and Pb but the former two are much smaller than the latter two.
3. The stability of the +2 oxidation state increases down the group.
4. Catenating ability decreases down the group.
5. Extent of π bonding decreases down the group.
6. Except for SnF_4 and PbF_4, which are predominantly ionic solids, all other Group 14 halides are rather volatile covalent compounds.
7. $C(Hal)_4$ are all stable to air or moisture but $Si(Hal)_4$ are not.

Group 15

	ELECTRONIC CONFIGURATION	NATURE	PAULING ELECTRONEGATIVITY
N	[He] $2s^2$ $2p^3$	diatomic gas	3.04
P	[Ne] $3s^2$ $3p^3$	solid, insulator	2.19
As	[Ar] $3d^{10}$ $4s^2$ $4p^3$	metalloid, semiconductor	2.18
Sb	[Kr] $4d^{10}$ $5s^2$ $5p^3$	metalloid, semiconductor	2.05
Bi	[Xe] $4f^{14}$ $5d^{10}$ $6s^2$ $6p^3$	metal	2.02

ALLOTROPY OF PHOSPHORUS There are three allotropes of P: white, red and black. White phosphorus is made up of P_4 tetrahedra held together by van der Waals' forces. The phosphorus atoms here can be regarded as approximately sp^3 hybridized and should, therefore, prefer P—$\hat{P}$—P bond angles of 109°, rather than the 60° that is found. Thus, the P_4 tetrahedra are very strained and white phosphorus is extremely reactive. Much of its chemistry can be rationalized by its desire to produce better bond angles, e.g.

$$P_4 + S \rightarrow P_4S_3$$

The interconversion of the allotropes also follows this rationale, i.e. white phosphorus, when heated, is converted to red phosphorus and then to black phosphorus, both of which have a more open structure where the P—$\hat{P}$—P bond angles tend more towards the ideal tetrahedral angle of 109°.

ATOMIC RADII AND IONIZATION ENERGIES As with Group 14, there is not a regular increase in the atomic size, again due to the interposition of the d and f block elements. This also leads to there not being a regular decrease in ionization energy.

IONIC BONDING Group 15 elements have the general valence electronic configuration

$$ns^2\ np^1\ np^1\ np^1$$

and hence could form 3+, 5+ and 3− ions.

It is possible to form cations of Sb and Bi, i.e. ionic bonding, but compounds with N and P in positive oxidation states involve mostly covalent bonding. Thus, BiF_3 is an ionic solid but NF_3 is a covalent gas.

Sb and Bi in oxidation state +3 form compounds with a high degree of ionic character because:

- the total ionization energy, which decreases down the group, is not excessively high and can be counterbalanced by exothermic terms, such as lattice energy, or hydration energy, in a Born–Haber cycle;
- the cation (quite large) produced will not be excessively polarizing.

As for Group 14, the real charges on the ions in these predominantly ionic compounds are going to be less than the formal oxidation state of the Group 15 atom.

M^{5+} ions do not exist because the exceedingly high total ionization energy required to remove five electrons would not be counterbalanced by hydration or lattice energies. An M^{5+} ion would also be too polarizing to exist as a distinct ion (i.e. the highly charged ion would pull electrons towards itself, reducing its real charge and introducing some covalency into the bonding), thus PF_5 is a covalent gas.

In our journey from left to right across the Periodic Table, Group 15 is the first point where formation of distinct negative ions becomes thermodynamically viable. Discrete E^{3-} ions are only found for nitrogen because:

- the electronegativity (which is related to the desire to form negative ions) increases across a period and decreases down a group;
- the N^{3-} ion is very small and highly charged, therefore there will be a very large lattice energy associated with its compounds.

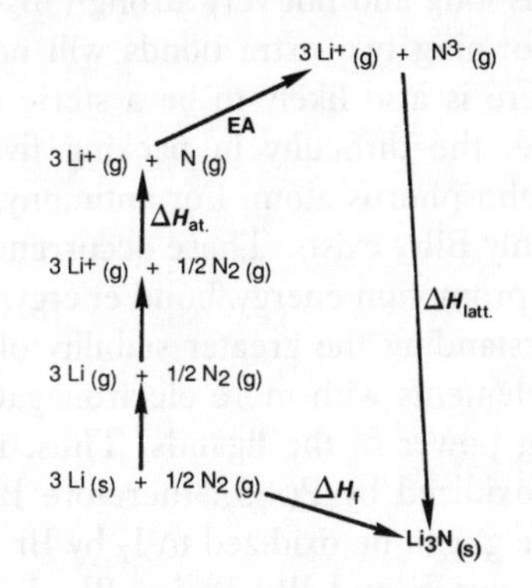

Two very highly endothermic processes are involved in this cycle, i.e. the energy required to dissociate N_2 and the energy needed to add three electrons to a nitrogen atom (even the first electron affinity is slightly endothermic, the second and third are very endothermic). The lattice energy for the formation of the salt must be very large to counterbalance these terms, therefore N^{3-} only occurs in salts with very small cations, i.e. Li^+, Mg^{2+} and Al^{3+} (Na^+, Ca^{2+}, etc., are too big – lattice energy not large enough to compensate for N_2 bond dissociation energy and electron affinity). The lattice energies in these salts are very high because of large charges on the ions, small inter-ionic distances and the compatibility of the radii of the anion and the cation (radius ratio approaching 1), resulting in high coordination numbers.

COVALENT BONDING The ground state electronic configuration is

$ns^2\ np^1\ np^1\ np^1$

which, with three unpaired electrons, allows the formation of three covalent single bonds, i.e. the +3/−3 oxidation states. Promotion of an s electron to the lowest vacant orbital will give

$ns^1\ np^1\ np^1\ np^1\ nd^1$

i.e. five unpaired electrons allowing the formation of five single bonds and the +5 oxidation state. (Note that these orbitals can be considered to hybridize to give five non-equivalent sp^3d hybrid orbitals.)

The stability of the +3 oxidation state increases down the group, in accordance with the inert pair effect, so that Bi exhibits the +3 oxidation state almost exclusively. Thus, it is possible to make PCl_5 but not $BiCl_5$. One of the few Bi^V compounds is the low melting point, extremely reactive solid, BiF_5.

The ease of formation of the +5 oxidation state depends on a balance between the energy required to give five unpaired electrons (promotional energy) and the energy released when two extra bonds are formed. When twice the bond energy is greater than the promotional energy, the +5 oxidation state can exist, e.g. PF_5, PCl_5 and PBr_5 all exist but PI_5 does not because the P—I bond is long and not very strong ($3p_\sigma$–$5p_\sigma$ bond), therefore the energy released in forming two extra bonds will not counterbalance the promotional energy. There is also likely to be a steric factor involved in the non-existence of PI_5, i.e. the difficulty in packing five large iodine atoms round a relatively small phosphorus atom. For antimony, $SbCl_5$ exists but not $SbBr_5$ and for bismuth only BiF_5 exists. These occurrences and omissions can similarly be argued from promotion energy/bond energy/steric considerations.

Another way of understanding the greater stability of the higher oxidation states of the Group 15 elements with more electronegative atoms is in terms of the oxidizing/reducing power of the ligands. Thus, F^- is very difficult to oxidize and will not be oxidized by P(+5): therefore PF_5 exists. I^- is more easily oxidized than F^- (e.g. can be oxidized to I_2 by Br^- or $S_2O_3{}^{2-}$) and will be oxidized by P(+5) to give I_2 and P(+3): i.e. PI_5 does not exist.

Oxidation state +5, with five single bonds, is not accessible to nitrogen. This is due to two factors:

1. There are no 2d orbitals, therefore an electron would have to be promoted to the 3s orbital to give five unpaired electrons. This requires a lot of energy, because the 3s orbital is a further quantum shell out, and involvement of this high energy orbital wouldn't be counterbalanced by the two extra bond energies.
2. The nitrogen atom is small and, similarly to the other elements in the second period, the coordination number cannot be extended beyond four.

However, nitrogen can reach the +5 oxidation state in some compounds, e.g. HNO_3, but these always involve π bonding and therefore have lower coordination numbers (four in HNO_3). The lack of d orbitals requires resonance forms of the type

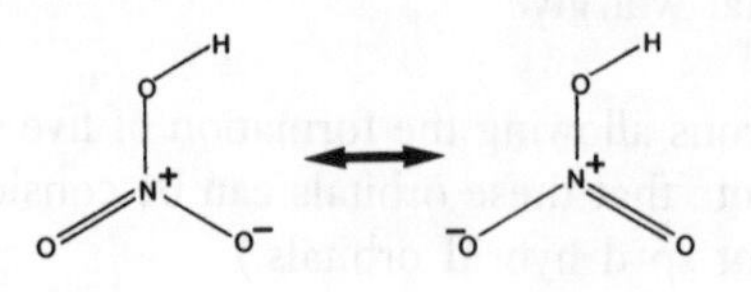

for N^V compounds.

COMPLEX FORMATION R_3E can act as an electron donor, i.e. a Lewis base

$R_3E: \rightarrow BCl_3$

The ability of the R_3E group to donate electrons depends on:

- the electronegativity of R;
- the nature of E.

For example $(Hal)_3N$ are not as good Lewis bases as Me_3N. This is due to the halogen atom being more electronegative than Me, therefore more electron density is withdrawn from the nitrogen, reducing its tendency to donate its electron pair.

Within Group 15, the Lewis basicity generally goes in the order $N < P > As > Sb > Bi$. This is because Lewis basicity depends mainly on two factors:

1. The electronegativity of the Group 15 atom. Nitrogen, as the most electronegative element in the group, holds on to its lone pair most strongly. The willingness to donate the lone pair increases down the group in accordance with the decreasing electronegativity.
2. The ability to form strong dative bonds to other elements. This is related to the diffuseness of the lone pair orbitals of the Group 15 atom – the more diffuse the orbital, the less strong the bonding. Therefore the strength of bonds to other elements decreases down the group.

These two factors act against each other in such a way as generally to give the above order of Lewis basicity (the order can sometimes depend upon the acceptor ability – steric and electronic effects – of the Lewis acid).

R_3N cannot act as a Lewis acid because it has no low-lying vacant orbitals available for the acceptance of electron density. However, the R_3E (E = P, As, Sb, Bi) molecules can act as Lewis acids because of the presence of low energy empty valence d orbitals, e.g.

$PBr_3 + Br^- \rightarrow PBr_4^-$

The $E(Hal)_5$ compounds cannot act as bases (no lone pair) but they can act as acids because they still have vacant orbitals. Thus, complexes such as PF_6^- and SbF_6^- can be formed by the reaction of EF_5 with F^-. EF_6^-, however, does not react with F^- to form EF_7^{2-} even though there are still vacant d orbitals (three of them) in EF_6^-. This is because of the difficulty in forcing together two negatively charged ions and the large steric repulsion that would exist in the resulting seven coordinate complex.

OXIDES The nitrogen oxides are all gases. The phosphorus oxides are solids but, unlike silicon oxide, some of them contain discrete molecular units, e.g. P_4O_{10}.

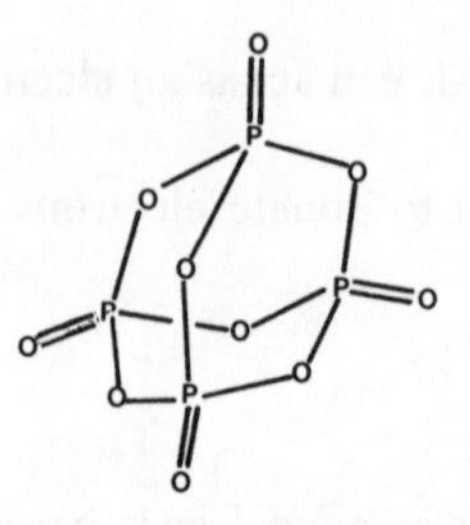

Most phosphorus oxides contain partial P—O π bonding. The greater tendency, than for Si, to form discrete molecules, rather than a macromolecular structure, comes from the desire to form double rather than single bonds to oxygen (e.g. $R_3P{=}O$ exists but $R_2Si{=}O$ polymerizes to give —O—Si—O— chains). P=O is stronger than Si=O for three reasons, which all arise from the increase in effective nuclear charge from left to right across a period:

1. P 3p orbitals are lower in energy than the Si 3p orbitals and therefore closer in energy to the O 2p orbitals.
2. The P 3p orbitals are more concentrated (less diffuse) than in silicon and therefore better π overlap is obtained with the O 2p orbitals.
3. P is smaller than Si, therefore P and O can get closer together, thereby enhancing the π side-on interaction.

As, like Si, forms a macromolecular oxide containing As—O single bonds – π bond strength increases across a period and decreases down a group, therefore there is an approximate 'diagonal relationship' for π bonding ability.

The degree of ionicity of the oxides increases down the group such that Bi_2O_3 possesses a predominantly ionic lattice structure.

CATENATION For Group 14, the catenating ability went in the order $C >> Si \approx Ge > Sn \approx Pb$. For Group 15, the order is $N < P > As > Sb > Bi$.

Why is this?

Let us look at single bonds: for carbon, the C—C single bond is quite strong (see page 67); however, for nitrogen, the N—N single bond is comparatively very weak. This is because there are lone pairs on N but not on C atoms. These lone pairs on adjacent atoms repel each other strongly, so that at the N—N single bond length there is a large electron–electron repulsion, which makes the bond quite weak. Thus, nitrogen does not form extended chains involving solely N—N single bonds.

N N N N

Another reason for nitrogen having a very low catenating ability is the incredible stability of N_2. The dissociation energy for N≡N is about five times the single bond dissociation energy and therefore all compounds involving N—N single bonds are very endothermic (positive ΔH_f), decomposing to give N_2 with the release of a very large amount of energy.

All nitrogen chains, containing more than two atoms, involve some π bonding between the nitrogens, e.g. the linear N_3^- ion, $[N—N—N]^-$.

Greater catenation occurs for phosphorus because of two factors:

1. The π bonding is not as strong between phosphorus atoms. This is because π bonding is a side-on overlap and is a lot weaker for longer bonds involving more diffuse (3p rather than 2p) orbitals.
2. The single bonds are stronger. P atoms are larger than N atoms, therefore a P—P bond is longer than an N—N bond, so that the repulsions between lone pairs on adjacent atoms are smaller for P.

The result of these two factors is that the dissociation energy of P≡P is a lot less than that of three P—P.

BOND	BOND ENERGY (kJ mol^{-1})
P≡P	493
P—P	209

From the table it can be seen that 3(P—P) bond energy = 627 kJ mol^{-1}. This means that P_2 is very unstable, in fact it only exists as a high temperature species in the gas phase. Phosphorus catenates tend to involve single rather than multiple bonds and are not as endothermic as their hypothetical nitrogen analogues.

After P the degree of catenation decreases down the group, reflecting the decreasing element–element bond strengths as we go to larger atoms (larger orbitals give decreased overlap).

Even though the phosphorus, and later Group 15, catenates are more extensive than those of nitrogen, these compounds are still very reactive and extremely unstable with respect to reaction with air and water (similar to their Group 14 analogues). The reactivity of these compounds is due to the availability of fairly low energy vacant d orbitals, ready to accept an electron pair from a Lewis base.

An understanding of the degree of catenation explains the states of the Group 15 elements, i.e. nitrogen exists as N_2 gas, whereas phosphorus exists in three allotropic forms, all of which contain P—P single bonds.

HYDRIDES The compounds EH_3, having a lone pair of electrons, can act as Lewis bases, e.g. $[Cr(NH_3)_6]^{2+}$.

The boiling points of the hydrides vary down the group as shown in the graph.

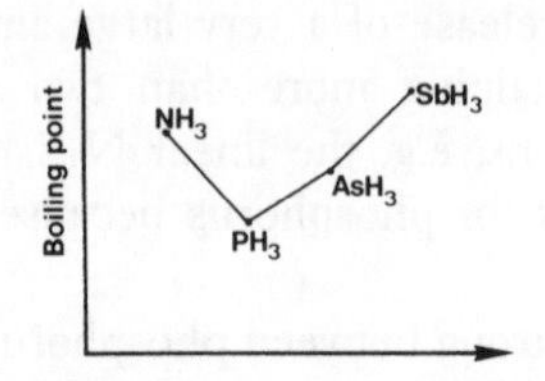

The general increase in boiling points from PH_3 to SbH_3 is mainly due to increasing van der Waals' forces as the central atom gets heavier, has more electrons and is more polarizable. The high boiling point of NH_3 is due to extensive hydrogen bonding; N, being the most electronegative element in the group, polarizes the N—H bond to a large extent, giving a $\delta+$ charge on the H. This positively charged hydrogen can interact with a lone pair on a nitrogen in an adjacent NH_3 molecule, giving quite a strong attractive force between the NH_3 molecules – a hydrogen bond. This is not just a simple dipole–dipole interaction but also involves a more directional, dative covalent, component, i.e. partial donation of the lone pair on N to an H on an adjacent molecule. The larger intermolecular forces in NH_3 mean that more energy has to be supplied (higher temperature) to completely separate the molecules to give a gas.

Summary of Group 15

1. Nitrogen is a gas, phosphorus is a covalent solid, arsenic and antimony are metalloid semiconductors, bismuth is a metal.
2. No discrete +5 ion exists for Group 15. However, essentially ionic compounds exist for Sb and Bi, in the +3 oxidation state, but not for the other Group 15 elements.
3. The stability of the +3 oxidation state with respect to the +5 oxidation state increases down the group from P to Bi.
4. Nitrogen is the only one of the Group 15 elements which is not known to be five coordinate.
5. Lewis basicity generally goes in the order, $N < P > As > Sb > Bi$.
6. Phosphorus oxides are molecular whereas arsenic oxides are macromolecular, but Bi_2O_3 is predominantly ionic.
7. Catenating ability goes in the order, $N < P > As > Sb > Bi$.
8. The boiling points of EH_3 compounds go in the order, $NH_3 >> PH_3 < AsH_3 < SbH_3 < BiH_3$.

Group 16

	ELECTRONIC CONFIGURATION	NATURE	PAULING ELECTRONEGATIVITY
O	[He] $2s^2$ $2p^4$	diatomic gas	3.44
S	[Ne] $3s^2$ $3p^4$	non-metallic solid, insulator	2.58
Se	[Ar] $3d^{10}$ $4s^2$ $4p^4$	metalloid, semiconductor	2.55
Te	[Kr] $4d^{10}$ $5s^2$ $5p^4$	metalloid, semiconductor	2.10
Po	[Xe] $4f^{14}$ $5d^{10}$ $6s^2$ $6p^4$	metallic	2.00

Group 16 has the general electronic configuration ns^2 np^4, which can give rise to oxidation states +6, +4, +2 and −2.

THE −2 OXIDATION STATE This occurs in both covalent and ionic compounds. The −2 anion is actually quite common in Group 16, whereas, in Group 15, only nitrogen can form a discrete anion, and then only in salts with small highly charged cations. This is because:

1. As we go from Group 15 to Group 16, the electronegativity, and hence the desire to form negative ions, increases.
2. The bond dissociation energy of O_2 is a lot lower than that of N_2, i.e. a double versus a triple bond. This will tend to make formation of negative ions relatively easier, i.e. the $\Delta H_{at.}$ term in the Born–Haber cycle is less endothermic for oxygen. For the other members of Group 16 the bond dissociation energies are not very different from those of their Group 15 counterparts.
3. For Group 16 we only have to form a −2 ion in order to give a full octet, as opposed to a −3 ion for Group 15, therefore the overall electron affinity involved in formation of the ion is less endothermic for Group 16.

Because the electron affinity term is less unfavourable, as we are only forming the −2 ion, we do not require such a large lattice energy for the ionic solid to compensate for this endothermic term. Thus, compounds containing O^{2-} can be formed with a much wider range of metal ions, hence we are no longer constrained by the need to have very small metal cations.

Sulphur, selenium and tellurium can also exist as discrete −2 ions but, because of the comparatively lower lattice energy associated with the larger ions, these are only formed with larger, electropositive metals, such as caesium. Caesium favours such an ionic compound because Cs to Cs^+ is relatively easy. Also, the sizes of Cs^+ and E^{2-} (E = S, Se) are quite similar, therefore the radius ratio Cs^+/E^{2-} is quite close to 1, which results in better packing being obtained in the lattice. This counteracts the decrease in lattice energy resulting from the larger inter-ionic distance (see Born–Landé equation – page 24).

POSITIVE OXIDATION STATES M^{6+} ions cannot be obtained for the same reason that we do not get M^{5+} in Group 15. Te and Po have relatively low ionization energies and are large enough so that the cation produced is not so polarizing as to induce a large degree of covalency in the bonding. This means that we can get predominantly ionic compounds containing these elements in the +4 oxidation state, e.g. TeO_2 and PoF_4, where they are in combination with very electronegative elements. Even though these compounds exhibit a high degree of ionic character it is unlikely that they contain discrete +4 cations (the polarizing power of such a cation means that the actual charge will be less than +4, i.e. some covalent character in the bonding).

For the elements above Te in Group 16, the ionization energy for the formation of the +4 cation would be too high to be counterbalanced by other terms ($\Delta H_{latt.}$, $\Delta H_{hyd.}$) and even compounds such as SF_4 (a gas at room temperature) involve predominantly covalent bonding and are molecular.

VARIATION IN OXIDATION STATE AND COVALENT BONDING The ground state electronic configuration for Group 16 is

$$ns^2\ np^2\ np^1\ np^1$$

allowing the formation of two single covalent bonds, i.e. the +2 or −2 oxidation state. The first promotion takes an electron from a p into a d orbital,

$$ns^2\ np^1\ np^1\ np^1\ nd^1 \quad \text{(four hybrid orbitals)}$$

generating four unpaired electrons, i.e. the +4 oxidation state. A second promotion takes an electron from an s orbital to another d orbital,

$$ns^1\ np^1\ np^1\ np^1\ nd^1\ nd^1 \quad \text{(six hybrid orbitals)}$$

giving six unpaired electrons, i.e. the +6 oxidation state.

Oxygen cannot exist in positive oxidation states higher than +2. This is because the lowest vacant orbital is the 3s and the energy to promote an electron to this orbital (a further quantum shell out) would not be counterbalanced by the extra bond energies obtained (i.e. promotion energy is greater than twice the bond energy).

For the elements sulphur to polonium, fairly low energy d orbitals are available and +4 and +6 oxidation states are observed. Similar to Group 15, the extent of formation of higher oxidation states depends on a balance between electronic promotion energies and bond energies/steric factors.

The covalent bonding will always be strongest for S because of the use of lower energy, less diffuse, orbitals and the formation of shorter bonds. Therefore the +6 oxidation state is quite available for S, as in SF_6 and H_2SO_4.

Why is SF_6 formed but not SCl_6?

Two factors need to be considered:

1. Steric factors: the inter-ligand repulsion is greater between six Cl atoms than between six smaller F atoms.
2. Electronic factors: the S—Cl bond energy is weaker than the S—F bond energy.

However, SF_5Cl and SF_4Cl_2 do exist because of the formation of strong S—F bonds and smaller inter-ligand repulsion compared with SCl_6. There are also many compounds involving sulphur in oxidation state +6 but with lower coordination numbers and therefore reduced steric interactions, e.g. SO_2F_2 and SO_3; here the stability will be determined almost totally by bond/promotion energy considerations (i.e. is promotion energy less than the energy released in the formation of extra bonds?).

As we go down Group 16, the +6 oxidation state becomes less stable as bond energies decrease (larger atoms) and do not counterbalance promotional energies. At the bottom of the group, the inert pair effect is again exhibited, thus SeF_6 and TeF_6 exist but PoF_6 does not. Also, TeF_6 is much less stable than SF_6 and readily decomposes to $TeF_4 + F_2$.

For compounds with the later halogens bond energies are lower and steric factors are larger, therefore there is no Group 16 bromide with an oxidation state greater than +4, i.e. EBr_6 does not exist.

As for Group 15, stabilities of the higher oxidation states can also be understood from a consideration of oxidizing/reducing power of the ligands. Thus, S(+6) is a strong enough oxidizing agent to oxidize Cl^- to Cl_2, being reduced to S(+4) in the process, but not to oxidize F^- to F_2.

There is also a kinetic term to be considered in the stability of the various oxidation states, i.e. SF_6 is stable to hydrolysis but SF_4 is not, although both reactions are thermodynamically favourable. In SF_6, although there are three vacant valence d orbitals available for attack by water, six F ligands around the S mean there is not enough room around the S for attack of a water molecule. Therefore the hydrolysis process would have to involve breaking of an S—F bond to generate a vacant coordination site, i.e. it would be a high activation energy process. However, for SF_4, there are four vacant low energy d orbitals and enough room (only four fluorines) for nucleophilic attack by H_2O. Hence, hydrolysis is kinetically as well as thermodynamically viable for SF_4.

COMPLEX FORMATION Oxygen is only ever a Lewis base because it has no low-lying vacant orbitals. The other elements can exhibit both Lewis acidity and basicity when in +2 and +4 oxidation states, e.g.

$$SCl_2 + Cl_2 \rightarrow SCl_3 + Cl^-$$

has SCl_2 acting as a Lewis base and

$$SF_4 + F^- \rightarrow SF_5^-$$

involves SF_4 as a Lewis acid.

In the +6 oxidation state there are no lone pairs of electrons available for donation, i.e. no Lewis base behaviour. The steric unfavourability of having the seven coordinate element also means that EX_6 compounds do not act as Lewis acids, even though there are vacant d orbitals available for attack.

CATENATION There is a similar trend to that seen in Group 15, i.e. O < S > Se > Te > Po. The position of oxygen is due to the weakness of the O—O single bond as a result of strong lone pair–lone pair repulsions. Inter-electron repulsion is greater than for Group 15 because of the presence of more valence electrons in, e.g. O as opposed to N.

BOND	BOND ENERGY ($kJ\ mol^{-1}$)
O—O	146
N—N	167
S—S	226
Se—Se	172

The bond energy in O_2 is more than twice the single bond energy due to reduced repulsion between lone pairs in sp^2 (O=O) rather than sp^3 (O—O) orbitals. So, the lone pairs in sp^2 orbitals are bent back further from the O—O bond resulting in less repulsion between the lone pairs on adjacent atoms.

Therefore, it is more favourable to just have O_2, which is incapable of further catenation without breaking the double bond. Other catenates, O_3, H_2O_2, O_2F_2 and O_4F_2 (decomposes at −183 °C!), tend to retain some degree of multiple bond character.

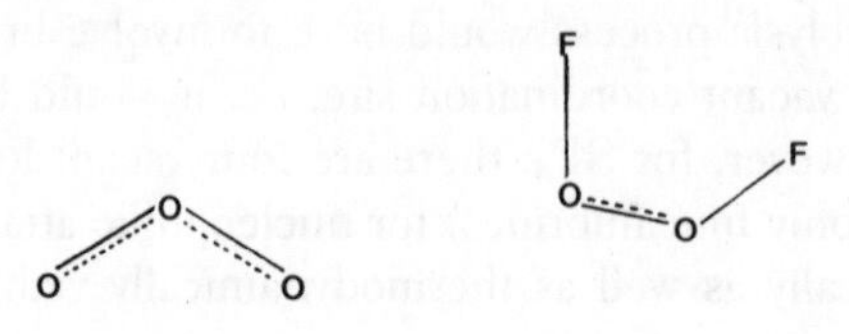

Sulphur forms many rings and chains containing S—S bonds, e.g. S_8. After sulphur the degree of catenation decreases down the group because of the weaker bond energies. Thus we can get,

H_2S_n n: 1–6
H_2Se_n n: 1–3
H_2Te_n n: 1 or 2

OXIDES The most important oxygen oxide is O_2. The structures of some other Group 16 oxides are:

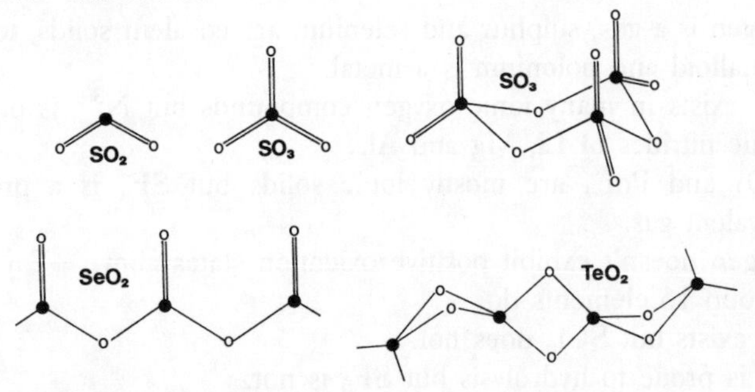

The difference in the structures of the oxides arises because of the decrease in the strength of the E=O bond relative to the E—O bond as the group is descended, i.e. the decrease in π bonding ability as we go down the group (see Group 14 and 15 oxides). Thus, once again, there is an approximate diagonal relationship in oxide structures, i.e. between C and S, and between P and Se. Carbon and sulphur both exhibit discrete molecular oxides which contain multiple bonding, i.e. SO_2 and CO_2. Phosphorus and selenium have less tendency towards E=O, i.e. P_4O_6 contains only P—O single bonds (with some strengthening due to d_π–p_π interaction) and P_4O_{10} contains a mixture of single and double bonds, SeO_2 in the gas phase has the same structure as SO_2 but forms a macromolecular solid with Se—O single bonds. The ionicity of the oxides increases down the group so that, like Bi_2O_3, PoO_2 is a predominantly ionic solid.

HYDRIDES Group 16 hydrides (EH_2, RSH) can act as Lewis bases due to the presence of lone pairs, e.g.

$$C_2H_5SH + TiCl_4 \rightarrow C_2H_5SH.TiCl_4$$

The basicity is lower than for their Group 15 analogues due to the higher electronegativity of the Group 16 atom, therefore the greater unwillingness to let go of their lone pairs.

The higher electronegativity of the Group 16 atom has two additional consequences:

1. The greater polarization of the E—H bond means that EH_2 can act as a protonic acid, i.e. it has a larger tendency to produce HE^- and H_3O^+ in water than its Group 15 analogue.
2. The boiling point of H_2O is very high due to the presence of substantial hydrogen bonding. Hydrogen bonding is stronger for H_2O than for NH_3 (O is more electronegative than N), thus the boiling point of H_2O, at 100 °C, is much higher than that for NH_3 (−33 °C).

Summary of Group 16

1. Oxygen is a gas, sulphur and selenium are covalent solids, tellurium is a metalloid and polonium is a metal.
2. O^{2-} exists in many ionic oxygen compounds but N^{3-} is only found in ionic nitrides of Li, Mg and Al.
3. TeO_2 and PoF_4 are mostly ionic solids but SF_4 is a predominantly covalent gas.
4. Oxygen doesn't exhibit positive oxidation states above +2 yet the other Group 16 elements do.
5. SF_6 exists but SCl_6 does not.
6. SF_4 is prone to hydrolysis but SF_6 is not.
7. The catenating ability of the Group 16 elements is: $O < S > Se > Te > Po$.
8. The structures of C and S oxides are similar, as are those of P and Se.
9. The Group 16 hydrides can act as both Lewis acids and bases.
10. H_2O has a higher boiling point than the other Group 16 hydrides and NH_3.

Group 17

	ELECTRONIC CONFIGURATION	NATURE	FIRST IONIZATION ENERGY ($kJ\,mol^{-1}$)	COVALENT RADIUS (pm)	PAULING ELECTRONEGATIVITY
F	[He] $2s^2\,2p^5$	diatomic gas	1681	72	4.00
Cl	[Ne] $3s^2 3p^5$	diatomic gas	1257	99	3.16
Br	[Ar] $3d^{10}\,4s^2\,4p^5$	diatomic liquid	1140	114	2.96
I	[Kr] $4d^{10}\,5s^2\,5p^5$	diatomic solid	1008	133	2.66

Unlike the groups to the left of the halogens, where the later elements had mostly metallic character, iodine is a completely covalent solid containing discrete I_2 molecules. Iodine is a solid because the van der Waals' forces between the molecules are large due to the large amount of electron density associated with iodine and its polarizability. Iodine is not metallic because formation of a metallic lattice involves 'ionization' of the atoms; the ionization energy is higher in Group 17 than for the groups to the left and, as such, cannot be counterbalanced by the energy released in formation of a metallic lattice.

As we have gone from Group 13 to Group 17, the difference between the third and fourth period elements has increased. For Group 13, which comes directly after the transition metal series, the elements are most influenced by the decreased shielding due to the d electrons, that is the increase in effective nuclear charge which occurs across the transition metal series. Because of the poor shielding due to the d electrons, the increase in size as we go from the third to the fourth period is cancelled out (see earlier), so that aluminium and gallium are very similar in size. Now, as we go from Group 13 to Group 17 we are adding electrons to the p shell and we get an increase in effective nuclear charge due to the p electrons not shielding each other perfectly. This increase in effective nuclear charge is different between Period 3 and Period 4, being larger for Period 3 because the outer electrons are closer to the nucleus and more influenced by changes in the nuclear charge. Thus, the difference between the size of Period 3 and Period 4 elements increases as we get further from the transition metals (i.e. go across a main group series), therefore chlorine and bromine are very different in size.

The ionization energies are all quite large because of the high effective nuclear charge at the right-hand side of the Periodic Table. These ionization energies decrease irregularly down the group and so $+1$ ions would be expected to be most likely at the bottom of the group. X^+ cations do not exist as discrete ions in ionic solids, the larger ionization energies not being counterbalanced by lattice energies (small for large ions such as I^+). However, Br^+ and I^+ can exist when complexed by donor molecules. Thus, $[I(pyridine)_2]^+$ and $[Br(quinoline)_2]^+$ both exist in predominantly ionic solids. Fluorine and chlorine do not form positive ions, under normal conditions, because of the high ionization energies involved.

OXIDATION STATES All of the halogens except F, which only exists in oxidation states -1 or 0, exhibit all the oxidation states -1, 0, $+1$, $+3$, $+5$, $+7$.

All the halogens can exist as discrete -1 anions. Because they are on the right-hand side of the Periodic Table, their electronegativities and electron affinities are high, they therefore have a strong tendency to gain an electron.

The degree of covalency increases down the group. The I^- ion, being the largest, is the most polarizable and therefore has the greatest tendency to, at least partially, give up its electrons to a cation (shared electrons = covalency). The most covalent compounds occur when the halogens are in combination with highly polarizing cations: AlF_3 is ionic; $AlCl_3$ forms a predominantly ionic solid but melts to give covalent Al_2Cl_6 dimers; and Al_2Br_6 and Al_2I_6 are covalent dimers.

POSITIVE OXIDATION STATES The ground state electronic configuration for Group 17 is

$$ns^2\ np^2\ np^2\ np^1$$

and this will give rise to the +1 oxidation state when the halogen is in combination with a more electronegative element.

The following promotions can produce +3, +5 and +7 oxidation states,

$ns^2\ np^2\ np^1\ np^1\ nd^1$	gives +3
$ns^2\ np^1\ np^1\ np^1\ nd^1\ nd^1$	gives +5
$ns^1\ np^1\ np^1\ np^1\ nd^1\ nd^1\ nd^1$	gives +7

The existence of these oxidation states depends on a balance between promotional energies, bond energies and steric factors, as for the other groups. For compounds just containing single bonds, the coordination numbers involved in the higher oxidation states are so high that steric factors will, in a lot of cases, outweigh any promotion/bond energy considerations.

Why does IF_7 exist while ICl_7 does not?

IF_7 has a fairly strong I—F bond (large ionic contribution) and the small size of the F atoms means that it is not too difficult to pack seven of them around the I. ICl_7 probably doesn't exist because the I—Cl bond is too weak to counteract the large amount of inter-ligand repulsion resulting from packing seven relatively large Cl atoms around the I.

Although IF_7 exists, ClF_7 and BrF_7 do not due to steric factors (more difficult to pack seven fluorines around a smaller atom). This appears to be contrary to what we have seen in other groups, in that here the stability of the higher oxidation states is increasing down the group (i.e. against the inert pair effect). However, when we go to multiply-bonded compounds, e.g. those involving oxygen, the coordination numbers are necessarily lower and the stability of the highest oxidation state tends to be higher for chlorine, which can form stronger bonds. Therefore ClO_4^- (oxidation state +7) is more stable than BrO_4^- and IO_4^- with respect to reduction.

The inert pair effect is seen for the halogens in that the +5 oxidation state becomes more stable relative to the +7 as the group is descended. This is as long as steric factors are not very important. Thus, the equilbrium,

$$2H^+ + 2e^- + ClO_4^- \rightleftharpoons ClO_3^- + H_2O$$

Oxidation state of Cl   +7   +5

lies mostly towards the left-hand side but, for iodine, the equivalent equilibrium lies further to the right.

E—E BOND ENERGIES

	F_2	Cl_2	Br_2	I_2
Bond energy (kJ mol^{-1})	158	242	193	151

For covalent bonding we normally expect bond energies to decrease down a group. Thus the F_2 bond energy is anomalously low. This is because of the large amount of repulsion between three lone pairs on one atom and three lone pairs on the other atom. This effect is most significant for F_2 where the F atoms are small and the lone pairs are close together.

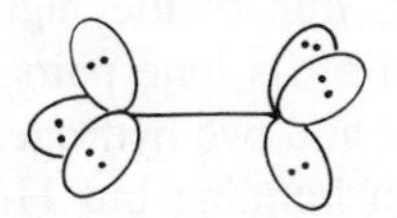

The bond energies from Cl_2 to I_2 follow the order expected for decreased overlap of the p orbitals forming the σ bond (i.e. 3p–3p > 4p–4p > 5p–5p).

Why is fluorine so reactive?

F_2 reacts with all elements except He, Ne and Ar. All reactions of F_2 involve breaking the F_2 bond.

The reactivity of fluorine is due to a combination of kinetic and thermodynamic factors, both of which are related to the weakness of the F—F bond (see above):

1. Kinetics: any transition state for a reaction involves at least partial breaking of the F—F bond and as relatively little energy is required to do this, the activation energy (energy to generate the transition state) is quite small.
2. Thermodynamics: overall, in any reaction, the small input of energy to break the F—F bond is more than offset by the energy released in the formation of strong bonds, ionic or covalent, to other elements.

In predominantly covalent compounds, the small size of F and its concentrated valence orbitals means that we get good overlap with orbitals on other atoms. Also, the electronegativity of F means that most of the bonds it forms are quite polar and are strengthened by an ionic contribution.

In predominantly ionic compounds, F^- is small for an anion and, in accordance with radius ratio rules and the Born–Landé equation, we get high coordination numbers and large $\Delta H_{latt.}$.

HYDRIDES Hydrogen fluoride has a relatively high boiling point due to the existence of hydrogen bonding in liquid HF. The other hydrogen halides do not show this behaviour to any great extent.

The boiling points of the Period 2 hydrides increase from NH_3 to H_2O, due to the increase in electronegativity from left to right, therefore it might reasonably be expected that the boiling point of HF (19.5 °C) would be higher again; in fact, it is lower than for H_2O (100 °C). This is because the polarity of the E—H bond is not the only thing that is important in determining the strength of hydrogen bonding; three additional factors must be considered:

1. As mentioned above, hydrogen bonding is not just a simple dipole–dipole interaction but also involves E partially donating its lone pair to a hydrogen atom, i.e. a partial dative bond. NH_3 and H_2O are stronger Lewis bases than HF, due to the high electronegativity of F, i.e. its unwillingness to let go of its lone pairs.
2. NH_3 and H_2O have three and two hydrogen atoms, respectively, which can participate in hydrogen bonding but HF has but one. Therefore NH_3 and H_2O can exhibit more multi-directional hydrogen bonding than can HF (more hydrogen bonds and therefore more energy is required to break apart molecules).
3. In the vapour phase HF is polymeric but gaseous NH_3 and H_2O just contain discrete molecules. Thus vaporization of liquid HF does not involve complete breaking of all hydrogen bonds and less energy is therefore required for vaporization, i.e. lower boiling point.

For NH_3 versus H_2O, the larger dipole wins out over factors (1) and (2) to give H_2O the higher boiling point.

We have said above that the other hydrogen halides do not exhibit hydrogen bonding to any real extent. Although Cl is more electronegative than N, this is not the only factor that determines whether hydrogen bonding is likely to be present. This is because the Cl lone pair is too diffuse to overlap efficiently with the $H^{\delta+}$ on an adjacent molecule (this confirms the picture of hydrogen bonding as not being just an electrostatic effect).

The high electronegativity of the Group 17 atoms means that, despite the presence of three lone pairs, HX compounds do not act as Lewis bases but rather as strong protonic acids, e.g.

$$HCl + H_2O \rightarrow H_3O^+{}_{(aq)} + Cl^-{}_{(aq)}$$

HF, unlike the other hydrogen halides, is not a strong acid in water; this is due to two factors, both reliant upon the hydrogen bonding ability of F.

1. We have the equilibrium

$$HF + H_2O \rightleftharpoons [H_3O]^+F^-{}_{(aq)} \rightleftharpoons [H_3O]^+{}_{(aq)} + F^-{}_{(aq)}$$

There is a large degree of association between the $[H_3O]^+$ and F^- which reduces the free concentration of $[H_3O]^+$ and hence the acidity. The strong interaction is the result of the small size of the two ions {cf. lattice energy where electrostatic interaction is $\propto 1/(r^+ + r^-)$} and hydrogen bonding.

2. There is also another equilibrium existing in solution,

$$a\text{HF} + b\text{F}^- \rightleftharpoons \text{HF}_2^-, \text{H}_2\text{F}_3^-, \text{H}_3\text{F}_4^-, \text{etc., i.e. } (\text{HF})_n\text{F}^-$$

where $a > b$

i.e. hydrogen bonding between F^- and one or more HF molecules generates a series of aggregates.

The strength of any acid is in essence judged by the position of the equilibrium,

$$HF + H_2O \rightleftharpoons F^- + H_3O^+$$

$$K_a = \frac{[F^-][H_3O^+]}{[HF]}$$

Formation of aggregates such as $[(HF)_2F]^-$, i.e. $H_2F_3^-$, lowers the concentration of HF more than that of F^-, forcing the equilibrium back to the left (Le Chatelier's principle), i.e. weakening the acid since the H_3O^+ concentration is reduced.

The acidity of HF decreases with increasing dilution, which is contrary to other aqueous acids. Increasing dilution should result in more dissociation of HF to give more F^- and more H_3O^+ (Le Chatelier's principle). However, these F^- ions can form complexes with more than one HF and so, the more F^- we produce, the more aggregation (i.e. removing more HF) we get and the more the equilibrium is forced back to the left (decreasing the concentration of H_3O^+).

HCl is a much stronger acid than HF because the larger size of Cl^- means that interaction between $[H_3O]^+$ and Cl^- is not so strong and extensive ion pairing does not occur. Also, HCl/Cl^- does not participate in very strong hydrogen bonding, therefore we do not get aggregation to form HCl_2^-, etc.

INTERHALOGEN COMPOUNDS Apart from a few ternary compounds (e.g. $IFCl_2$), these are all of the general empirical formula XY_n, where n is an odd number so that there are no unpaired electrons. For $n \geq 3$, Y is always the lighter atom, e.g. we get IF_3 but not FI_3. This is in accord with size and oxidation state considerations:

1. The heavier atoms are bigger and it is relatively easy to pack lots of little atoms around a big atom but not lots of big ones around a small one.
2. In IF_3 we have +3 and −1 oxidation states for I and F respectively; the

inverse compound would involve F(−3) and I(+1). Although the halogens, except F, can exist in positive oxidation states up to +7 there are no negative oxidation states above −1.

The bond energies are related to the electronegativity difference between the component atoms such that the strongest bond is in IF. This then suggests a large degree of ionic character in the bonding, i.e. we have resonance forms,

$$\underset{1}{X{-}Y} \longleftrightarrow \underset{2}{X^+Y^-} \longleftrightarrow \underset{3}{X^-Y^+}$$

If X is the heavier halogen then form **3** will be of negligible importance but **2** will be a major contributor to the overall bonding picture.

This also applies to the other interhalogens, XY_n, with $n \geq 3$. The bond energies here are lower than for their diatomic counterparts, since with more ligands around there is greater inter-ligand repulsion and therefore the ligands have to move further from the central atom.

The tendency to form XY_n, with higher n values, depends on three factors:

1. Size.
2. Ease of oxidation of the central halogen.
3. Electronegativity difference between the halogens.

Factors (2) and (3) are related.

IF_7 exists but BrF_7 does not. This is because:

- Br is smaller than I and so it is harder to pack more ligands around.
- Br^- is harder to oxidize than I^-, i.e. $2I^- + Br_2 \rightarrow 2Br^- + I_2$, i.e. it is harder to obtain Br in the +7 oxidation state.
- the Br—F bond is not as strong as the I—F (more polar = more ionic contribution) bond and forming two more of them will not release enough energy to offset the promotion energy.

POLYHALIDE IONS In accordance with radius ratio rules, the stability of these ions in the solid state is increased by having large cations (r^+/r^- close to 1 allows for maximum coordination number). The stability is greater for more symmetrical polyhalide ions and those with large central atoms, i.e. stability goes: $I_3^- > IBr_2^- > ICl_2^- > I_2Br^- > Br_3^- > BrCl_2^- > Br_2Cl^- > Cl_3^-$; there is no F_3^-.

These trends can be understood by looking at the bonding. Employing a valence bond approach we can draw resonance forms

$$I{-}I\ I^- \longleftrightarrow I^-\ I{-}I \longleftrightarrow I^-\ I^+\ I^-$$

The bonding in the first two forms involves polarization of the I_2 by the I^-, which induces a dipole (forces the electrons in the I_2 bond away from it), and therefore an attractive force, i.e.

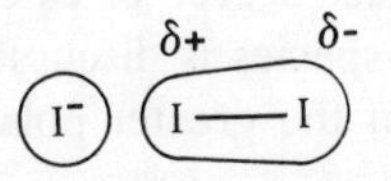

The bonding in the third form is purely electrostatic. The actual situation involves a combination of all three forms.

This explains the non-existence of F_3^- in that, with regard to the first two forms, F_2 is not very polarizable at all and, with regard to the third form, F is unknown in a positive oxidation state.

In general, the stability of these compounds follows the polarizability of the X—Y group in XYX^- and the ease with which Y can attain a positive oxidation state.

Nuclear quadrupole resonance studies (see page 314) show that in all these XY_2^- compounds there is more negative charge localized on the outer atoms, in accordance with the above resonance forms.

THE POLYIODIDES We only get compounds X_n^-, $n > 3$, for iodine, e.g. I_5^-.

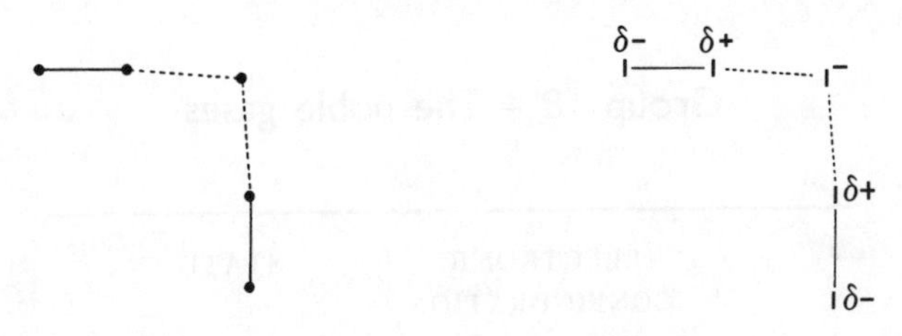

The formation of such compounds relies on the ease of polarizability of I_2.

POLYCATIONS Many interhalogens undergo autoionization, e.g.

$$ICl_3 \rightarrow ICl_2^+ + ICl_4^-$$

Alternatively, we can have reaction with a Lewis acid, e.g.

$$SbF_5 + BrF_5 \rightarrow [SbF_6]^-[BrF_4]^+$$

For ions of the form XY_2^+ (where Y can be X), I_3^+ is the most stable and F_3^+ does not exist. The reasons for this are similar to those discussed for the polyhalide anions, the resonance forms for I_3^+ are

$$I—I\ I^+ \longleftrightarrow I^+\ I—I \longleftrightarrow I^+\ I^-\ I^+$$

CATENATION For other groups in the Periodic Table the degree of catenation of the bottom element is the lowest in the group (e.g. Pb versus C). However, for the halogens, I is the only one which forms an extensive catenated series – the greatest degree of catenation is seen in the polyiodide ions. The bonding in these species is discussed above and the greater degree of catenation is the result of the greater polarizability of I_2.

Summary of Group 17

1. None of the Group 17 elements can be thought of as metallic under normal conditions.
2. Similarity between Cl and Br is no greater than between Br and I.
3. Fluorine does not exhibit positive oxidation states.
4. IF_7 exists but ClF_7 does not although ClO_4^- is more stable than IO_4^-.
5. Fluorine is far more reactive than the other halogens.
6. HF is a high boiling point liquid and is a weak acid in aqueous solution.
7. IF_3 exists but FI_3 does not.
8. The I—F bond is stronger than the Cl—F or the Br—F bond.
9. I_3^- is more stable than Cl_3^- and F_3^- does not exist.
10. Catenation is only exhibited to any great extent by iodine.

Group 18 – The noble gases

	ELECTRONIC CONFIGURATION	STATE
He	$1s^2$	monatomic gas
Ne	[He] $2s^2\ 2p^6$	monatomic gas
Ar	[Ne] $3s^2\ 3p^6$	monatomic gas
Kr	[Ar] $3d^{10}\ 4s^2\ 4p^6$	monatomic gas
Xe	[Kr] $4d^{10}\ 5s^2\ 5p^6$	monatomic gas
Rn	[Xe] $4f^{14}\ 5d^{10}\ 6s^2\ 6p^6$	monatomic gas

The elements are all gases at room temperature, the boiling points increasing down the group due to increasing numbers of electrons, greater polarizability and, therefore, stronger van der Waals' interactions.

Radon has more electrons and a greater mass than Br_2 and might also be expected to be a liquid. However, radon is a gas and this is because the bromine *molecule* is more polarizable than the radon *atom*, therefore the van der Waals' forces are larger in Br_2. The greater polarizability of Br_2 comes from the fact that the highest energy electrons are in antibonding molecular orbitals (see Chapter 2) and, being held less strongly by the nucleus, are able to be displaced more easily by charge density on an adjacent molecule. (An example of the large polarizability of Br_2 can be seen in its reaction with ethene.)

$$CH_2{=}CH_2 \cdots \overset{\delta+}{Br}{-}\overset{\delta-}{Br} \longrightarrow [CH_2CH_2Br]^+ \; + \; Br^- \longrightarrow CH_2Br{-}CH_2Br$$

ELECTRONIC CONFIGURATION AND UNREACTIVITY The elements are extremely unreactive – helium, neon and argon will not react with anything and krypton, xenon and radon only react with extremely electronegative species. Thus, virtually all compounds of the Group 18 elements are oxides, fluorides or oxofluorides.

The unreactivity of the Group 18 elements is a consequence of the closed shell electronic configuration. The 'magical stability' of this electronic arrangement arises from the difficulty in forming either positive or negative ions and having no unpaired electrons available for covalent bonding.

POSITIVE IONS These are difficult to obtain because the Group 18 elements, being at the extreme right of the Periodic Table, have the highest effective nuclear charge in a particular period and therefore the highest ionization energy. Thus, the noble gases do not form discrete cations but xenon can form complex ions such as $[Xe{-}F]^+$, where the energy released in formation of the Xe—F bond goes some way to counterbalancing the large ionization energy (cf. formation of I^+ and Br^+).

NEGATIVE IONS None of the noble gases forms a discrete anion.

Electronegativity increases across a period (higher effective nuclear charge), therefore the electronegativities of the Group 18 elements should be the highest in each period. However, electronegativity refers to the attraction of an atom for electron density in bonds and if an element does not form bonds (He, Ne, Ar), then it is not really possible for it to have an electronegativity. So far we have observed a rough correlation between electronegativity and the tendency to form negative ions, i.e. the electron affinity. This correlation is not valid for the noble gases because electrons must be added to orbitals which are a further shell out, therefore the attraction between the added electron and the nucleus is lower because they are further apart and the electron is well shielded by the full inner shells (cf. the alkali metals which have low ionization energies because the process involves the removal of an electron from an outer shell; such an electron is not tightly held). Thus, although the electron affinity for the noble gases is still going to be exothermic, it is not going to be nearly as exothermic as for the halogens. The electron affinity for the noble gases is probably just the positive side of zero.

We have explained that the electron affinities for the noble gases are quite low. However, this still does not really explain why negative ions should not be formed, since some elements (e.g. S to S^{2-}) have total electron affinities which are endothermic but they are still able to form anions. To fully understand this let us consider a specific question:

Why does CsI exist while CsKr does not?

A possible explanation may be found by considering the terms in a Born–Haber cycle that are contributed by the anion:

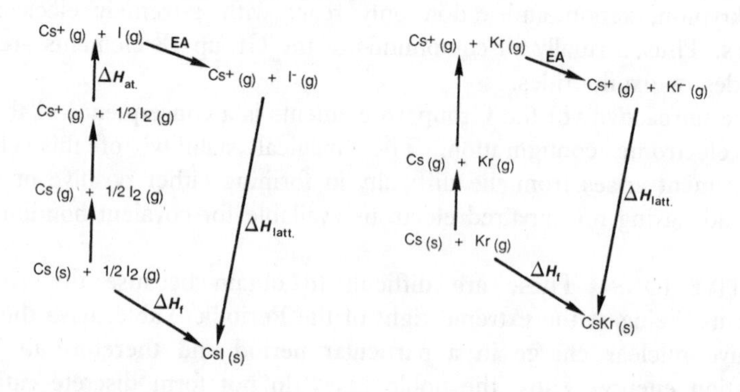

For iodine we have an endothermic $\Delta H_{at.}$ of $107\,\text{kJ mol}^{-1}$ and then an exothermic electron affinity of $-314\,\text{kJ mol}^{-1}$. Thus, the overall energy change involved in formation of $I^{-}_{(g)}$ is $-207\,\text{kJ mol}^{-1}$. The enthalpy of formation of $CsI_{(s)}$, from a Born–Haber cycle is $-337\,\text{kJ mol}^{-1}$.

For krypton, there is no $\Delta H_{at.}$ since it is already a monatomic gas. The electron affinity will be significantly lower than for iodine but still an exothermic term. So, the overall energy change when $Kr^{-}_{(g)}$ is formed is small and exothermic.

The difference between the energies required to form $I^{-}_{(g)}$ or $Kr^{-}_{(g)}$ is, thus, very small. The lattice energies will be comparable, since Kr^{-} and I^{-} are of a similar size, and since all energy terms just involving Cs are the same for both systems, there appears to be no energetic reason why CsI should exist but CsKr should not. In fact, even if the electron affinity of Kr is zero and the lattice energy is about $100\,\text{kJ mol}^{-1}$ less than that of CsI, the overall enthalpy of formation for CsKr will still be exothermic.

However, it is ΔG, and not ΔH, which determines whether a reaction is thermodynamically favourable or not (if ΔG is positive then a reaction is unfavourable),

$$\Delta G = \Delta H - T\Delta S$$

For the reaction between Cs and I_2, there is only a small entropy change because we are starting with two solids and ending up with a solid. (Also, iodine consists of diatomic molecules so that, even if the the reaction between gaseous iodine and Cs is considered, the entropy change is not that unfavourable because of the further constraint of the two I atoms being joined together.) Therefore the ΔG value for formation of CsI will be governed by ΔH. However, for the reaction involving Kr and Cs, we start with a monatomic gas (no constraints on the atoms) and a solid and finish with a solid. The entropy of a monatomic gas is very high (a great deal of randomness), therefore there is going to be a very large decrease in entropy (ΔS is negative) upon formation of CsKr. This is going to more than cancel out any favourable ΔH term, giving a positive ΔG and, therefore, making the overall reaction thermodynamically unfavourable. The noble gases, therefore, do not form negative ions in the solid state.

OXIDATION STATES AND REACTIONS Oxidation states +2, +4, +6 and +8 are observed, e.g. XeF_2, XeF_4, XeF_6 and XeO_6^{4-}. The absence of XeF_8 is likely to be due to steric factors, with the +8 oxidation state only being formed in multiply-bonded (lower coordination number) oxygen compounds.

The noble gas–fluorine bonds have a large amount of ionic character, which is very important in strengthening the bond. The electronegativity difference between Xe and F is greater than between Kr and F, therefore Xe—F bonds tend to be stronger. Thus, KrF_2 is much less stable than XeF_2.

Although noble gas chemistry is dominated by the formation of bonds to F and O, it is also possible to form some compounds containing bonds to N or C. These compounds contain $(FO_2S)_2N$ and CF_3 ligands which are both very electronegative, fluorine-like, 'pseudohalogens'.

Summary of Group 18

1. Noble gases are generally very unreactive.
2. He, Ne and Ar are not known to react with anything and Kr and Xe are only found in combination with very electronegative ligands.

Across the Periods

In Chapter 1 we saw how quantities such as size, electronegativity, etc., varied within the Periodic Table. In the first part of this chapter we looked at trends down the groups. We are now going to draw these together and make some generalizations going across the Periodic Table.

Structure and bonding in the elements

Metallic character decreases from left to right and increases from top to bottom, thus:

- All Group 1 elements are metallic.
- In Group 14 C is a non-metal, Si and Ge are metalloid and Sn and Pb are metals at room temperature.
- In Group 17 only I can be said to have any leaning towards metallic character but even this is best regarded as a non-metal.

Metallic bonding involves positive ions within a sea of electrons. The tendency towards metallic character depends on two factors:

- Ionization energy, which increases across a period and decreases down a group.
- The availability of electrons for, and the strength of, E—E covalent bonds.

The Group 1 elements have low ionization energies, therefore they readily form M^+ plus delocalized electron. As these elements only have one valence electron they could, and do, form M_2 but the dimers are highly electron deficient and the bonds are weak because of the diffuseness of the valence atomic orbitals. The available electrons and orbitals are used most efficiently by the adoption of a delocalized metallic structure.

As we go across the period, Group 14 elements have higher ionization energies, more concentrated valence orbitals and sufficient electrons to form four strong 2c–2e bonds to give a full outer shell. Therefore the elements at the top of Group 14 form 3-D macromolecular, covalent structures. As we descend Group 14, covalent bonds become weaker as the valence orbitals become more diffuse and the ionization energy decreases, so that it is easier to form metallic structures with delocalized electrons and less favourable to form structures involving covalent bonds.

In Group 17, where the electronegativities are high, the valence orbitals are concentrated (strong covalent bonds), the ionization energies are high ($E \rightarrow E^+ + e^-$ is unfavourable, therefore metallic bonding is not observed, even for I) and only one electron is needed for a full outer shell, therefore all elements exist as distinct E_2 units.

The structure of the elements and the tendency towards metallic character is rather closely mirrored by the electrical conductive properties of the elements (Chapter 9):

1. All Group 1 elements are metallic conductors.
2. In Group 14, of the two common allotropes of C, diamond is an insulator

and graphite conducts electricity (although only in 2-D); Si and Ge are semiconductors; Sn and Pb are metallic conductors at room temperature.

3. All Group 17 elements are insulators under standard conditions.

Ionic character of compounds

Ionic character is greatest at the left-hand and right-hand sides of the Periodic Table; at the left-hand side it increases down the group but at the right-hand side it decreases:

- *Group 1*: compounds are mostly ionic, with the alkali metal forming a cation, M^+. Li has the greatest tendency towards covalency.
- *Group 14*: only Sn and Pb have any tendency towards ionic character.
- *Group 17*: these form a great number of ionic and covalent compounds; ionic character is greatest for F (exists as F^-).

At the extremes of the periods we only have to form singly (Groups 1 and 17) or doubly (Groups 2 and 16) charged ions to produce a full outer shell; this requires less energy (exothermic for formation of E^- in Group 17) and is easily offset by exothermic lattice energies ($\Delta H_{\text{latt.}}$).

At the left-hand side positive ions are formed – this gets easier down the group (decreasing ionization energy). Also, the polarizing power of M^+ decreases down the group (i.e. the power of M^+ to attract electrons back to itself from an anion, inducing covalency in the bonding). These two factors lead to ionic character increasing down the group.

At the right-hand side negative ions are formed – the electron affinity (exothermic for $E + e^- \rightarrow E^-$) generally decreases down the group. Also, the polarizability of M^- is highest at the bottom of the group (the tendency of M^- to let go of its electrons leading to covalency). Therefore, ionic character is highest at the top of the group.

Elements at the right-hand side often participate in covalent bonding since the orbitals are contracted (high effective nuclear charge), giving good overlap, i.e. strong covalent bonds.

In the middle, Group 14, at the top the electronegativity is mid-range, the orbitals are quite concentrated (good overlap in covalent bonds), the ionization energy/electron affinity to form +4 or −4 ions (to give full outer shell) is excessively high, therefore compounds containing predominantly covalent bonding are formed. As the group is descended the ionization energy decreases (making cation formation more favourable) the valence orbitals become more diffuse (poor overlap in covalent bonds), therefore the ionic character increases so that Sn and Pb can form predominantly ionic compounds in combination with highly electronegative elements, as in, e.g. PbF_4 and PbO.

Reactivity

This is a difficult concept to make generalizations about because it depends on so many factors, both kinetic and thermodynamic. It is influenced by bond energies, electron affinities, heats of atomization ($\Delta H_{at.}$), etc. But, as a rough guide, reactivity of the elements peaks at the right-hand and left-hand sides of the Periodic Table and is lowest in the middle. At the left-hand side it increases down the group; at the right-hand side it decreases down the group.

The reactivity of the Group 1 elements is mainly a consequence of the low heats of atomization (very open structures with weak metallic bonding) and the low first ionization energy (ease of formation of M^+). These factors contribute to both the activation energy for a reaction (kinetics) and the overall stability of the compound formed (thermodynamics). As the ionization energy and $\Delta H_{at.}$ decrease down the group the reactivity increases. Thus, Cs explodes in contact with water but Li just fizzes.

At the right-hand side, in Group 17, the formation of E^- is exothermic and lattice energies are large with the small halide ions, therefore there is a strong driving force towards the formation of ionic compounds. The high electronegativity means that the valence orbitals are concentrated, resulting in strong covalent bonds with other elements, and a strong driving force towards formation of covalent compounds. The exceptional reactivity of fluorine has already been discussed (see halogens, page 87). The decrease in reactivity down the group comes from increasing size, lower electron affinities and lattice energies, decreasing electronegativity and more diffuse valence orbitals.

In the middle, Group 14, $\Delta H_{at.}$ is high (e.g. macromolecular structure for diamond), ionization energies are large, electron affinities are strongly endothermic ($C \rightarrow C^{4-}$) and reactivity is generally low.

The Oxides and π bonding

Looking at the oxides around the Periodic Table, the following generalizations can be made:

1. The degree of ionicity decreases from left to right across a period and increases down a group.
2. For the more covalent oxides, the tendency to form discrete molecules containing E=O increases from left to right and decreases down a group.

3. The tendency towards macromolecular oxide structures increases down a group.

For Group 1, the oxides are all ionic.

In Group 14, there is multiply-bonded CO_2 at the top, then macromolecular SiO_2 and predominantly ionic PbO_2 at the bottom.

In Group 17 all the oxides are discrete molecules and even iodine has some tendency towards E—O π bonding.

The ionicity was discussed above. The tendency towards π bonding depends on electronegativity and size (see Chapter 4). Electronegativity and size influence the diffuseness of orbitals (more concentrated gives better overlap), compatibility with the p orbitals of O (the smaller the energy difference between the orbitals, the better the overlap) and the E—O bond length (the longer the bond, the weaker the π interaction). Electronegativity decreases down a group and increases from left to right. Thus, diffuseness decreases from left to right and increases from top to bottom and compatibility of p orbitals with those of O decreases down a group. Size, and hence E—O bond length, increases down a group and decreases from left to right. Therefore, overall, π bonding in the oxides increases from left to right and decreases down a group.

Hydrides

Basicity decreases from left to right and increases down a group. At the left-hand side we have basic hydrides $[M^{n+}(H^-)_n]$, which react with water to give MOH and H_2, e.g.

$$2NaH + H_2O \rightarrow 2NaOH + H_2$$

As we move across the table the hydrides become less basic, so that at Group 14 they are essentially neutral at the top of the group but slightly basic at the bottom. As we move past Group 14 we get hydrides with some acidic character; the acidity increases as we keep going across, so that the halogen hydrides are really quite acidic, e.g.

$$HCl_{(aq)} \rightarrow H^+{}_{(aq)} + Cl^-{}_{(aq)}$$

This trend across the Periodic Table is related to the electronegativity difference between H and the elements to which it is bound. At the left-hand side the elements are less electronegative than H and so the bonds are polarized $M^{\delta+}$—$H^{\delta-}$, i.e. hydridic; as we get to Group 14 the electronegativity is very similar to that of H and the bonds are essentially non-polar; beyond Group 14 the elements are generally more electronegative than H and we get $E^{\delta-}$—$H^{\delta+}$, i.e. protonic H and the hydride can be acidic. As we go down a group E becomes less electronegative and so H becomes more hydridic and less protonic, therefore acid strength should decrease down a group. However, it is not quite this simple and acidity depends on several factors, so that HI is more acidic than HF (see halogen hydrides, page 88).

Questions

1 Rationalize the following data:

	NaF	MgF_2	AlF_3	SiF_4	PF_5	SF_6
m.p. (°C)	988	1266	1291	−90	−94	−50
			InF_3	SnF_4	SbF_5	TeF_6
m.p. (°C)			1170	705	8	−36

Answer

Ionic salts consist of an infinite array of anions and cations held together by strong electrostatic forces. On melting an ionic solid this structure is broken down to produce mobile ions. Breaking all the ionic bonds requires a lot of energy, and therefore a high temperature.

Most covalent compounds contain discrete molecules. The interactions between these molecules are relatively weak (dipole–dipole, van der Waals'). On melting a covalent molecular compound only the weak intermolecular, and not covalent, bonds are broken, which does not require very much energy, i.e. a low melting point.

For Na, Mg, Al, In and Sn the electronegativity difference between the element and F is very large so that the fluorides have mostly ionic character and are high melting point solids (infinite ionic lattice). All these elements have relatively low ionization energies, which can be easily compensated for by the large amount of energy released in the formation of a mostly ionic lattice.

The electronegativity differences between F and Si, P, S, Sb and Te are relatively small and these compounds are predominantly covalent molecular, with low melting points. The elements are not going to form ionic fluorides because the ionization energies (higher at the right-hand side of the Periodic Table) are too high and the energy released in formation of the lattice will not cancel out this endothermic term. Also, the valence orbitals of these elements are fairly contracted (high effective nuclear charge) so that they can form strong covalent bonds.

The melting point of NaF is lower than that of MgF_2. Electrostatic interaction is proportional to the product of the charges and inversely proportional to the inter-ionic distance. Thus, the electrostatic interactions in MgF_2 $(+2 / -1)$ are larger than those in NaF $(+1 / -1)$ and more energy is required to break apart the lattice of MgF_2, therefore the melting point is higher. Reinforcing this trend is the much smaller size of the Mg^{2+} ion (Mg^{2+} is approximately the same size as Li^+), i.e. the smaller inter-ionic distance.

These considerations would lead to a prediction that the melting point of AlF_3 should be substantially higher than that of MgF_2 – $(+3 / -1)$ better than $(+2 / -1)$ and Al^{3+} is very small. The small size and high charge of the Al^{3+} ion, however, make it very polarizing – Al^{3+} pulls electrons back from the F^- – so that the bonding has a significant amount of covalent character and the melting point is not as high as would be expected if the compound were totally ionic, (i.e. the charges on the ions are not really $+3$ and -1. Within the first coordination shell of a particular Al cation this loss of some electrostatic attraction is paid back by an increase in covalent interactions with neighbouring F anions. However, there will be no such compensating covalent contribution for the decrease in electrostatic interaction beyond the first coordination shell, i.e. longer range attractive forces, which are purely electrostatic are also in operation in a lattice, and it is these which hold the 3-D lattice together.) The covalent character of the AlF_3 can be understood from a different angle – the ionization energy for formation of the Al^{3+} ion is too high so that, in reality, the Al is going to be incompletely ionized in the solid, i.e. there is incomplete transfer of negative charge to the F and therefore covalency in the bonding.

The larger size of In, as compared to Al (atomic radius increases down a group), means that the inter-ionic distances in the fluoride are longer, hence the electrostatic interactions between ions are weaker and the melting point is lower for InF_3.

SnF_4 has a high degree of covalent character (Sn^{4+} would be excessively polarizing) so that the actual charge on the Sn cation will be significantly lower than $+4$. The reduced charge and the relatively large size of the Sn cation results in SnF_4 having the lowest melting point of the 'ionic' compounds.

There is a large decrease in the melting point between AlF_3 and SiF_4. This is due to the change from an ionic lattice structure to a covalent molecular one. The difference between the type of bonding shown by Si and Al is due to combination of ionization energy and size effects. The ionization energy to generate an Si^{4+} ion is very large and the ions so produced would be extremely polarizing. The Si ion would also be significantly smaller than the Al ion so that its coordination number in an ionic compound would be expected to be lower – lower coordination number, fewer ionic bonds, less stable structure. Si has fairly concentrated valence orbitals (higher effective nuclear charge towards the right-hand side of a period) and is therefore able to form strong covalent bonds (formation of covalent bonds is more favourable than formation of an ionic lattice for this compound).

Si and Sn are in the same group but one forms a covalent fluoride while the other forms a mostly ionic one. The difference is the result of the decreasing ionization energy as we descend a group, thus the ability to form cations, is much greater for Sn.

The variation in the melting points of the 'covalent compounds' is due to a combination of factors. Van der Waals' forces increase with increasing relative molecular mass so that there is a general increase along the series SiF_4, PF_5, SF_6 and down the groups (e.g. from SF_6 to TeF_6). The melting point of SiF_4 is larger than that of PF_5 because of the more spherical shape of SiF_4 (tetrahedral) so that the molecules can pack together better in the solid.

SbF_5 has the highest melting point among this group of compounds because it is a tetrameric (F bridges) structure and therefore has a high molecular mass (large van der Waals' forces).

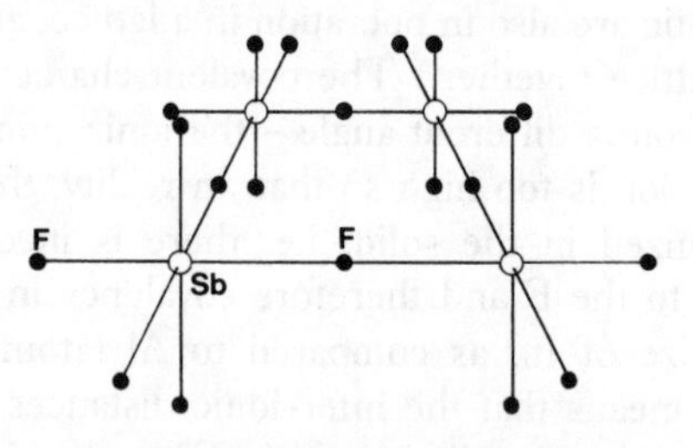

2 Alkyl substitution in NH_3 decreases the tendency towards complex formation in the order, $NH_3 > RNH_2 > R_2NH > R_3N$. The order for phospines is the opposite way around. Discuss.

Answer

Electronic and steric factors have to be considered. Replacement of H by electron releasing R groups causes an increase in the amount of electron density on the donor atom, therefore it should make it a better donor. This is the case for phospines but not for nitrogen. Nitrogen is a small atom, therefore E—N bonds are short and the ligands are closer together within the coordination sphere of the E. As more R groups are attached to the N then the amount of steric repulsion within the complex increases and this outweighs any favourable electronic effect. E—P bonds are longer, therefore steric effects are smaller and electronic effects dominate.

CHAPTER 4

Main Group π-Bonded Systems

This chapter will be concerned with π bonding between main group elements and the factors that affect the stability of compounds formed.

π Bonds can be categorized as:

second period–second period
second period–third period
third period–third period

↓ decreasing stability of π bonds

Factors that affect the stability of π bonds

1. π Bonding is a side-on interaction and depends critically on how diffuse/concentrated and how far apart the interacting orbitals are: *the more diffuse and further apart the interacting orbitals, the weaker the π bonding.*

   In a particular group diffuseness of valence orbitals increases from top to bottom as the orbitals are further away from, and less influenced by, the nucleus. More electronegative atoms pull in their outer orbitals more closely so that they are more concentrated.
2. *π Bonding decreases in strength with increasing electronegativity difference between the interacting atoms.* A large electronegativity difference generally means a large energy difference between interacting orbitals and, therefore, poor overlap (Rule 4, page 19).
3. Kinetic factors: the availability of vacant low energy orbitals on the interacting atoms will greatly reduce kinetic stability by making π bonds more susceptible to nucleophilic attack – donation of lone pairs of

electrons into these orbitals. Greater polarity in the π bond will increase the susceptibility towards nucleophilic/electrophilic attack – nucleophiles will be attracted by δ+ in a bond, electrophiles by the δ−.

It is often the case that the bond dissociation energy of E═E is less than that of 2 × E—E, i.e. polymerization is usually thermodynamically favourable. This may not occur if the activation energy for the process is too high, i.e. a 'kinetic block' – see third period–third period π bonds.

The above order for the stability of π bonds can now be rationalized:

- *Second period–second period.* This produces the best interaction because the orbitals employed are more concentrated and the distances between atoms are small, giving good side-on interaction. The absence of low energy d orbitals means that they are not readily able to extend their coordination number beyond four and so there is not generally a low activation energy pathway for attack: they are kinetically stable. These factors lead to π bonding being much more extensive in the second period of the Periodic Table than anywhere else.
- *Second period–third period.* These π bonds are generally stronger than those between two third period atoms and there are many examples of compounds containing second period–third period π bonds (see below). This might not be expected from (2) above since there is likely to be a larger electronegativity difference (energy difference between orbitals) between a second and a third period atom than between two third period atoms. However, (1) outweighs this – the bonds are shorter and one of the atoms employs a more concentrated 2p orbital, consequently 2p–3p overlap is better than 3p–3p.
- *Third period–third period.* The orbitals involved in π bonding are more diffuse, the bond distances are longer (because the atoms are bigger) and there are vacant valence d orbitals available to be attacked by nucleophiles (third period atoms are able to extend their coordination number beyond four, e.g. PF_5 exists but NF_5 does not, and there is thus a low activation energy pathway for nucleophilic attack). Therefore π bonds between third period atoms are far less stable than those between second period atoms and are not commonly found.

In summary, *the strength of π bonding follows the order* $2p_\pi$–$2p_\pi$ > $2p_\pi$–$3p_\pi$ > $3p_\pi$–$3p_\pi$.

Second period–second period

N_2 This has an exceptionally strong triple bond, with a dissociation energy about five times that for an N—N single bond. Therefore it is much more thermodynamically favourable to have a triple bond rather than three single bonds and elemental nitrogen exists as N≡N.

Why is it such a strong bond?

1. The orbitals involved in π bonding are 2p orbitals (which are concentrated/contracted).
2. The interacting orbitals are of the same energy.
3. The bond distance is short and so the p orbitals are close together, leading to a good side-on interaction.

Part of the stability of the N≡N triple bond comes from the weakness of the N—N single bond (weak because of lone pair–lone pair repulsions – see page 76), i.e. the thermodynamic unfavourability of any reaction which involves a triple bond going to a single bond (many N—N compounds are explosive; decomposition leads to formation of gaseous N_2 with the evolution of a great deal of energy).

As we form a triple bond via a single and two double bonds, the atoms get closer together and the π interaction increases. This would seem to apply equally well to P_2 as to N_2. However, P_2 does not exist under normal conditions (P≡P bond energy is only about 2.5 times the P—P bond energy so that P≡P going to 3 × P—P is thermodynamically favourable). This is because, as we force two atoms together, we increase the inter-electron and inter-nuclear repulsion between them. For N, the concentrated nature of the 2p orbitals means that, as the atoms approach, there is more than enough increase in π bonding to overcome increasing repulsions. For P, the repulsions between core electrons are going to be larger (there are more electrons) and use of the less concentrated 3p orbitals will not result in sufficiently strong π bonding to offset these increased repulsions. Therefore the predominantly singly bonded P_4 structure is preferred (page 77).

ALKYNES The dissociation energy of the C≡C triple bond is about 2.5 times that of the C—C single bond; why does it differ from N_2? Three factors are important:

1. N is more electronegative than C, therefore its p orbitals are more concentrated, giving better π overlap.
2. N is smaller than C, therefore the p orbitals on adjacent atoms are closer together, resulting in a stronger π interaction.
3. A comparison of the N—N and C—C single bonds is slightly artificial as the N—N bond is very weak due to lone pair–lone pair repulsions, which are not present in C—C bonds (see Chapter 3).

BF_3 π Bonding involves donation of electrons from full p orbitals on the fluorines into a vacant p orbital on the sp^2 hybridized B.

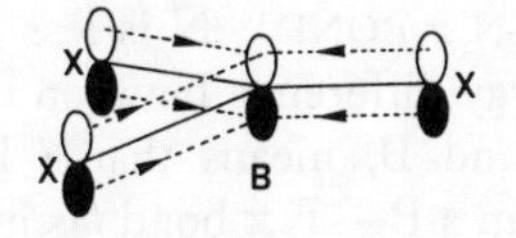

Although π bonding is not very strong in BF_3, because of the large electronegativity difference between B and F, it still has a major influence on the physical and chemical properties of BF_3, i.e. the B—F bond is much shorter than the sum of the covalent radii and BF_3 is not as strong a Lewis acid as would have been expected from electronegativity considerations (B should be quite δ+). Also, the loss of π bonding upon dimerization (four coordinate B cannot participate in π bonding) means that BF_3 does not dimerize (cf. BH_3 – no π bonding – dimerizes to give B_2H_6).

BORAZINES

$$B_2H_6 + NH_3 \rightarrow [BH_2NH_2]^+[BH_2]^- \xrightarrow{\text{Heat}} B_3N_3H_6$$

Borazine has a planar hexagonal ring structure. The nitrogens and borons can be considered to be sp^2 hybridized, leaving the N with a filled, and the B with a vacant, p orbital perpendicular to the ring. The valence bond forms are:

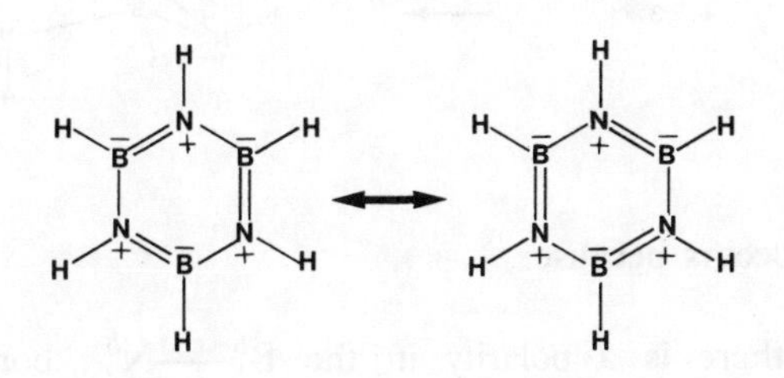

The actual structure is a resonance hybrid of these two forms. Although both resonance forms show positive charges on the N, calculations and reactivity show that the B—N bond is polarized $B^{\delta+}$—$N^{\delta-}$. This is due to the greater electronegativity of the N; N → B donation is counterbalanced by N's desire to pull both σ and π electrons back to itself.

The bonding can also be described using an MO treatment, which would regard it as involving delocalization of the six available π electrons over the whole ring, i.e. similar to benzene (see below).

THE STRENGTH OF THE B—N π BOND N is less electronegative than F and the resultingly smaller energy difference between the p orbitals of N and B, than between those of F and B, means that a B—N π bond (as in borazine) should be stronger than a B—F π bond (as in BF_3). This is seen in, e.g. $B(NMe_2)_3$, which is a weaker Lewis acid than BF_3: π bonding is lost upon coordination of a fourth ligand (a Lewis base). The loss of π bonding, which occurs when BF_3 forms an adduct, is more than compensated for by the formation of an extra bond. However, for $B(NMe_2)_3$, where the π bonding is stronger, this is not the case. (This could be, at least in part, due to steric factors – with three large NMe_2 groups around the B there is little room for attack of another group.)

BORAZINE VERSUS BENZENE B—N is isoelectronic with C—C and therefore $(BN)_3H_6$ (borazine) is isoelectronic with $(CC)_3H_6$ (benzene); borazine has been called 'inorganic benzene'. The physical properties of borazine and benzene are quite similar (they are both liquids) but the chemical properties are very different. Borazine is far more reactive than benzene, e.g. it reacts with HCl whereas benzene does not.

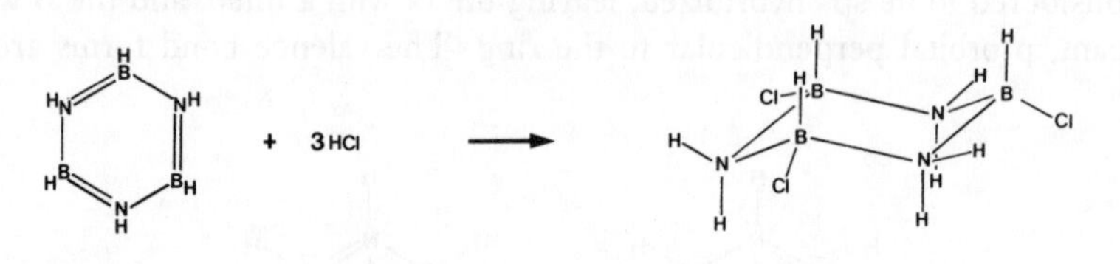

The difference occurs because:

1. in borazine there is a polarity in the $B^{\delta+}$—$N^{\delta-}$ bond due to the electronegativity differences between B and N;
2. the p orbitals are of different energies and delocalization is not as complete in borazine as in benzene, i.e. in borazine there is a 'lumpy' π cloud.

H—Cl is polar and initial attack is going to be facilitated by the polarity of the B—N bond – $H^{\delta+}$ attracted to $N^{\delta-}$ (the C—C bond in benzene is non-polar). The reaction begins with addition of H^+ to give partial breakdown of the aromatic system, i.e.

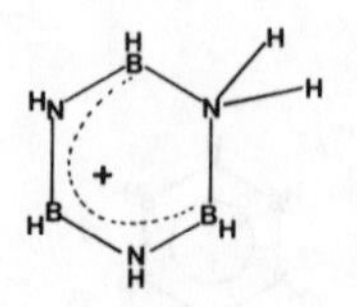

Aromatic character is lost upon formation of the intermediate/transition state. Breakdown of the aromatic system in benzene requires more energy (greater aromatic character in benzene).

These two factors result in the activation energy for the process being much lower for borazine than for benzene (high activation energy means the reaction is immeasurably slow for benzene; $K \alpha e^{-E_{act.}/RT}$).

Benzene tends to undergo electrophilic substitution reactions in which the aromatic system is retained, illustrating the greater stability of the aromatic system in benzene.

Benzene and borazine derivatives can undergo similar reactions, e.g. they will both coordinate to transition metals, although the ring is non-planar in the borazine complex.

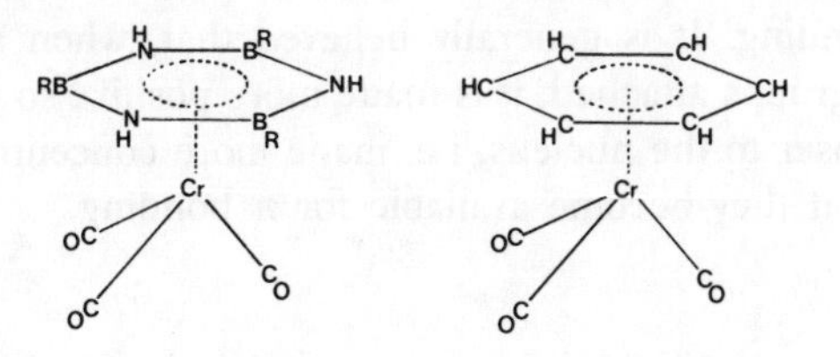

Second period–third period

$2p_\pi$–$3p_\pi$ BONDING The order of stability of multiple bonds between second and third period atoms is, e.g.

$$Si{=}C < P{=}C < S{=}C$$

e.g. silabenzene, $SiHC_5H_5$, is less stable than phosphabenzene, PC_5H_5, which is less stable than thiophen, C_4H_4S. This is due to a combination of kinetic and thermodynamic factors:

1. Kinetic: the bonds are polarized as follows:

   $$C^{\delta-}{-}Si^{\delta+} \qquad C^{\delta-}{-}P^{\delta+} \qquad C^{\delta+}{-}S^{\delta-}$$

   The electronegativity difference (0.8) between C and Si is relatively large, making the bond quite polar ($C^{\delta-}{-}Si^{\delta+}$). This, together with the availability of vacant d orbitals on the Si, means that these bonds are susceptible to nucleophilic attack at the Si.

   There is also a significant (but smaller, 0.4) electronegativity difference between P and C, so that P=C is also readily attacked but not as readily as Si=C.

   The electronegativities of C and S are virtually identical so that the C—S bond is almost non-polar. What polarization there is causes the

C to be the δ+ atom, i.e. the end of the bond that is attacked by nucleophiles. Attack at the C is more difficult than attack at the third period atom since the C has no low energy vacant orbitals into which a lone pair of electrons can be donated by a nucleophile.

2. Thermodynamic: a large electronegativity difference between the π bonded atoms results in a large energy difference between the interacting p atomic orbitals, which causes poor overlap and weaker π bonding.

p_π–d_π BONDS The degree of involvement of 3d orbitals in π bonding is a matter of great debate. The argument is that the 3d orbitals are too high in energy and too diffuse to participate in significant π interaction. However, there are many differences in behaviour between compounds involving second and third period elements that are difficult to explain without the admission of some p_π–d_π bonding. It is generally believed that, when an atom has very electronegative groups attached, it is made more positive so that the d orbitals are pulled in closer to the nucleus, i.e. made more concentrated and lowered in energy, so that they become available for π bonding.

EXAMPLES

NMe_3 versus $N(SiMe_3)_3$

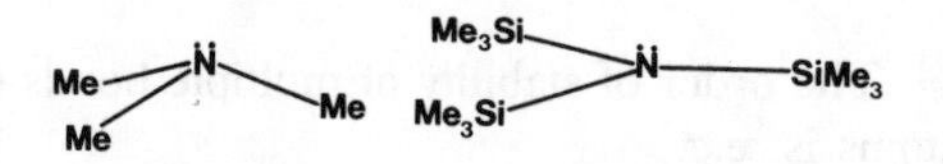

VSEPR would predict a pyramidal shape in both cases and it is generally believed that the planarity of the Si compound comes from a d_π–p_π interaction,

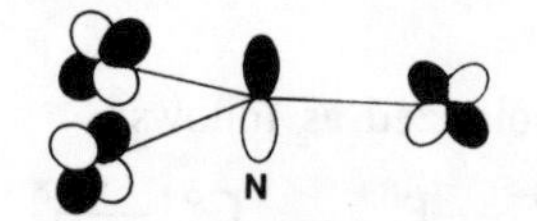

The planarity maximizes the π interaction.

No low energy vacant d orbitals on C means that there is no π bonding at all in NMe_3, therefore the shape is governed by normal VSEPR rules, i.e. it is pyramidal.

$P(SiMe_3)_3$ is pyramidal – π bonding here involves a 3p–3d (as opposed to 2p–3d) interaction and is just not strong enough to overcome the electron pair–electron pair repulsion terms trying to make the structure pyramidal.

Opponents of d_π–p_π bonding would argue the observed facts on steric grounds, i.e. $SiMe_3$ is an extremely bulky group and three of these around a small N atom would force a planar structure (pushing the large groups as far apart as possible). P is larger, the ligands are further apart and a pyramidal structure is possible.

Phosphacumulene ylids, e.g. $Ph_3P{=}C{=}PPh_3$

These are the two extreme ways of viewing the bonding in this structure; π bonded A would be expected to have a PCP bond angle of 180°, whereas in B it would be expected to be 109°. The actual bond angle of about 130° indicates at least some contribution from the π bonded structure.

R_3PO and R_3NO

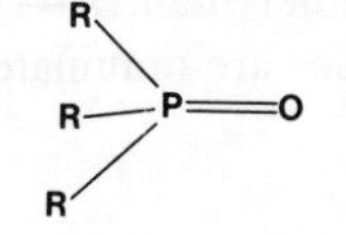

COMPOUND	$\theta_{P—O}(cm^{-1})$
F_3PO	1404
Cl_3PO	1295
Me_3PO	1176

P—O infra-red stretching frequencies provide a guide to P—O bond strengths; a higher stretching frequency corresponds to a stronger bond (page 319).

The P—O stretching frequency is largest for R = F. A plausible explanation of this is that the F is very electronegative and pulls electron density away from the P, causing the d orbitals on the P to be sufficiently lowered in energy so that they can participate in p_π–d_π bonding with the O 2p orbital. This effect decreases from F to Cl to Me as the electronegativity of the R group decreases. Me is actually an electron releasing group and so will force the P 3d orbitals to higher energy.

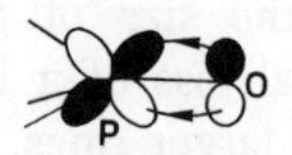

An alternative explanation of this would describe the bonding as R_3P^+—O^- (with a high degree of ionic character), so that, as R becomes more electronegative, the positive charge on the P is increased and there is greater ionic contribution to the P—O bonding, making the bond stronger and the P—O stretching frequency higher. Also, increasing the positive charge on P lowers the energy of the σ orbitals making them more compatible with those of O. However, these factors also apply to R_3NO where the difference in stretching frequencies is much smaller and so an explanation in terms of d_π–p_π bonding is more acceptable (although, the greater electronegativity of N, compared with P, means that the charge on the N atom is not going to be so influenced by changing the R group).

Siloxanes and phosphazenes

Neither SiO nor PN exists at room temperature. This is because they are both very unstable with respect to oligomerization or polymerization due to the weakness of the d_π–p_π bonds in any such monomer, i.e. the favourability of having 2Si—O or 2P—N rather than Si=O or P=N (page 69).

Their very similar chemistries are dominated by the formation of rings and chains, e.g.

$R_2SiO \rightarrow (R_2SiO)_n$
siloxanes

$R_2PN \rightarrow (R_2PN)_n$
phosphazenes

Siloxanes

An example is $Me_6Si_3O_3$

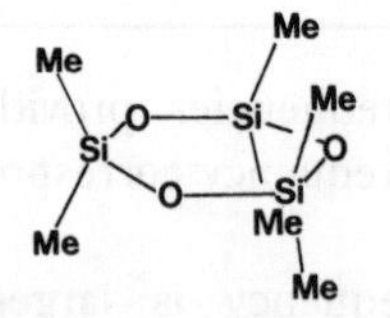

A puckered ring is predicted on steric grounds (cf. C_6H_{12}) but the ring is actually planar due to the necessity of maximizing d_π–p_π bonding between a lone pair on O and a vacant d orbital on Si (cf. benzene, where the ring is planar to maintain good p_π–p_π bonding). As we go to larger rings there is some puckering due to steric and entropy effects (a rigid planar ring is more ordered than a flexible puckered ring – the difference in entropy between these structures increases as the size of the ring increases). Puckering, however, does not result in total loss of π bonding and there is still some present in eight-membered and larger rings.

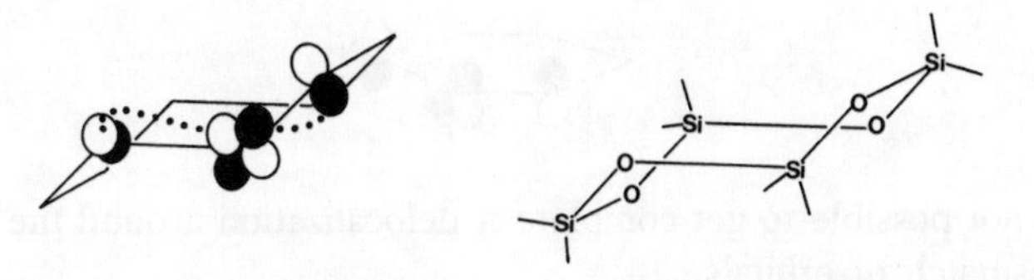

The O—Ŝi—O bond angle in $Me_8Si_4O_4$ is significantly larger than the 109° found for the C—Ĉ—C bond angle in cyclooctane – the benefit of maintaining π bonding in the siloxane is preventing full puckering to give tetrahedral bond angles.

A good indicator of bond strength is the Si—O infrared stretching frequency in these compounds, which show the same trends as observed for R_3PO compounds; the same arguments can be applied.

Phosphazenes (Phosphonitriles)

$PCl_5 + NH_4Cl \rightarrow P_3N_3Cl_6$ + other products

$P_3N_3Cl_6$ is a planar six-membered ring, with all P—N distances equal and shorter than the sum of the covalent radii.

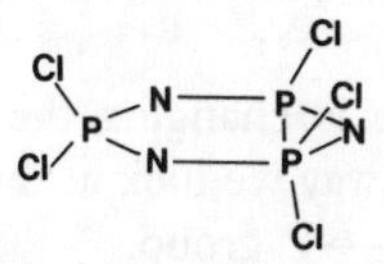

There are two components to the bonding:

1. In addition to the basic σ bonding framework further in-plane bonding utilizing d orbitals on P is possible.

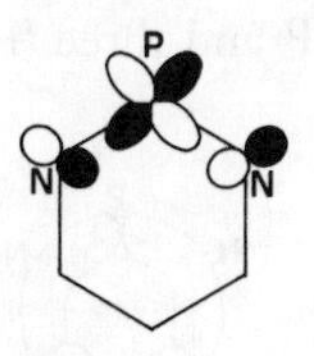

2. Out-of-plane bonding. For P to be able to form four single bonds we need to generate four unpaired electrons, which can be achieved by promotion of an electron from the P 3s orbital to a 3d orbital,

$$3s^2\ 3p^1\ 3p^1\ 3p^1 \rightarrow 3s^1\ 3p^1\ 3p^1\ 3p^1\ 3d^1$$

We can now get d_π–p_π bonding involving overlap of the half-filled d orbital on the phosphorus atoms with half-filled p orbitals on the nitrogens.

It is not possible to get complete π delocalization around the ring because of a mismatch of orbitals,

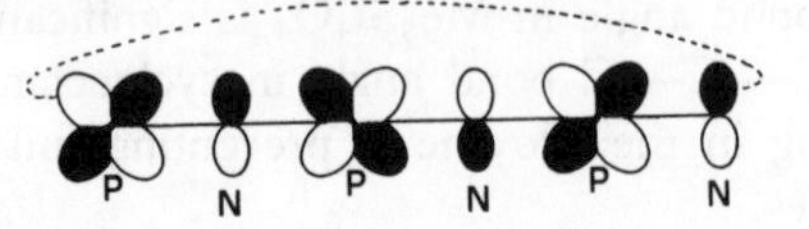

There is a bonding interaction all the way from P(1) to N(3) but then there is a mismatch between the orbitals of P(1) and N(3), which are joined together when a ring is formed.

An alternative view of the bonding is to consider rehybridization of the out-of-plane d orbitals to give hybrid d orbitals pointing towards the nitrogens.

(All we are really doing here is changing the axis system, thus we are not changing anything except the way we look at the system.) Thus, we get 3c–2e π bonding about each P—N—P group,

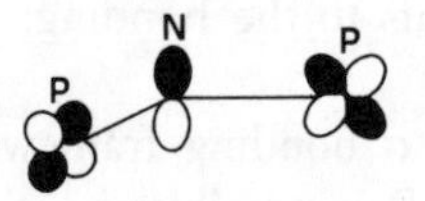

This leads to a node at each P and three 'islands' of π bonding rather than complete delocalization,

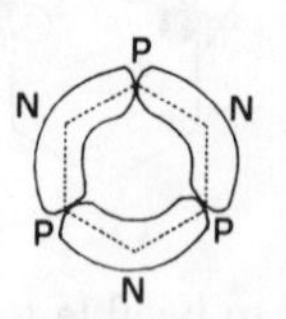

The exact nature of the bonding in the cyclic phosphazenes is a matter of great debate, but everybody agrees that there is not complete delocalization around the ring.

Why is $P_4N_4Cl_8$ puckered while $P_4N_4F_8$ is planar?
F, being more electronegative than Cl, causes a greater lowering of the energy

of the 3d orbitals on P, making them more energetically compatible with the N 2p orbitals, therefore there is strong enough π bonding to warrant a completely planar structure. In $P_4N_4Cl_8$ the lowering of the d orbitals, and hence the π bonding, is not sufficient to warrant a planar structure. However, the d orbitals are quite adaptable and (as for the siloxanes) a puckered structure does not result in complete loss of π bonding in $P_4N_4Cl_8$.

As with siloxanes and R_3PO, the P—N stretching frequencies in the phosphazenes provide a guide to the strength of P—N π bonds. For compounds $(PNR_2)_n$ the stretching frequencies decrease along the series $R = F > Cl > Br > Me$.

Reactions

If $P_3N_3Cl_6$ is heated a chain polymer $(Cl_2PN)_n$ is obtained, which is just as reactive as the ring compound. This is because there is still π bonding in the polymer.

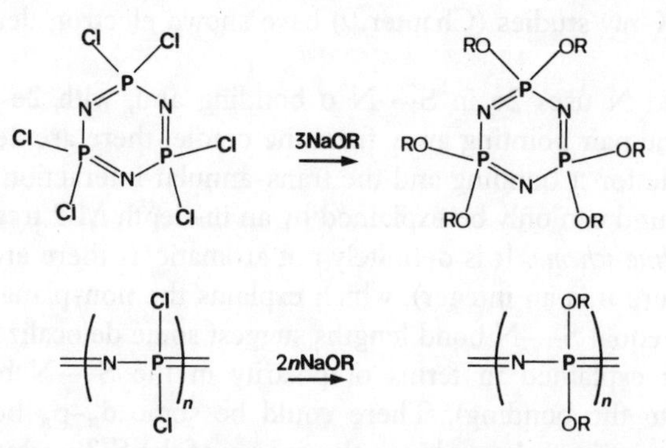

Thus, both cyclic and chain compounds react with NaOR to give replacement of Cl by OR. The reactivity of these polymers can be contrasted with organic polymers, which are generally much less reactive than their monomers.

SULPHUR–NITROGEN CHAINS, RINGS AND CAGES SN π bonded compounds are considerably more stable than their PN and SiO analogues. This is because the electronegativity difference between S and N is quite small, therefore the π bonding is quite strong (thermodynamics), also the bond is not very polar, i.e. not as susceptible to attack by nucleophiles or electrophiles (kinetics).

Compare the following bond lengths (Å):

SN	FSN	F_3SN
1.49	1.45	1.42

A plausible explanation of this is that the electronegative fluorines withdraw electron density from the S, which lowers the energy of the S 3p and 3d orbitals, making them more concentrated and more energetically compatible with the N 2p orbitals, resulting in better π overlap. However, this can also be explained in terms of bond polarities (ionic contribution), without recourse to d_π–p_π bonding, as for R_3PO, above.

S_4N_4 This has a cradle-like structure,

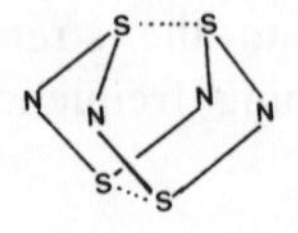

All S—N bond lengths are equal and shorter than the sum of the covalent radii. There is interaction between the trans-annular S atoms – the S · · · S distance is substantially less than the sum of the van der Waals' radii and low temperature X-ray studies (Chapter 9) have shown electron density along the S · · · S edge.

Each S and N uses 2e in S—N σ bonding and, with 2e on each atom acting as a lone pair pointing away from the cradle, there are 2e per S and 1e per N available for π bonding and the trans-annular interaction. The bonding in this compound can only be explained by an in-depth MO treatment: *there is no simple bonding scheme.* It is definitely not aromatic as there are not $(4n+2)\pi$ electrons (where n is an integer), which explains the non-planarity. However, the short and equal S—N bond lengths suggest some delocalization, although this could be explained in terms of polarity in the S—N bond (an ionic contribution to the bonding). There could be some d_π–p_π bonding – N is fairly electronegative and may lower the energy of the S 3d orbitals sufficiently so that they can participate in π bonding. The situation is made even more complicated by the trans-annular S · · · S bond and which orbitals and electrons are involved in this.

$[S_4N_4]^{2+}$ The situation in $[S_4N_4]^{2+}$ is much more simple:

Total number of valence electrons	$= 4 \times 6 + 4 \times 5 - 2 =$ (S, N, charge)	42e
Eight S—N bonds		−16e
Eight lone pairs (one on each atom)		−16e
Number of electrons for π bonding		10e

We have 10 electrons, i.e. $(4n+2)$, with $n = 2$, for π bonding. Thus, the system is aromatic and planar (to maximize the π bonding), with all S—N bond lengths equal and shorter than the sum of the covalent radii.

$[S_3N_3]^-$

Number of valence electrons = 3 × 6 + 3 × 5 + 1 =	34e
Six S—N bonds	−12e
Six lone pairs	−12e
Number of electrons for π bonding	10e

Thus, this is another $(4n+2)$ planar aromatic benzene analogue.

In sharp contrast to planar aromatic hydrocarbons, aromatic S—N ring compounds are all coloured; this is a consequence of the weaker π bonding in the SN compounds. The MO diagrams for $[S_3N_3]^-$ and C_6H_6 are of the form,

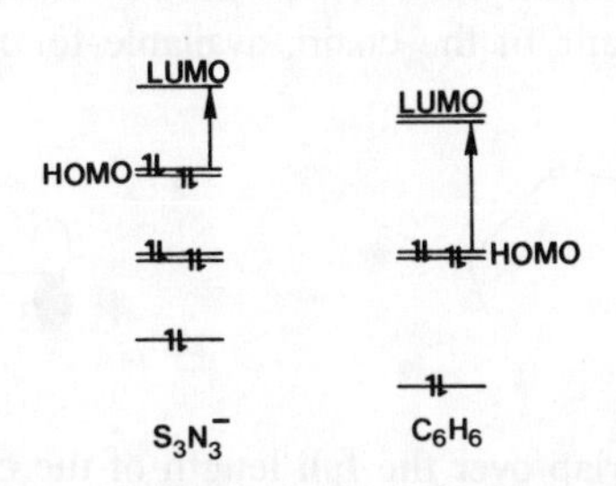

The colour of a compound depends on how much energy is required to promote an electron from the HOMO to the LUMO; a large HOMO–LUMO gap corresponds to absorption of high frequency radiation. The HOMO–LUMO gap depends on how strong the π bonding is, since between each set of orbitals two more antibonding interactions (two more nodes) are introduced. Therefore, for benzene, where there are strong π bonds (second period atoms, same electronegativity), the gap is large enough so that it absorbs in the ultraviolet (i.e. it is colourless). In $[S_3N_3]^-$, where the π bonding is weaker (second period–third period, different electronegativities) the gap is smaller and it absorbs in the visible region (lower frequency radiation and it is therefore coloured).

S_2N_2

Total number of valence electrons = 2 × 6 + 2 × 5 =	22e
Four S—N bonds	−8e
Four lone pairs	−8e
Number of electrons for π bonding	6e

Therefore, this is a $(4n+2)$, $n = 1$, system and S_2N_2 is aromatic and planar,

S—N
N—S

In this and all the aromatics discussed above, the delocalization cannot be complete because of the electronegativity differences between the atoms, i.e. we get 'lumpy' π clouds. The '$(4n+2)$ rule' for aromaticity only applies to p_π–p_π systems; it does not apply to d_π–p_π systems, e.g. the phosphazenes. Thus $P_4N_4F_8$, with 8π electrons is planar with all equal bond lengths, although the $(4n+2)$ rule would predict it to be antiaromatic and to exhibit a puckered structure with alternating bond lengths.

When S_2N_2 is heated to 20 °C we get $(SN)_x$, a linear chain polymer which has a metallic lustre. The chain is near planar and the bond lengths are almost equal (small differences probably being the result of packing forces in the solid), indicating a high degree of delocalization along the chain. All the atoms can be considered to be sp^2 hybridized and with two SN bonds (2e) and one lone pair (2e) per atom each S has 2e and each N 1e, in a p orbital perpendicular to the plane of the chain, available for π bonding.

These p orbitals overlap over the full length of the chain to generate a full band and a half-filled band (see Chapter 9 – a band is filled by $2n$ electrons). The electrons in the partially filled band are free to move under the influence of an applied potential difference and thus we get conduction along the chain, i.e. $(SN)_x$ is a 1-D conductor. A partially filled band means that this is a metallic conductor (page 288), i.e. conduction increases with decreasing temperature and, indeed, $(SN)_x$ becomes a superconductor below 0.26 K.

Bromine doped derivatives of $(SN)_x$, with stoichiometries in the range $(SNBr_{0.25})_x$ to $(SNBr_{1.5})_x$, can be prepared. These have conductivities higher than the parent polymer (page 288).

Third period homonuclear multiple bonds

There are two major problems in synthesizing this type of bond:

1. 3p–3p π bonds are weak and so quite susceptible to polymerization. This is the result of the dissociation energy of E=E being much less than twice that of E—E (thermodynamic instability).
2. Unlike their second period counterparts, they have fairly low energy vacant d orbitals and so are prone to attack by nucleophiles (kinetic instability).

To make these compounds we must inhibit the possibility of polymerization and nucleophilic attack, which is best achieved by coordinating very bulky groups to the π bonded centres. This is a kinetic effect, blocking the approach of attacking groups, whether they be nucleophiles or other π bonded molecules.

e.g.

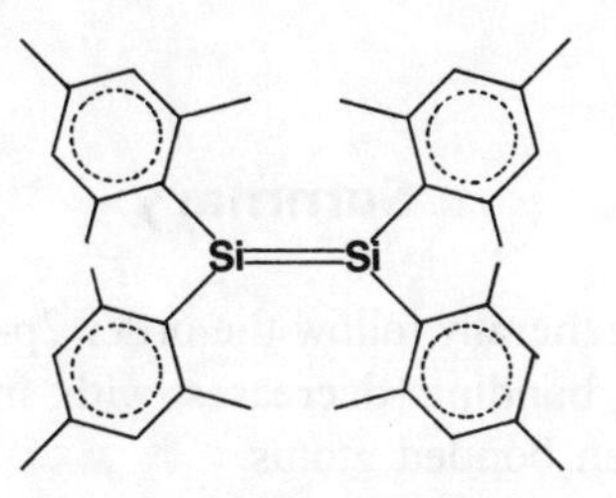

Polymerization is very effectively inhibited, due to the large repulsive interaction between the mesityl groups, but nucleophilic attack, although made more difficult, can still occur, e.g.

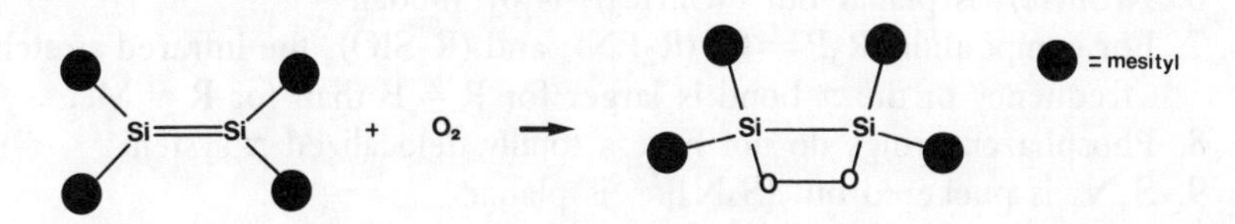

The driving force here is the production of strong Si—O bonds, strengthened by partial $2p_\pi$–$3d_\pi$ interactions,

Compare this reaction with

$C_2H_4 + 3O_2 \rightarrow 2CO_2 + 2H_2O$

There is no reaction at room temperature because π bonding is stronger in C_2H_4 and there are no vacant orbitals for initial attack, i.e. the activation energy for the reaction is high so that it does not occur at room temperature despite being thermodynamically favourable.

As we proceed from Si to P to S the elements become more electronegative; this results in:

- stronger π bonds for S – the p orbitals are more concentrated (contracted); and
- bonds to C, N, O and F which are less polar – the susceptibility to nucleophilic attack is reduced, e.g.

$$S + AgF \rightarrow FS\text{—}SF \rightarrow F_2S{=}S \text{ which is stable at room temperature}$$

Note that large bulky groups are not required to prevent polymerization/attack by nucleophiles in $F_2S{=}S$ because the π bond is strong ($S{=}S \rightarrow 2 \times S\text{—}S$ is not as thermodynamically favourable) and the $S{=}S$ bond is not very polar.

Summary

1. π Bond strengths generally follow the order 2p–2p > 2p–3p > 3p–3p . . .
2. The extent of π bonding decreases with increasing electronegativity difference between bonded atoms.
3. N_2 exists under normal conditions but P_2 does not.
4. Borazine is more reactive than benzene.
5. The strength of third period–carbon π bonds is $S{=}C > P{=}C > Si{=}C$.
6. $N(SiMe)_3$ is planar but $P(SiMe_3)_3$ is pyramidal.
7. For compounds, $R_3P{=}O$, $(R_2PN)_n$ and $(R_2SiO)_n$ the infrared stretching frequency of the π bond is larger for R = F than for R = Me.
8. Phosphazene rings do not have a totally delocalized π system.
9. S_4N_4 is puckered but $[S_4N_4]^{2+}$ is planar.
10. Third period homonuclear diatomics can usually be stabilized if they have large bulky ligands attached to the π bonded unit.

Questions

1 $S_4N_4H_4$ and $S_4N_4F_4$ have the structures

In $S_4N_4H_4$ the ring is puckered (crown structure – similar to S_8) and all the S—N bond lengths are equal at 1.65 Å. In $S_4N_4F_4$ the ring is also puckered but this time there are alternating double (1.55 Å) and single (1.65 Å) S—N bonds. Discuss the differences between the bonding and structures of these two compounds.

Answer

In $S_4N_4F_4$ the fluorines are joined to the S atoms. The fluorines being very electronegative withdraw electron density from the S, causing the 3d orbitals on S to be lowered in energy and contracted so that they become available for p_π–d_π bonding with the N.

Each N uses two electrons in forming two σ bonds with adjacent sulphurs, leaving it with one unpaired electron in a hybrid orbital and a lone pair. The lone pair points out of the ring, away from the sulphurs and is not involved in bonding.

Each S uses three electrons for three single covalent bonds (two to N and one to F) – generation of three unpaired electrons requires promotion of an electron to a 3d orbital, i.e. $3s^2\ 3p^2\ 3p^1\ 3p^1 \rightarrow 3s^2\ 3p^1\ 3p^1\ 3p^1\ 3d^1$.

This leaves the S with one lone pair not involved in bonding and a single electron in a 3d orbital. Overlap between this unpaired electron on the S and the unpaired electron in the hybrid orbital on N gives rise to d_π–p_π bonding and the short (1.55 Å) S—N distances.

The S—F and N—F bond energies are approximately equal and so the desire to gain a π bonding interaction dictates that the fluorines should become attached to the S atoms.

In $S_4N_4H_4$ the S forms two single bonds to nitrogen and is left with two lone pairs; the N forms three single bonds (two to S and one to H) and is left with one lone pair. There is thus no capacity for p_π–p_π bonding since here it would have to involve overlap of two filled orbitals.

The H would not be electronegative enough to lower the S 3d orbitals sufficiently so that they could become involved in d_π–p_π bonding. An N—H bond is sufficiently stronger than an S—H bond – 2p–1s gives better overlap than 3p–1s and, with a greater electronegativity between N and H than between S and H, there is a larger ionic contribution to the bonding. Thus, as there is no π bonding to be gained by H becoming attached to the S, the H bonds to the N, releasing more energy in the formation of stronger N—H bonds.

No π bonding in the structure means that all bond lengths are going to be equal and about the same length as the single (long) bonds in $S_4N_4F_4$.

2 Explain the following:
In the molecule,

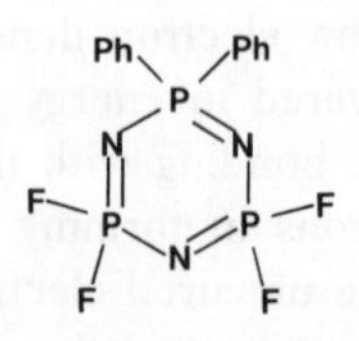

the three nitrogen atoms and the fluoro-substituted phosphorus atoms are coplanar, but the phenyl-substituted phosphorus atom lies 0.2 Å above this plane.

Answer

In the structure above there is P=N π bonding (3d–2p). The strength of this π interaction is dependent upon the energy separation between the N 2p and P 3d orbitals (i.e. the lower in energy the P 3d orbitals the stronger the π bond). Fluorine is more electronegative than a C in a phenyl (Ph) group. Hence the P bonded to two fluorines is more $\delta+$ than the P bonded to two Ph groups, and consequently its 3d orbitals are pulled lower in energy and made more available for π bonding.

Planarity maximizes π bonding because the interacting orbitals are better alligned for the side-on overlap. The P=N π bond is strongest with F substituents and hence a planar configuration is adopted to maximize this interaction.

In a planar structure, there is greater steric repulsion between substituent groups than in a puckered one (cf. cyclohexane, which has only σ bonding, is puckered). For the Ph_2P group P=N π bonding is not strong enough to overcome the large steric repulsions associated with the planar configuration and it puckers out of the plane.

CHAPTER 5
Main Group Organometallics

What is a main group organometallic compound?

The definition we shall use here is that a main group organometallic compound is one containing a bond between the C of an organic group and a main group metal. Also included in this are bonds to semi-metals such as Si or As.

Stability of main group organometallic compounds

Thermodynamic stability

There are two components to the bonding in organometallic compounds: an ionic and a covalent one. The importance of each contribution depends on the electronegativity difference between the metal, M, and the C of the organic ligand. Thus, with very electropositive elements (Groups 1 and 2), the bonding is mostly ionic; this is especially true with organic ligands which form stable negative ions as in, e.g $Na^+(C_5H_5)^-$, where the $C_5H_5{}^-$, cyclopentadienyl (Cp^-), ligand contains a 6e π aromatic system. With more electronegative 'metals' the bonding is much more covalent as in, e.g. $SiMe_4$. In most cases the bonding lies somewhere in between the two extremes.

The strength of the ionic component will depend on the actual charges (rather than the formal oxidation states) on atoms/ions and inter-ionic distances, as we have seen before (small highly charged ions give rise to a strong electrostatic interaction – Chapter 2). The strength of the covalent part of a bond will depend on the diffuseness of, and energy difference between, interacting orbitals (concentrated orbitals of comparable energy give a strong covalent interaction – Chapter 2).

Although some compounds are stable with respect to the elements, all main group organometallics are thermodynamically unstable with respect to formation of M—O bonds (e.g. hydroxides, oxides, etc.). M—O bonds are stronger than M—C bonds.

The favourability of the oxidation reaction leads to most organometallic compounds having to be handled under dry nitrogen, or some other inert gas. However, there are some compounds which are stable in the presence of oxygen (e.g. SiR_4) and this must be a case of kinetic stability, i.e. a high activation energy barrier to the oxidation reaction.

Kinetic stability

Probably the most important pathway for decomposition is β-hydrogen eliminat'

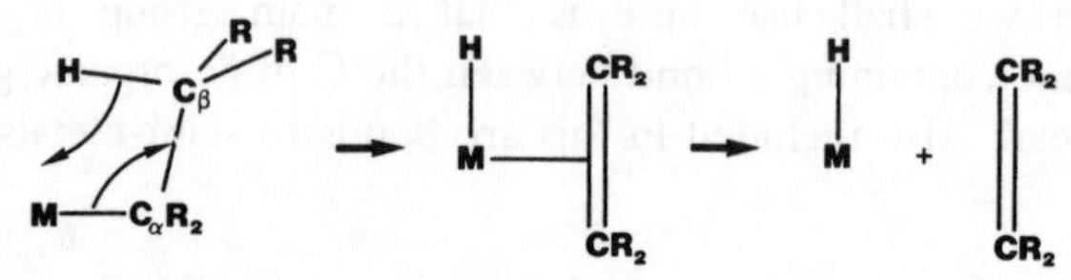

This can only occur when the ligand possesses a β-H, so stability is generally increased when none is present, e.g. $PbMe_4$ (no β-H) boils without decomposition at 110 °C but $PbEt_4$ (β-H) decomposes at 110 °C.

The lability of main group organometallic compounds has been considered here as a rather negative property, but it is actually what makes them most useful, i.e. they act as sources of organic groups, e.g.

$$MeLi + CH_3CH_2Br \rightarrow CH_3CH_2CH_3 + LiBr$$

Stability with respect to attack by air and water (nucleophilic attack) generally depends on two factors:

1. The polarity of the M—C bond. The greater the electronegativity difference between M and C, the more positive M will be and the more susceptible it will be to nucleophilic attack, e.g.

   Me_3In ($\Delta_{electroneg.} = 0.8$) but Me_4Si ($\Delta_{electroneg.} = 0.7$)
   pyrophoric in air — inert to air and water

2. The availability of vacant valence orbitals on the metal and the ability of the metal to extend its coordination number.

Metals which have vacant low energy orbitals and are readily able to extend their coordination number are more likely to be attacked by nucleophiles. This is because, if there is an orbital available for initial attack by the nucleophile, it can occur without the need for dissociation of other ligands. This is a lower activation energy process than that involving ligand dissociation, i.e. an example of kinetic instability.

Group I

Organolithiums

Lithium alkyls form aggregates, $(RLi)_n$. In the absence of donor ligands, tetramers and hexamers are obtained. There are two extreme views of the bonding in organolithiums, i.e. predominantly ionic or predominantly covalent; reality probably lies between the two extremes.

COVALENT BONDING If the bonding is assumed to be predominantly covalent, these compounds can be considered as electron deficient (with respect to 2c–2e bonding), that is comprising of RLi units joined together by four centre–two electron bonds (4c–2e), e.g. Li_4Me_4 whose structure is

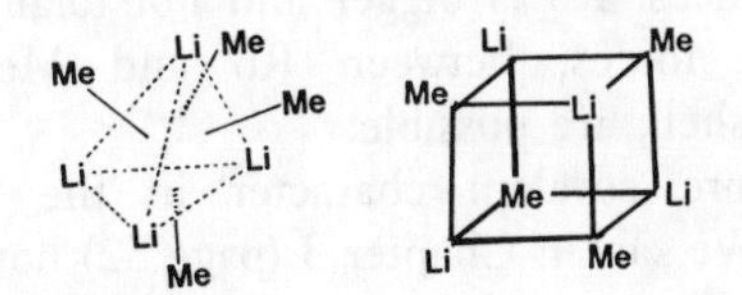

There are *no direct Li—Li bonds* and interaction between metal centres occurs via the methyl groups.

The structure, as drawn here, appears to have 12 Li—C bonds but there are only eight electrons (one each from the lithiums and one each from the methyls), four electron pairs, available for bonding. The compound is thus electron deficient in the sense that there are not enough electrons for 2c–2e bonds between all adjacent atoms.

The bonding can be represented as involving multi-centre interactions. We assume that each Li is sp^3 hybridized, with one hybrid orbital pointing into the centre of each face of the tetrahedron. There is overlap between the sp^3 hybrid orbitals of the three lithiums around the triangular face and the sp^3 orbital on the Me sitting above the face, i.e.

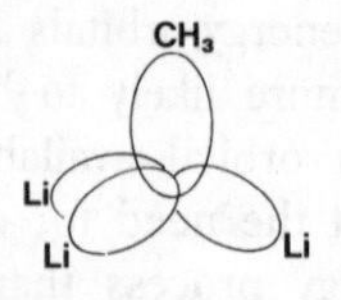

There is, thus, a four centre bond on each face of the tetrahedron and the four electron pairs fill up these four bonding MOs, i.e. there are four 4c–2e bonds and Li_4Me_4 is not electron deficient with respect to having all bonding MOs filled, but it is electron deficient with respect to 2c–2e bonding.

The fourth sp^3 orbital on Li contains no electrons and is not involved in bonding within the tetrahedron; it points directly outward from the vertex so that, in the solid state, we get interaction between the lithiums and methyl groups on adjacent tetrahedra.

How does covalent bonding explain the formation of Li aggregates?

1. Li has only one valence electron but four valence orbitals – by forming aggregates it uses its orbitals as efficiently as possible in an attempt to make up its electron deficiency.
2. If LiMe is compared with RbMe, LiMe contains tetrahedral aggregates but RbMe has an infinite ionic lattice structure; this is the result of two factors:
   (a) Li is a smaller atom – almost total envelopment of the Li_n group by the ligand polyhedron prevents association to give infinite polymers/ionic type lattices. Rb is bigger and not totally enveloped so that longer range forces, between Rb and Me beyond the first coordination shell, are possible.
   (b) There is more covalent character in the bonding within Li compounds. We saw in Chapter 3 (page 52) how Li, because of the high polarizing power of the very small Li^+ ion, differs from the other members of Group 1 in having a much greater tendency towards covalent bonding. Rubidium has a greater tendency towards ionicity (infinite lattice structure).

The degree of aggregation in Li alkyls depends on presence/type of donor solvent.

Toluene is a non-coordinating solvent and will not bond to Li, thus LiR groups aggregate to get the best bonding between themselves. Ethoxyethane is a weakly coordinating solvent; in solution, breaking up of the hexamer to generate the tetramer results in the release of an sp^3 orbital for bonding to the solvent. The Li can form stronger bonds to the solvent (2c–2e) than it can to itself in the hexamer. Ethoxyethane is not such a good solvent that it causes total break up of the structure; the bonds to the solvent are not *that* strong.

Dependence of the degree of aggregation of $LiCH_3$ on the solvent:

SOLVENT	OLIGOMER
toluene	hexamer
Et_2O	tetramer
TMEDA	monomer

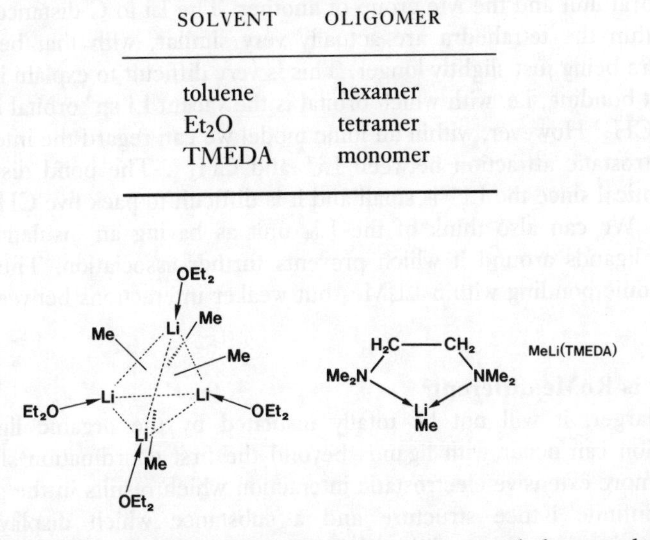

Tetramethylethylenediamine (TMEDA) is a very good donor solvent and the Li atom can form stronger bonds to the N donor atoms in TMEDA than it can to itself (2c–2e Li—N bonds are stronger than 4c–2e Li—C bonds). (TMEDA is a bidentate ligand and so we get some extra stability due to the chelate effect – see page 192.)

IONIC BONDING

Are lithium alkyls better represented as Li—R or $Li^+ R^-$?

There is a great deal of debate as to the answer to this question. The explanations given above are widely accepted but it is also possible to rationalize the structure, bonding and general properties of the compounds by considering the bonding as almost totally ionic. We still consider Li_4Me_4 as being made up of molecules but with ionic bonding within the molecules, i.e.

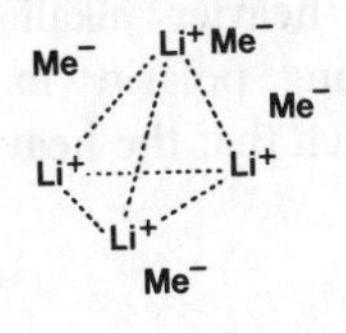

The aggregate is held together by electrostatic interaction between Li^+ and CH_3^-.

Why doesn't LiMe form an infinite ionic lattice?

As mentioned above, in the solid state, there is interaction between Li in one tetrahedral unit and the Me group of another. The Li to C distances between and within the tetrahedra are actually very similar, with that between the terahedra being just slightly longer. This is very difficult to explain in terms of covalent bonding, i.e. with which orbital is the vacant Li sp^3 orbital interacting on the CH_3? However, within an ionic model we can regard the interaction as an electrostatic attraction between Li^+ and CH_3^-. The bond distances are not identical since the Li^+ is small and it is difficult to pack five CH_3^- groups around. We can also think of the Li_4 unit as having an insulating coat of organic ligands around it which prevents further association. Thus there is strong ionic bonding within Li_4Me_4 but weaker interactions between Li_4Me_4 units.

Why is RbMe different?

Rb is larger, it will not be totally insulated by the organic ligands and interaction can occur with ligands beyond the first coordination shell. Thus we get more extensive electrostatic interaction which results in the generation of an infinite lattice structure and a substance which displays all the macroscopic properties traditionally associated with an ionic compound: high melting point, etc.

However, the properties normally associated with ionic or covalent compounds (ionic compounds have high melting points and melt to form ions; covalent compounds melt at lower temperatures, retaining the molecular integrity) are really a property of the gross structure of the compound, rather than the nature of the bonding (page 27). Thus a compound which was made up of 'ionic molecules' would be expected to have similar properties to a 'traditional' covalent compound, i.e. it would melt at relatively low temperatures to generate molecules.

Heavier alkali metals

Ionic character increases down the group as the electronegativity of the alkali metal atom increases. The heavier alkali metal alkyls are ionic lattice compounds and the increasing polarity in the bonds leads to reactivity increasing down the group, such that the heavier Group 1 alkyls are extremely reactive.

Group 2

Organoberylliums – how ionic are they?

The higher electronegativity of Be (Be 1.6, Li 1.0), the large amount of ionization energy required for formation of Be^{2+} and the high polarizing power of Be^{2+} lead to organoberylliums having more covalent character than their lithium analogues. (The high polarizing power of the small highly charged Be^{2+} ion causes a large amount of electron density to be pulled away from the anion, introducing substantial covalency into the bonding.) The structures of Be organometallics can be rationalized using a covalent multi-centre bonding scheme.

$Be{}^{Me}_{Me}Be$ is polymeric in the solid state,

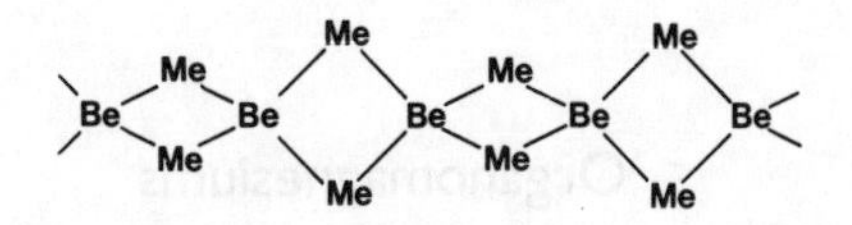

the bonding involving 3c–2e alkyl bridges,

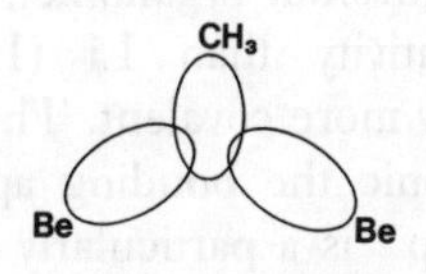

Each Be has two electrons and so in each $Be{}^{Me}_{Me}Be$ unit there is 1e from each Be and 1e from each Me, i.e. 4e in total, enough electrons for two 3c–2e bonds. So the compounds are not electron deficient in the sense that they have enough electrons to fill all the bonding MOs, but they are electron deficient in that there are not enough electrons for 2c–2e bonds between all atoms (see boranes, Chapter 8). The polymerization is an effort to use all valence orbitals as efficiently as possible, given the limited number of electrons available for bonding (i.e. in a monomeric structure each Be would only have a total of 4e in its valence shell, due to formation of two 2c–2e bonds).

With larger alkyl groups, polymerization is less favourable (four large groups around the small Be atom), e.g. $Be(Bu^t)_2$ is monomeric and linear.

R_2Be compounds are very strong Lewis acids and monomeric compounds such as $R_2Be(OEt_2)_2$ are readily made. In this compound the Be has a full outer shell of electrons.

Mixed Be alkyl hydrides can be formed and in all cases H bridges are favoured over alkyl bridges. This is due to a combination of steric and electronic factors:

1. The small size of the H atom means that it can more easily adopt a bridging position.
2. The Be—H—Be bridge is a closed three centre unit but the Be—CH_3—Be bridge is an open unit. So, with the Be—H—Be bridge the H nucleus lies more between the Be atoms, which gives a more direct interaction between the H nucleus and the electrons in the Be valence orbitals. In the open unit this interaction is less direct.

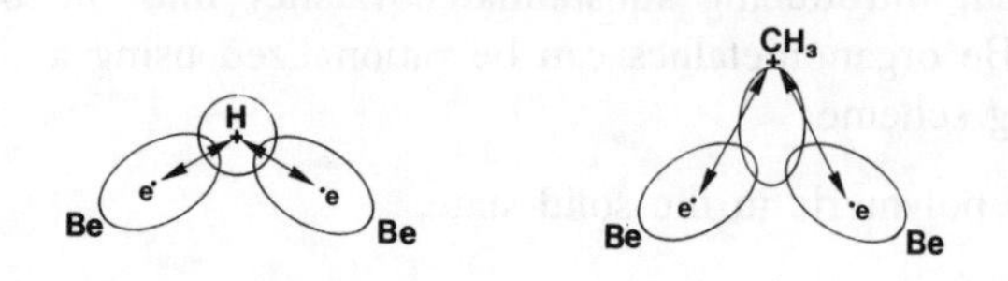

Organomagnesiums

As for lithium (diagonal relationship – see page 55), there is a great deal of debate as to how ionic magnesium organometallics are. However, Mg does have a higher electronegativity than Li (1.3 versus 1.0) and so its organometallics are generally more covalent. The nature of the R group has a strong influence on how ionic the bonding appears to be, e.g. $MgCp_2$ is regarded as mostly ionic (Cp^- is a particularly stable anion, page 123).

In solution there are the equilibria,

$$2RMgX \rightleftharpoons R{-}Mg(\mu\text{-}X)_2Mg{-}R \rightleftharpoons MgR_2 + MgX_2 \rightleftharpoons Mg(\mu\text{-}R)(\mu\text{-}X)_2Mg$$

The positions of equilibria depend on various factors, including the nature of R, X and the solvent.

In all polymeric structures of RMgX it is the halide, rather than the alkyl group, which bridges between the magnesiums. An Mg—R—Mg bridge is a 3c–2e bond whereas in Mg—X—Mg there are two 2c–2e bonds.

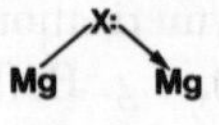

Heavier Alkaline Earth Organometallics

The preparation of these compounds is extremely difficult and they have not been widely studied. The degree of ionicity and reactivity increases down the group, as was observed for Group 1. The increase is much more marked than for Group 1 – electronegativity difference between Li and Cs is 0.2 and that between Be and Ba is 0.7.

Group 3

Organoborons

The electronegativity difference between B and C is quite small (0.5), so that the B—C bond is not very polar; this gives rise to fairly high kinetic stability. Most alkyls and aryls are thus stable in water, although they are fairly air sensitive (cf. Li alkyls, which can be spontaneously flammable in water – electronegativity difference between Li and C is large, 1.5). Because of the high ionization energy for the formation of the (unknown) B^{3+} ion and the small electronegativity difference between B and C, B organometallics are essentially covalent. These compounds are electron deficient with respect to the formation of 2c–2e bonds (Chapter 8).

BH_3 dimerizes to give diborane (Chapter 8) but BR_3 exists as a monomer; this is due to two factors:

1. *Steric terms*: the difficulty in packing four large R groups, rather than four small H atoms, around the small B atom.
2. *Hyperconjugation*: overlap between a C—H σ bonding orbital and the vacant p orbital on B partially alleviates the electron deficiency of B.

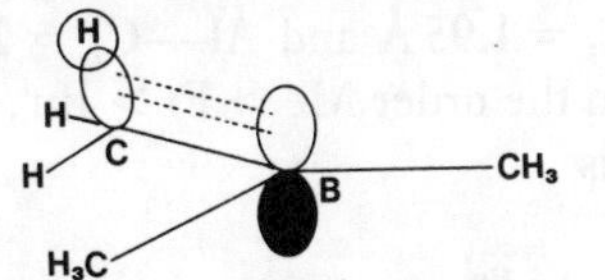

This hyperconjugation is lost upon dimerization (no longer a vacant p orbital on B).

π BONDING Strong π bonding between B and C is expected ($2p_\pi$–$2p_\pi$) and boron heterocycles, containing multiple bonding, can be prepared, e.g.

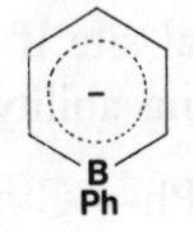

which is a 6e $(4n+2)\pi$ aromatic system (Chapter 4).

Organoaluminiums

The greater electropositivity of Al compared to B means that the E—C bond is more polar. This, together with the availability of fairly low energy d orbitals on Al, leads to AlR_3 compounds being extremely air and water sensitive (pyrophoric in air or water). AlR_3 is a much stronger Lewis acid than BR_3; this is due to a combination of steric and electronic factors:

1. *Steric*: it is easier to pack more ligands around the larger aluminium atom.
2. *Electronic*: Al is more positive than B in compounds MR_3, i.e. Al is more attractive to electron donors. There is little hyperconjugation in AlR_3 (because of the worse energy match between 3p on Al and C—H σ bond) and therefore little alleviation of the electron deficiency of Al.

In hydrocarbon solution and the solid state there is a tendency for AlR_3 to dimerize; this is very dependent on the size of the R group, e.g. Al_2Me_6 is a dimer but $Al(Bu^i)_3$ is a monomer.

The bonding in Al_2Me_6 can be considered as involving 'normal' 2c–2e bonds to the terminal methyls and two 3c–2e bridges (cf. boranes, Chapter 8),

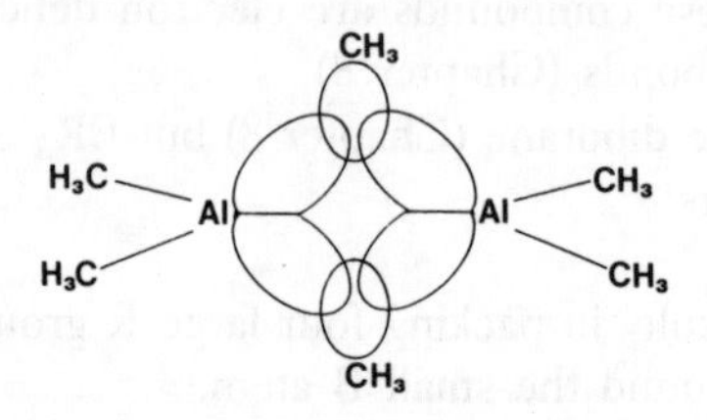

This picture is confirmed by the bond lengths to the terminal methyls being significantly shorter than those to the bridging methyls, e.g. Al_2Me_6; bond lengths are $Al—C_t = 1.95$ Å and $Al—C_b = 2.12$ Å.

Bridging ability goes in the order Me > Et > Bu^t, in accordance with steric factors, e.g. $MeBu^t{}_5Al_2$ is

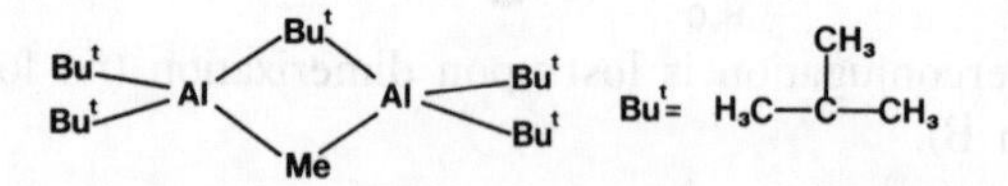

This is not in accordance with statistical prediction – there are twice as many terminal sites as bridging ones and therefore there would be a higher chance of the methyl occupying a terminal site if all things were equal.

In compounds $(AlR_2X)_2$ bridging ability goes in the order,

X = R_2N > RO > Cl > Br > Ph—C≡C > Ph

The bridge unit can be drawn as,

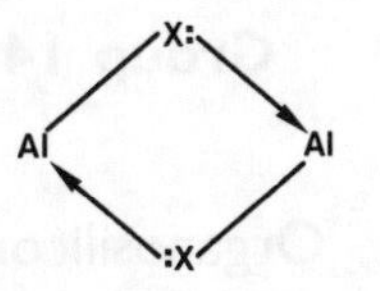

and the above order mirrors the Lewis basicity of the X group, i.e. R_2N is the strongest Lewis base.

If we consider the monomer–dimer equilibrium in these mixed species,

$$2R_2AlX \rightleftharpoons (R_2AlX)_2$$

then the favourability for the dimer follows the order,

$$H > Cl > Br > I > CH_3$$

This order is due to a balance between the competing factors of π bonding in the monomer (π bonding is lost upon dimerization) and Al—X—Al bond strengths in the dimer.

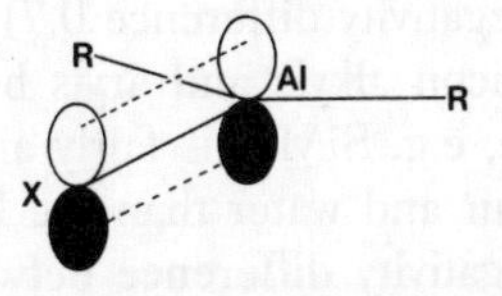

H cannot participate in π bonding and gives rise to reasonably strong 3c–2e bonds (see above). Cl forms stronger bridging bonds (2c–2e) but there is π bonding in the monomer which is lost upon dimerization. Br has weaker π bonding ($4p_\pi$–$3p_\pi$), which would tend to favour dimerization but weaker bridge bonds counteract this.

Heavier Group 13 Organometallics

These have no tendency to form dimers; there would be a great deal of repulsion between the large metal atoms,

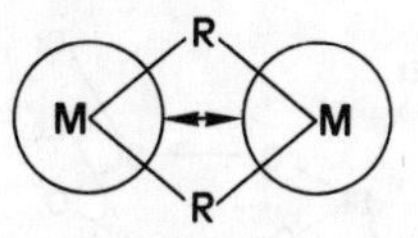

and reasonably weak (long) M—C bonds.

The MR_3 compounds are not as strong Lewis acids as AlR_3, and the Lewis acidity decreases down the group. This is a consequence of the acceptor orbitals becoming more diffuse, such that bonds to Lewis bases are not strong. The overall order of Lewis acidity in Group 13 organometallics is

$$B < Al > Ga > In > Tl$$

Group 14

Organosilicons

All SiR_4 molecules are pyramidal (VSEPR) with four 2c–2e covalent bonds and an octet in the outer shell of the Si, i.e. they are electron precise.

The Si—C bond is quite strong.

Bond energies (kJ mol^{-1})	C—C	C—Si
	358	311

This is a consequence of the reasonably high electronegativity of the Si, which leads to contracted valence orbitals and a good energy match with the C orbitals.

The low polarity (electronegativity difference 0.7) and the high strength of the Si—C bond leads to silicon alkyls and aryls being generally kinetically and thermodynamically stable, e.g. $SiMe_4$ is fairly air and water stable. They are generally more stable to air and water than are B organometallics; this is despite the smaller electronegativity difference between B and C, i.e. lower polarity in the B—C bond. This is because of the existence of a low activation energy pathway for attack at B by donation of electron density into a vacant 2p orbital; for Si, attack must occur at the relatively high energy 3d orbitals.

However, there is some polarity in the Si—C bond, with the Si being δ+. This polarity and the presence of fairly low energy vacant 3d orbitals on the Si makes it susceptible to nucleophilic attack, e.g.

$$Me_3SiCR_3 + OR^- \rightarrow Me_3SiOR + CR_3^-$$

A large number of organometallic silicon compounds contain Si also bonded to other atoms, such as O, N, S. Siloxane compounds such as

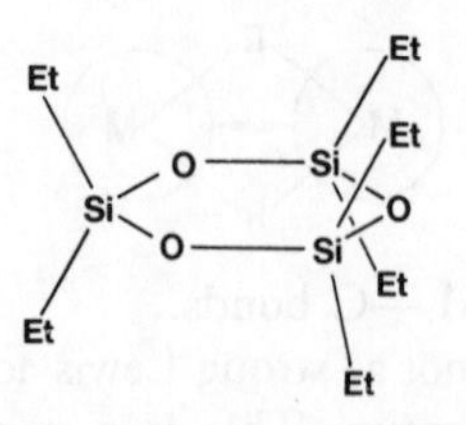

can be formed by hydrolysis of silicon alkyl halides. These siloxanes always exist as rings or chains containing formally Si—O single bonds. In this respect Si should be compared with C, which forms a large number of compounds containing C=O. This difference is due to $3p_\pi$–$2p_\pi$ bonding being less favourable than $2p_\pi$–$2p_\pi$. However, the Si—O bond is probably strengthened by some d_π–p_π interaction (see page 113).

COMPOUNDS CONTAINING MULTIPLE BONDS

Si=C

Simple compounds such as $HMeC{=}SiMe_2$ only exist at very low temperatures (kinetic stability); the Si=C π bond is much weaker than a σ bond ($2p_\pi$–$3p_\pi$; see Chapter 4) and there is a strong tendency to dimerize, or react, to form single bonds,

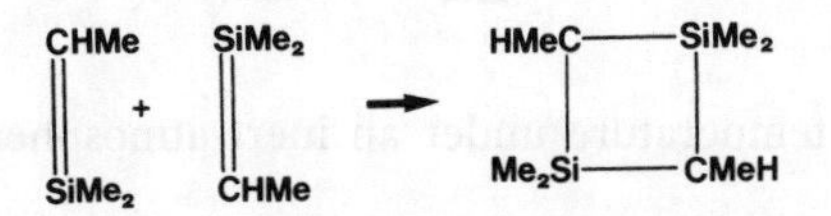

The dimers always involve just Si—C rather than C—C and Si—Si bonds:

Bond energies (kJ mol^{-1})	4×Si—C	4×311	C—C	358
			+Si—Si	222
			+2×Si—C	2×311
		1244		1202

Dimerization can, however, be inhibited by the presence of very bulky groups on Si and C, such that the close approach of Si=C units is prevented by large repulsions between the bulky groups on adjacent molecules, i.e. kinetic stabilization.

Silabenzene, an analogue of benzene,

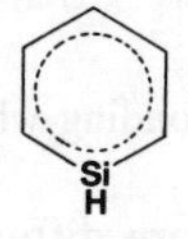

exists at low temperatures, but polymerizes as the temperature is raised.

Si=Si

Si=Si π bonds ($3p_\pi$–$3p_\pi$) are even weaker than Si=C π bonds and, once again, only exist at very low temperatures unless polymerization is inhibited by the presence of bulky groups. Thus,

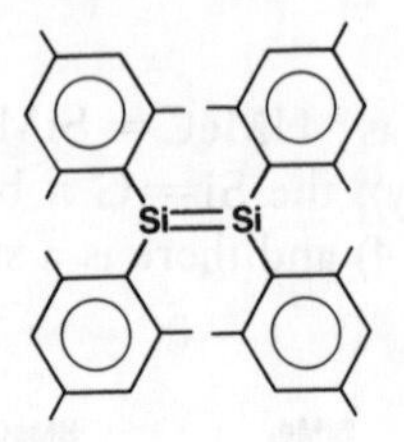

is stable at room temperature under an inert atmosphere.

Organogermaniums

The GeR_4 compounds, like their Si analogues, are quite thermally stable and fairly inert to air and water. This is the result of a reasonably strong Ge—C bond (213 kJ mol^{-1}) and the low polarity of the Ge—C bond (electronegativity difference 0.5). As for Si these compounds are susceptible to nucleophilic attack because of the availability of valence d orbitals. However, the lower polarity in the bond and the higher energy and greater diffuseness of the d orbitals makes Ge not so prone to nucleophilic attack.

Dimeric germoxanes, $R_3GeOGeR_3$, can be formed and, unlike their siloxane counterparts, they are bent,

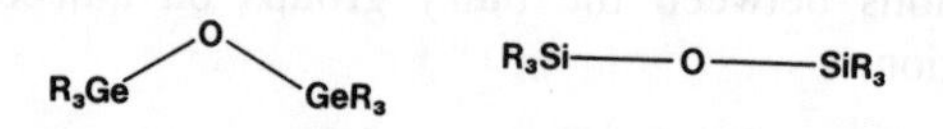

The linear structure of the siloxane can be rationalized by the use of 3d orbitals in d_π–p_π bonding with O, i.e.

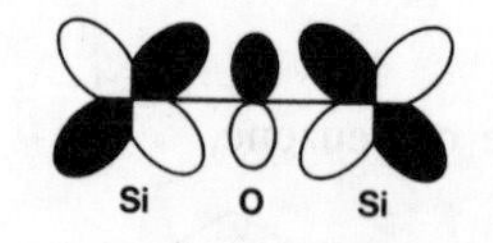

Maximum π bonding when group is linear

The 4d orbitals on Ge are more diffuse than Si 3d orbitals, therefore less good overlap is obtained with the O 2p orbitals. The weaker π bonding means that the molecule prefers a bent structure, which reduces steric interaction. However, the bond angle of about 140° (rather than the ideal tetrahedral angle of about 109°) does suggest that there is some d_π–p_π bonding present.

This less favourable use of d orbitals is also seen in the haloalkylgermanes, R_nGeX_{4-n} being less readily hydrolysed than their Si analogues; initial attack occurs at d orbitals on Si/Ge and the higher energy and greater diffuseness of the Ge orbitals renders them less susceptible to attack.

MULTIPLY BONDED COMPOUNDS

Ge=C

This π bond ($4p_\pi$–$2p_\pi$) is even weaker than the Si=C bond and compounds containing this group, e.g. germabenzene, only exist as transient intermediates in polymerization reactions (produce compounds containing all single bonds).

Ge=Ge

Similarly to Si, such a double bond can be stabilized by the presence of extremely bulky groups, which prevent close approach of Ge=Ge units in adjacent molecules, e.g.

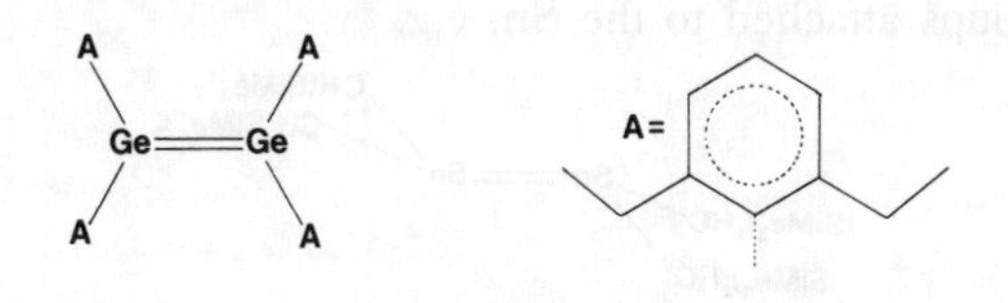

Organotins

Tetraalkyls and -aryls tend to be fairly stable (both kinetically and thermodynamically – reasons as above), e.g. $SnMe_4$ is inert to air and water and stable to about 400 °C.

The tetraalkyls react readily with halogens,

$$Me_4Sn + 2Br_2 \rightarrow \underset{\text{2 Sn—C broken}}{Me_2SnBr_2} + 2MeBr$$

compare with

$$Me_4C + Br_2 \rightarrow \underset{\text{No C—C broken}}{CMe_3CH_2Br} + HBr + MeBr$$

and

$$SiMe_4 + Br_2 \rightarrow \underset{\text{1 Si—C broken}}{SiMe_3Br}$$

This series of reactions illustrates the decreasing strength of E—C bonds down the group (interacting orbitals further apart in energy and valence orbitals are more diffuse).

Tin differs from lighter members of Group 14 in that its larger size (Si and Ge are about the same size but Sn is significantly larger – see Chapter 3) leads to an ability to very readily extend its coordination number beyond four, such that five coordinate adducts and polymers are readily formed, e.g. Me_3SnF

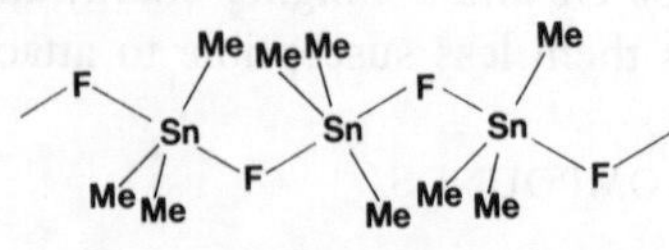

Stannoxanes are five coordinate cross-linked polymers,

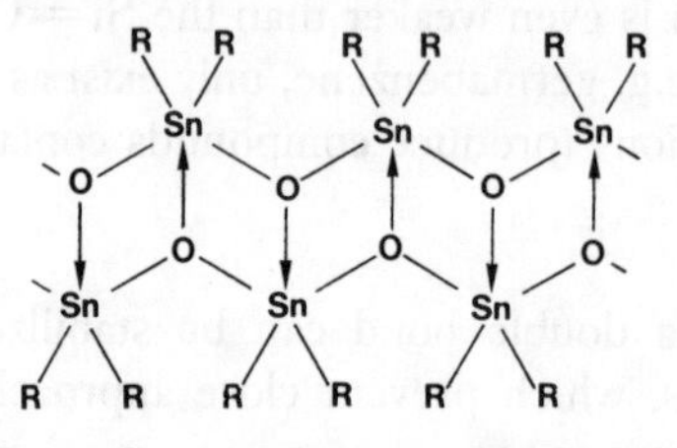

MULTIPLE BONDS Sn=Sn bonds can only be obtained when there are very bulky groups attached to the Sn, e.g.

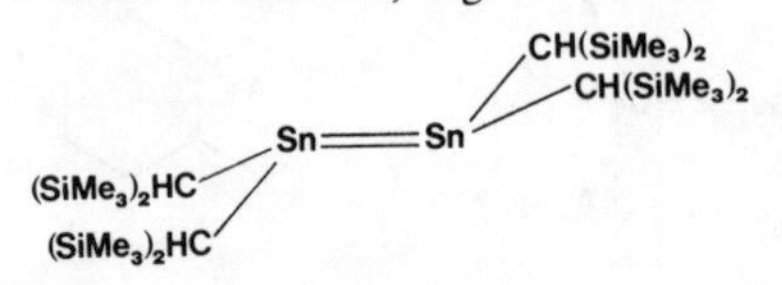

Organoleads

PbR_4 compounds are the least thermally stable in this group (weak Pb—C bond) but $PbMe_4$, for example, is still stable at room temperature and is not attacked by air or water.

The alkyls and aryls are readily cleaved, e.g.

$$PbEt_4 \xrightarrow{110\,°C} Et_3Pb\cdot + Et\cdot$$

With halogens all the Pb—C bonds are broken,

$$PbMe_4 + 4Cl_2 \rightarrow PbCl_4 + 4MeCl$$

which shows the weakness of the Pb—C bond (6p–2p σ bonds).

The organometallic chemistry of lead is broadly similar to that of tin. The large size of the lead atom leads to its having the greatest tendency towards formation of compounds with coordination number greater than four, e.g.

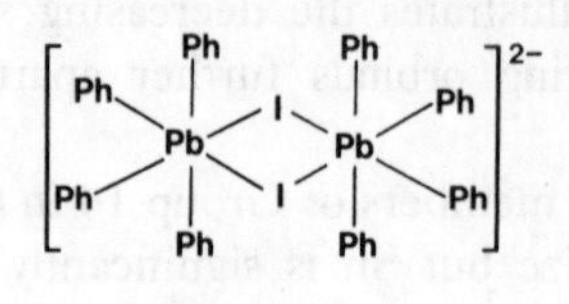

Although the Pb—Pb bond is weak (6p–6p σ interaction – approximately $100\,kJ\,mol^{-1}$), compounds containing this unit can be made, e.g. $Ph_3Pb—PbPh_3$.

The expected weakness of Pb=Pb ($6p_\pi$–$6p_\pi$) and Pb=C ($6p_\pi$–$2p_\pi$) π bonds is shown in that these units have not so far been observed.

Group 15 (As, Sb, Bi)

There are two classes of Group 15 organometallics: MR_3 and MR_5 compounds.

MR_5 molecules are trigonal bipyramidal (in accordance with VSEPR rules); the bonding has been discussed in Chapter 2. Low polarity in the M—C bond leads to these compounds being fairly water stable.

MR_5 can act both as a Lewis acid and a Lewis base, e.g.

$$Ph_3SbCl_2 + 3PhLi \rightarrow 2LiCl + Li^+[SbPh_6]^-$$

Lewis acid (accepts Ph^-)

$$SbPh_5 + I_2 \rightarrow [Ph_4Sb]^+I^- + PhI$$

Lewis base (donates Ph^-)

MR_3 molecules are pyramidal (cf. NH_3); the bonding involves exclusively 2c–2e bonds and results in an octet in the valence shell of M. Incomplete envelopment of the M by R groups leads to the MR_3 compounds being considerably more air sensitive than their MR_5 counterparts. Air sensitivity for MR_3 compounds goes in the order:

$$R_3Bi > R_3Sb > R_3As$$

most air sensitive — least air sensitive

This is the result of Bi being the largest M (more room for O_2 to attack) and the Bi—C bond being the weakest (longest, involving diffuse Bi 6s/p orbitals) and most polar (Bi most δ+ since it has the lowest electronegativity) of the three.

MR_3 can act as a Lewis base, there being a particularly varied chemistry involving coordination to transition metals, as in, e.g. $Cr(CO)_5(AsR_3)$. Donor ability (Lewis basicity) goes in the order:

$$R_3As > R_3Sb > R_3Bi$$

which mirrors the electronegativity of M and the concentration of the lone pair orbital. The greater electronegativity of the As means that it is more attractive to Lewis acids (although this also means that it would have a greater tendency to hold on to its lone pair); the less diffuse (more concentrated) lone pair results in a stronger bond to the Lewis acid.

The three coordinate alkyl halides undergo hydrolysis to give oligomers/polymers containing M—O single bonds (cf. nitrogen oxides, which contain N—O multiple bonds – $3p_\pi$–$2p_\pi$ bonds are weaker than $2p_\pi$–$2p_\pi$ bonds), e.g.

$$RAsX_2 \xrightarrow{H_2O} RAs(OH)_2 + 1/n(RAsO)_n$$

As—O—As angles of about 140° are consistent with partial use of d orbitals in bonding (cf. Si and Ge above).

CHAINS AND RINGS M—M bond strength goes in the order As > Sb > Bi, which is reflected in the stability of compounds containing these bonds. Thus, Me_2As—$AsMe_2$, with small R groups, is stable. In order to get air-stable compounds containing Sb—Sb bonds we require large ligands coordinated to the Sb, which inhibit the attack of nucleophiles, thus Me_2Sb—$SbMe_2$ is pyrophoric whereas $(Bu^t)_2Sb$—$Sb(Bu^t)_2$ is air stable.

The Bi—Bi bond is very weak and no organometallic compounds containing this unit are known to exist (although clusters such as $[Bi_9]^{5+}$ do exist; see page 279).

MULTIPLE BONDS

M=C

Stable compounds containing the M=C unit can be obtained by the use of bulky groups, incorporation into aromatic systems or coordination to transition metals, e.g.

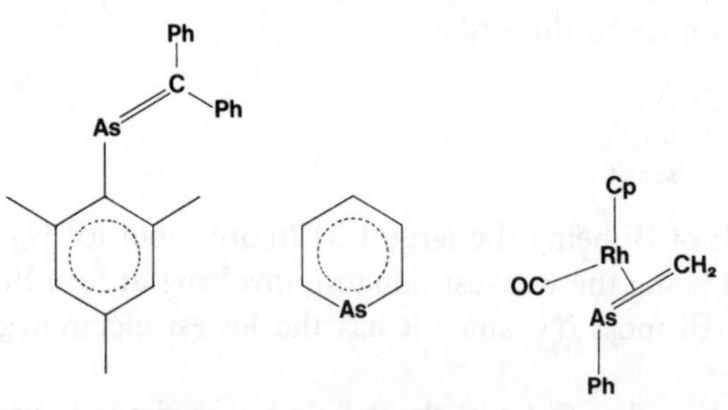

The stability of the benzene derivatives decreases down Group 15, in accordance with the weaker π bonding; thus phosphabenzene and arsabenzene are stable at room temperature in the absence of air but bismabenzene only exists as a transient species.

The Group 15 benzene derivatives are more stable than their Group 14 counterparts – a consequence of the increase in electronegativity across the Periodic Table. With a more electronegative element the orbitals used for π bonding are lower in energy and more concentrated, i.e. they match the C π orbitals better. Also the polarity in the M—C bond is lower, i.e. M is less δ+ and less prone to nucleophilic attack. Thus, silabenzene is unstable above about 10 K.

M=M

To get isolable compounds containing M=M we need to use very bulky ligands, e.g.

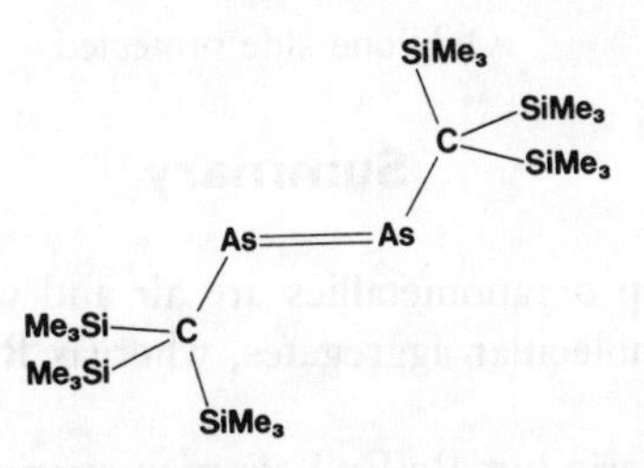

Although π bonding is stronger than for Group 14 (less diffuse orbitals for Group 15 give better overlap), larger ligands are generally required to stabilize multiple bonds involving Group 15 atoms. This is because the two coordinate Group 15 atom is more exposed than a three coordinate Group 14 atom to attack by, e.g. O_2.

Weaker π bonding between Sb atoms, compared to As, requires the use of big R groups *and* coordination to a transition metal,

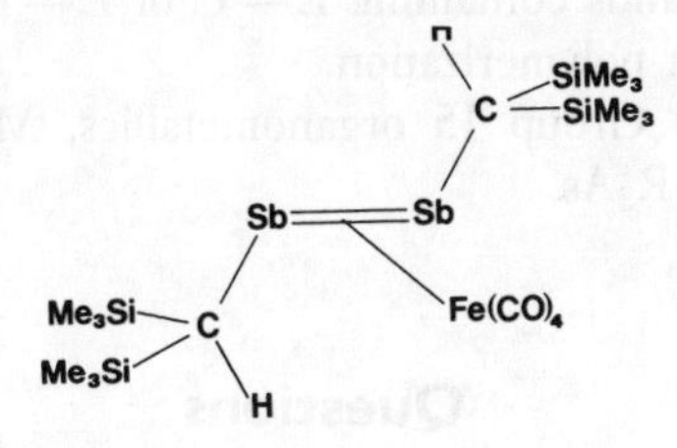

Group 16 (Se and Te)

Angular MR_2 compounds containing 2c–2e bonds are formed. There is very little tendency to form compounds in oxidation states above two; a manifestation of the inert pair effect (cf. S, which can form SR_4).

C=Se and Se=Se are quite stable. This is because Se orbitals are quite compact and reasonably low in energy and the electronegativity difference between C and Se is virtually zero, making the Se not at all susceptible to nucleophilic/electrophilic attack. Therefore excessively large groups are not required to stabilize these units, e.g.

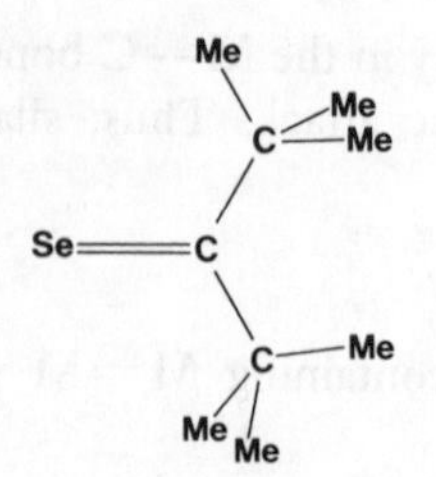

Only one side protected

Summary

1. Most main group organometallics are air and water sensitive.
2. Li alkyls form molecular aggregates, whereas Rb alkyls are infinite ionic solids.
3. $BeMe_2$ is polymeric but $Be(Bu^t)_2$ is monomeric.
4. Bridging ability generally goes in the order: halogen > H > alkyl.
5. Most boron alkyls and aryls are stable in water, but organoaluminiums are extremely water sensitive.
6. Lewis acidity for Group 13 MR_3 compounds goes in the order: M=B < Al > Ga > In > Tl.
7. Organosilicons are generally air and water stable.
8. Germoxanes, $R_3GeOGeR_3$, are bent; siloxanes, $R_3SiOSiR_3$, are linear.
9. Group 14 compounds containing E=C or E=E bonds require large R groups to inhibit polymerization.
10. Air sensitivity for Group 15 organometallics, MR_3, goes in the order: $R_3Bi > R_3Sb > R_3As$.

Questions

1 Compare and contrast:
(a) Lewis acidity,
(b) Lewis basicity,
(c) ability to act as a carbanion reagent
of the following:

LiPh, ClMgPh, BPh_3, $GePh_4$, $AsPh_3$, $AsPh_5$.

Answer

(a) Lewis acidity relates to the ability of a compound to accept an electron pair into its valence shell. For a compound to act as a Lewis acid it must have acceptor orbitals and a desire to gain electrons (i.e. be electron deficient).

If the bonding of LiPh ClMgPh and BPh_3 is considered as covalent, then they are all electron deficient with respect to forming 2c–2e bonds, and all three have vacant orbitals into which an electron pair can be donated. They are therefore strong Lewis acids, e.g.

$$BPh_3 + Ph^- \rightarrow BPh_4^-$$

$GePh_4$ and $AsPh_3$ are electron precise compounds, where both the Ge and the As have eight electrons in their outer shell. However, both Ge and As have five vacant valence 4d orbitals which are capable of accepting electrons from an electron pair donor, e.g.

$$AsPh_3 + Ph^- \rightarrow AsPh_4^-$$

Hence $GePh_4$ and $AsPh_3$ are both Lewis acids, although somewhat weaker than LiPh, ClMgPh, and BPh_3.

$AsPh_5$ also has vacant valence 4d orbitals (four of them) which can accommodate an electron pair, and so is also a Lewis acid, e.g.

$$AsPh_5 + Ph^- \rightarrow AsPh_6^-$$

$AsPh_5$ is a weaker Lewis acid than $GePh_4$ or $AsPh_3$ because it is more difficult to pack six relatively large ligands around a central atom than it is to pack five or four respectively.

(b) Lewis basicity relates to the ability of a compound to donate an electron pair. Of the compounds in the list only $AsPh_3$ has a lone pair of electrons (five valence electrons; three used in bonding and a lone pair) that can be donated (i.e. not used in covalent bonding).

(c) Carbanions are C^- species. In order for a compound to generate carbanions, the bonding within the molecule must have a large degree of ionicity, i.e. the resonance form $M^+ R^-$ must be a major contributor to the bonding. Ionic bonding will be more prevalent when the electronegativity difference between the metal and the C of the organic ligand is larger. This electronegativity difference is largest for Li and Mg (electronegativities 1.5 and 1.2 respectively), but much smaller for B, Ge and As. Therefore the bonding in Li and Mg organometallics is quite ionic whereas that in B, Ge and As organometallics is much more covalent in character, e.g.

$$LiPh + ClML_n \rightarrow PhML_n + LiCl$$

whereas,

$$GePh_4 + ClML_n \rightarrow \text{no reaction}$$

2 Dimethylphosphine reacts with diborane with the elimination of hydrogen and the formation of $\{(CH_3)_2PBH_2\}_3$. This compound is exceptionally stable; even the B—H bonds are very difficult to hydrolyse. Dimethylstibine, on the other hand, yields the relatively reactive monomer, $(CH_3)_2SbBH_2$. Discuss these observations.

Answer

In $\{(CH_3)_2PBH_2\}_3$ the P and B both have full outer shells,

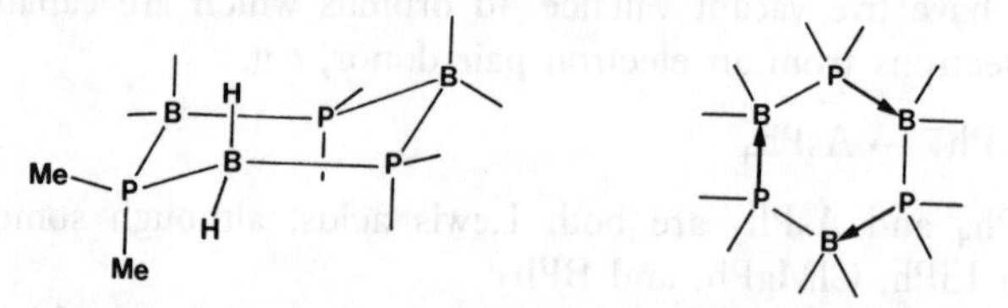

Note that there is no π bonding in this compound as B uses all its four valence orbitals in σ bonds. The B therefore has 8 electrons in its valence shell and is not electron deficient. The unreactivity of $\{(CH_3)_2PBH_2\}_3$ is attributable to two factors:

- The B is not sufficiently electronegative to cause the P 3d orbitals to be significantly lowered in energy, i.e. they are not very available for attack by nucleophiles.
- The electronegativity difference between P and B is very small (P is just slightly more electronegative) so that the P—B bond is essentially non-polar and therefore not susceptible to attack by nucleophiles/electrophiles.

The stability of $\{(CH_3)_2PBH_2\}_3$ is thus the result of kinetic factors.

$(CH_3)_2SbBH_2$ does not trimerize due to a combination of weak 4p–2p σ bonds and large steric effects associated with trying to pack two large Sb atoms into the coordination sphere of the small B. The B in this compound, forming three covalent bonds, is electron deficient (6e in its outer shell). The presence of a vacant p orbital on the B makes this compound very susceptible to nucleophilic attack – there is a low activation energy pathway for the process involving donation of a lone pair into the vacant p orbital on the B. The availability of the p orbital on B is reduced slightly by p_π–p_π bonding,

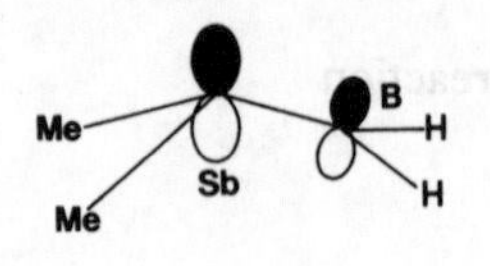

But this is not a large effect as the π bonding is between a 4p orbital on the Sb and a 2p orbital on the B and is therefore very weak.

CHAPTER 6
Transition Metals

Section A

Sc Ti V Cr Mn Fe Co Ni Cu Zn

First row transition metal atoms have the general electronic configuration $3d^n$ $4s^2$ (there are two slight exceptions, Cr and Cu, with just one electron in the 4s orbital) and the 4s orbital is lower in energy than the 3d. However, transition metal +2 ions all have the electron configuration $3d^n$ $4s^0$, i.e. it appears that lower energy s electrons, rather than higher energy d ones, have been removed.

There is actually no real problem here since it is not the energies of the individual electrons that have to be considered but, rather, the overall energy of the final ion; removal of electrons from the s orbital results in a more stable +2 ion than if the electrons had been removed from the d orbital. This is because the s electrons are able to shield the 3d electrons quite effectively so that, when they are removed, the d electrons feel a significantly higher effective nuclear charge ($z_{eff.}$) and are pulled in closer to the nucleus and hence stabilized. The 3d electrons do not shield the 4s electrons at all well (order of shielding is s > p > d > f – see page 3) and their removal would do little to stabilize the 4s electrons, i.e. to stabilize the overall ion.

Bonding in Transition Metal Complexes

The d orbitals are as follows:

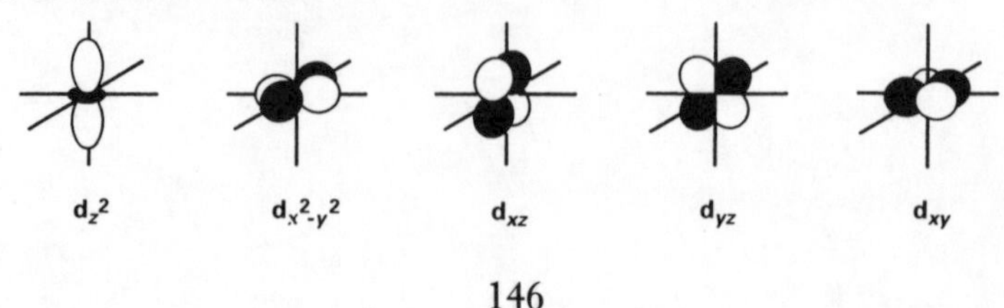

Transition metal complexes, ML_n, consist of a central metal atom surrounded by other atoms or groups, called ligands, e.g. $[Ti(H_2O)_6]^{3+}$. The bonding can be considered in exactly the same way as for main group complexes, such as SF_6, discussed on page 38. The most important case for consideration is an ML_6 octahedral complex.

If σ bonding is considered, then each ligand will have a σ bonding orbital pointing towards the centre of the octahedron. This could be a lone pair, as in H_2O, or a half-filled orbital, as in Cl.

The six ligands are considered as a single unit and the same six group orbitals, as in SF_6, can be drawn.

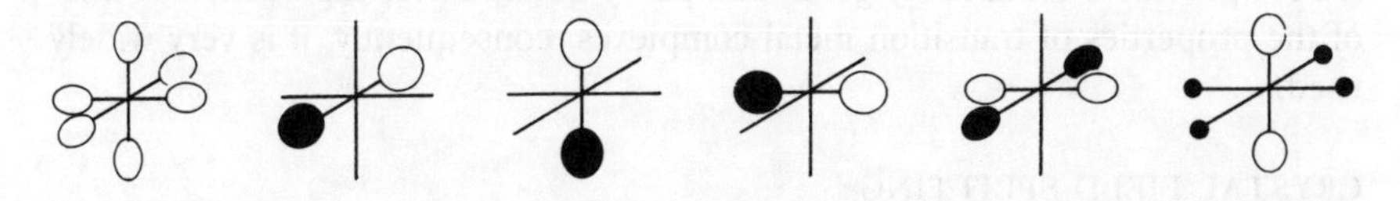

A first row transition metal atom uses its 3d, 4s and 4p orbitals for bonding (although in a transition metal atom the 4s orbital is lower in energy than the 3d this is not necessarily the case in a complex where the effective nuclear charge in the metal atom is different and the 3d is lower than the 4s – see later). The ligand group orbitals will in general be lower in energy than the metal 3d orbitals because of the greater electronegativity of the ligands (O is more electronegative than Ti).

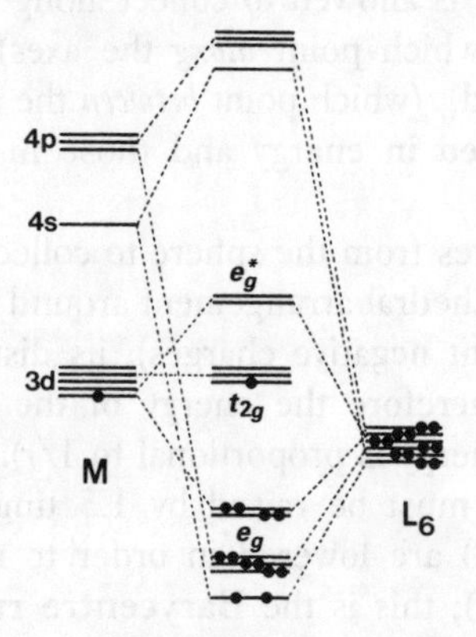

Only orbitals of the same symmetry interact, therefore just as for SF_6, the s orbital on M interacts with the s type group orbital, the p with the p type and the d with d type. There is no group orbital of the same symmetry as d_{xy}, d_{xz}

and d_{yz} orbitals on M and these metal orbitals remain non-bonding.

In filling orbitals, e.g. for $[Ti(H_2O)_6]^{3+}$, there are four electrons in the d and s orbitals of the Ti, 6×2 from each of the waters then -3 for the $+3$ charge. This gives 13 electrons available for occupying the MOs and these go in as shown.

Crystal Field Theory (A repulsive theory!)

This is a purely electrostatic approach to the bonding within transition metal complexes. This is not a very realistic approach, e.g. it would consider MnO_4^- as containing the Mn^{7+} cation (the ionization energy required to produce a $+7$ cation would be phenomenal, and the ion produced would be extremely polarizing so that it would pull electron density away from O^{2-}, inducing a large amount of covalency in the bonding) and four O^{2-} ions, but it does provide a remarkably good, and partly quantitative, explanation of a lot of the properties of transition metal complexes; consequently, it is very widely used.

CRYSTAL FIELD SPLITTING

Octahedral Complexes

The way the theory was originally developed considered a metal ion surrounded by a uniform sphere of negative charge, of some radius *r*. Electrons in the metal d orbitals will be repelled equally by this negative charge and therefore they are raised in energy but remain degenerate (equal energies).

If the negative charge is allowed to collect along the axes, electrons in the d_{z^2} and $d_{x^2-y^2}$ orbitals (which point *along* the axes) are repelled more than those in the d_{xz}, d_{yz} and d_{xy} (which point *between* the axes). Electrons in the d_{z^2} and $d_{x^2-y^2}$ are thus raised in energy and those in the d_{xz}, d_{yz} and d_{xy} are lowered.

When the charge moves from the sphere to collect at the axes (i.e. as if we had six ligands in an octahedral arrangement around the metal ion; the ligands are regarded as just point negative charges), its distance from the metal ion does not change and therefore the energy of the system does not change (electrostatic potential energy is proportional to $1/r$). This means that the two orbitals (d_{z^2} and $d_{x^2-y^2}$) must be raised by 1.5 times as much as the three orbitals (d_{xz}, d_{yz} and d_{xy}) are lowered in order to maintain the balance (no overall change in energy); this is the **Barycentre rule**.

The energy between the two sets of orbitals is called $\Delta_{oct.}$ (sometimes Δ_o or 10 Dq).

The d_{xz}, d_{yz} and d_{xy} orbitals are equal in energy (degenerate), because they

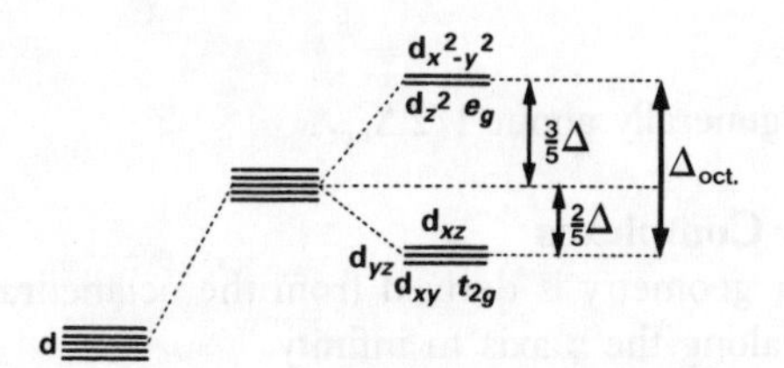

are the same shape and positioned identically with respect to the axes of an octahedron (which are all equivalent), but why should the d_{z^2} and $d_{x^2-y^2}$ orbitals be degenerate? The d_{z^2} orbital is actually a linear combination of two components,

It is made up of equal contributions from these two components. These are the same shape as $d_{x^2-y^2}$, point directly along the axes and will therefore be repelled to the same extent as the $d_{x^2-y^2}$. Another way of thinking about it is that the d_{z^2} actually has about 30 per cent of its electron density in the *xy* plane (due to the 'doughnut') and will therefore not just be repelled by ligands along the *z* axis, but also by those in the *xy* plane.

Tetrahedral Complexes

The tetrahedral geometry can be considered as derived from a cube,

This time the negative charge collects *between* the axes and electrons in the d_{xz}, d_{yz} and d_{xy} orbitals are repelled more than those in the d_{z^2} and $d_{x^2-y^2}$ orbitals.

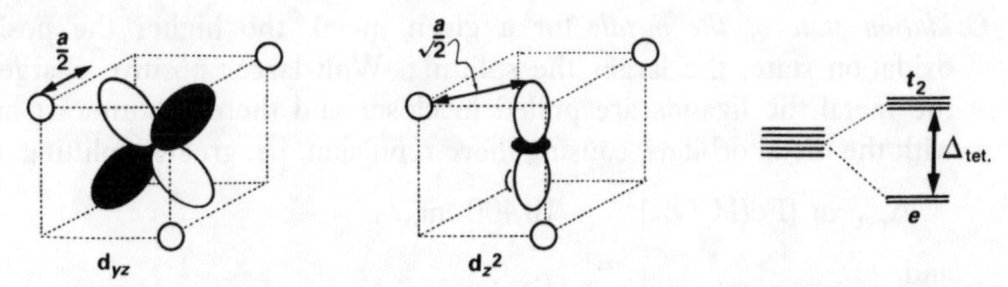

None of the orbitals points directly at the negative charge and therefore the stabilization/destabilization is not as great as for the octahedral case and the energy separation between the two sets of orbitals is smaller.

For a purely point charge model, with bond lengths and charges equal between the two geometries,

$\Delta_{tet.} = 4/9\Delta_{oct.}$

in reality, $\Delta_{tet.}$ is generally about $1/2\Delta_{oct.}$.

Square Planar Complexes

The square planar geometry is derived from the octahedral one by removing electronic charge along the z axis to infinity.

The $d_{x^2-y^2}$ orbital is highest in energy because it points directly at all four regions of electronic charge. The d_{xy} orbital, which does not point directly at any regions of negative charge but is the only other orbital to lie totally in the xy plane (which contains all the negative charge), is next highest. The separation between $d_{x^2-y^2}$ and d_{xy} remains as $\Delta_{oct.}$ since they are unaffected by the presence or absence of electronic charge along the z axis, and so for the xy plane the situation is the same as for the octahedral splitting.

The d_{z^2} orbital is the next highest below the d_{xy} because it has some electron density in the xy plane (the 'doughnut'). d_{xz} and d_{yz} remain degenerate and lowest in energy. It is important to realize that this is not a constant ordering.

(The splitting diagrams given above are all for ideal geometries and when deviations from these geometries occur it can be very difficult to predict the ordering of d orbitals. The splitting diagrams are then generally worked out from empirical data, e.g. electronic spectroscopy.)

FACTORS AFFECTING CRYSTAL FIELD SPLITTING

1. *Oxidation state of the metal*: for a given metal, the higher the positive oxidation state, the larger the splitting. With larger positive charge on the metal the ligands are pulled in closer and therefore interact more with the metal orbitals, causing more repulsion, i.e. greater splitting, e.g.

   $\Delta_{oct.}$ in $[Fe(H_2O)_6]^{2+} = 10\,400\,cm^{-1}$

   and

   $\Delta_{oct.}$ in $[Fe(H_2O)_6]^{3+} = 17\,400\,cm^{-1}$

2. *Geometry about the metal*: the spatial distribution of the ligands about the metal affects the splitting of the d orbitals, as discussed above, i.e. $\Delta_{tet.} \approx 1/2\Delta_{oct.}$.

3. *The position of the transition metal in the Periodic Table*: down a transition metal triad the splitting increases:

COMPOUND	$\Delta_{oct.}$ (cm^{-1})
$[Co(NH_3)_6]^{3+}$	22 900
$[Rh(NH_3)_6]^{3+}$	34 100
$[Ir(NH_3)_6]^{3+}$	41 000

This is because the 5d orbitals (third row) are more diffuse than their 4d (second row) and 3d (first row) counterparts. They therefore extend out further towards the ligands and there is greater repulsion between the metal d and ligand orbitals. (Remember that the interaction between metal and ligand is an electrostatic one and the repulsion is between non-bonding electrons on the anion and cation – the further out, towards the anion, that the metal electrons extend, the greater the repulsion.)

4. *The nature of the ligand*: this is incorporated into the spectrochemical series.

THE SPECTROCHEMICAL SERIES For a given metal ion the ligands can be arranged in order of increasing Δ, e.g.

$$I^- < Br^- < Cl^- < F^- < OH^- < O^{2-} < H_2O < NH_3 < PR_3 < CN^- \approx CO$$

Ligands such as CO, which cause a large value of Δ are termed **strong field**. Ligands such as I^-, which cause only a small value for Δ, are called **weak field**. This order is not absolute and minor reversals in order can occur.

From crystal field theory it would be expected that F^- would be a strong field ligand; being small and charged it would be expected to repel metal electrons strongly, causing large splitting of the metal d orbitals; however, the opposite is true. To explain this phenomenon, crystal field theory was adapted to incorporate covalent bonding – this is ligand field theory.

The difference between crystal and ligand field theories is that crystal field theory adopts a purely electrostatic approach to the bonding in transition metal complexes whereas ligand field theory, though it incorporates the same basic principles, i.e that the bonding is mostly ionic, also allows for some covalent contribution to the bonding in the form of localized σ and π interactions with the ligands.

STRONG AND WEAK FIELD LIGANDS Strong field ligands are able to accept electron density from the metal into high energy empty orbitals by the formation of π bonds with metal d orbitals of the correct symmetry. This is termed π acidity (a Lewis acid being an electron acceptor), e.g. for CO

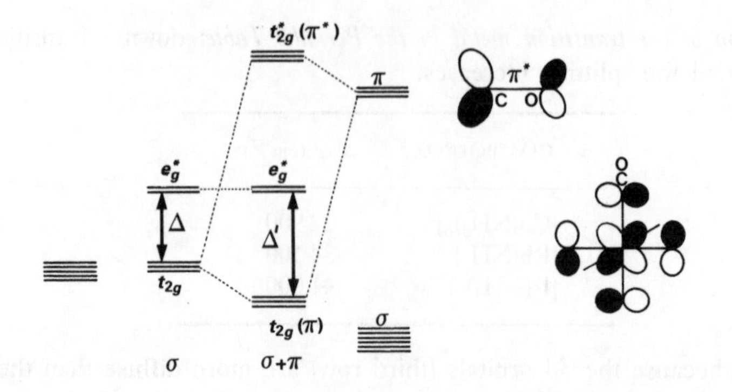

So we have C—O π^* antibonding orbitals interacting with the t_{2g} set of the metal producing a π bonding and a π antibonding set. The CO π^* orbitals are well above the t_{2g} orbitals so that the π bonding set produced is very similar to the metal t_{2g} in orbital character (page 22); the π^* set is very similar to the CO π^* orbitals. CO contributes no electrons (the π^* set is empty), therefore we have essentially purely metal electrons in the π bonding set. Δ is between the π bonding set and the e^*_g and is therefore larger than for just σ bonding.

Weak field ligands are able to donate electrons to the metal from low lying orbitals of π symmetry, e.g. for Cl^-

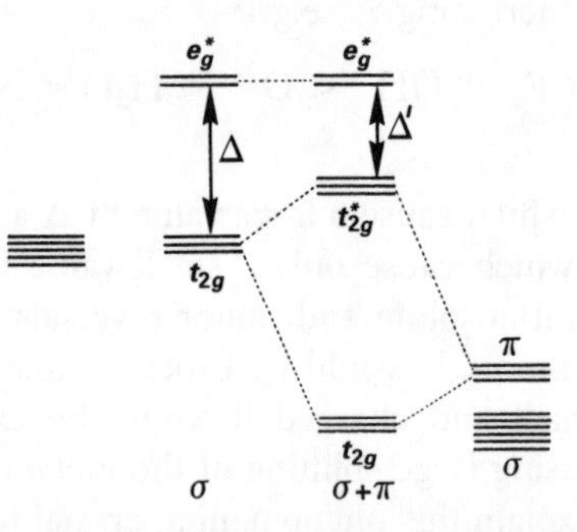

Once again the t_{2g} set of ligand orbitals interacts with the metal t_{2g} set to produce three bonding and three antibonding orbitals. In this case the lower, t_{2g} bonding, set is occupied by ligand electrons and the metal electrons go into the t^*_{2g} set. Δ is now the gap between the t^*_{2g} set and the e^*_g set and is therefore lower than for a purely σ bonding scheme.

Intermediate ligands, e.g. NH_3, have neither π donating nor π accepting orbitals and participate only in σ bonding.

LOW SPIN VERSUS HIGH SPIN COMPLEXES

Octahedral

For each of the electronic configurations d^3 to d^7 ions two configurations are possible, one having the maximum number of unpaired electrons, termed **high spin**, and the other having the minimum number of unpaired electrons, **low spin**. Which electronic configuration is obtained depends on two factors:

- The value of $\Delta_{oct.}$.
- The energy required to pair up electrons in the same orbital, **pairing energy** (*P*).

For an octahedral geometry when Δ is bigger than the pairing energy *P* it takes more energy to promote an electron to the higher energy e_g set than to pair two electrons in the lower energy t_{2g} set and a low spin electronic configuration results. When Δ is smaller than *P* the converse occurs and a high spin configuration results. In general *strong field ligands cause large Δ and usually give rise to low spin configurations; weak field ligands cause small Δ and usually give high spin configurations.*

Tetrahedral

Both high and low spin complexes are theoretically possible but because of the small value of Δ, i.e. $1/2\ \Delta_{oct.}$ it is always energetically more favourable to adopt the high spin configuration.

Square Planar

These are always low spin with respect to the occupation, or not, of the $d_{x^2-y^2}$ orbital (i.e. for a d^8 electronic configuration). This is due to the very large splitting between the the d_{xy} and the $d_{x^2-y^2}$ orbitals,

Low spin

CRYSTAL FIELD STABILIZATION ENERGY (CFSE) This is the energy by which a particular d electronic configuration is stabilized by the splitting of the d orbitals. The splitting energy is relative to the hypothetical case in which the d orbitals are unsplit, i.e. if the metal ion were just in a spherical field of charge. For an octahedral complex, an electron in a t_{2g} orbital contributes $-2/5\Delta_{oct.}$ to the crystal field stabilization energy whereas each e_g electron destabilizes the system by $+3/5\Delta_{oct.}$. For a tetrahedral complex this is the other way around, with the *e* electrons stabilizing by $-3/5\Delta_{tet.}$ and the t_2 electrons destabilizing by $+2/5\Delta_{tet.}$.

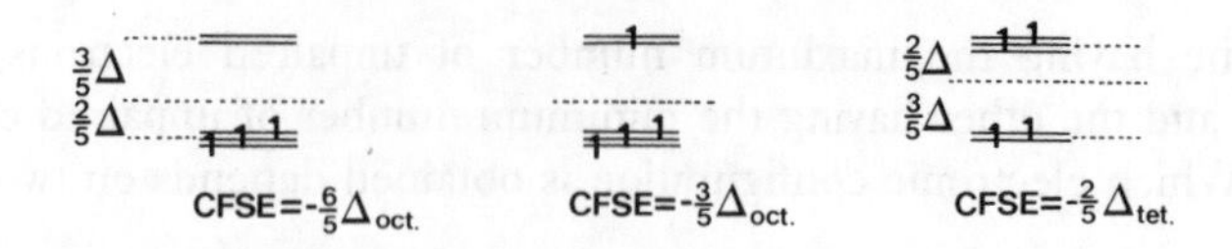

APPLICATIONS OF CFSE CFSE is usually only a small energy term (rarely more than 10 per cent of the total energy in the formation of any complex) but it does make its presence felt in a number of cases.

Hydration energies of the M^{2+} transition metal ions

The hydration energy is the energy change for the process,

$$M^{2+}_{(g)} + 6H_2O_{(l)} \rightarrow [M(H_2O)_6]^{2+}_{(aq)}$$

The diagram shows a plot of hydration energy against d electron configuration. As we go across the transition metal series the effective nuclear charge ($z_{eff.}$) increases so that the outer electrons are pulled in more closely, the size of the transition metal ion decreases, and so the charge to radius ratio (ionic potential) increases. The larger the charge to radius ratio, the more strongly the water molecules are pulled in closer to the cation and the higher the hydration energy. The smooth increase in hydration energy (the lower line) is what might be expected from the decreasing radius of the M^{2+} ion as we go across the period; the 'double-humped' curve is what is observed experimentally.

These complexes all involve high spin metal ions and if the calculated CFSE for each ion is subtracted from the double-humped curve, the smooth curve is obtained. The hydrated metal ions are thus 'more stable than expected' and this extra stability is the crystal field stabilization energy. Note that there is no CFSE for d^0, d^5 (high spin) and d^{10} electronic configurations, and these all lie on the hypothetical smooth curve.

The lattice energies of MX_2 (X = halide) salts

These salts consist of the transition metal +2 ion octahedrally surrounded by six halide ions in an infinite 3-D lattice. A double-humped curve is obtained for a plot of lattice energy against d electron configuration, with the 'extra stabilization' being due to the crystal field stabilization energy.

In the above two examples we are comparing like with like, i.e. the only real difference between the systems is in the electronic configuration of the metal ion, and hence the CFSE. It is usual that CFSE only generally becomes significant when other terms, such as steric effects, bond energies, cancel each other out.

Octahedral versus tetrahedral coordination

Three main factors determine whether a complex adopts an octahedral or tetrahedral geometry:

1. It is always more favourable to form six bonds instead of four {it is also more favourable to form 23 bonds instead of six but steric factors (inter-ligand repulsions) and the lack of suitable low energy orbitals generally dictate against this}.
2. Large and/or highly charged ligands will suffer large inter-ligand repulsions and so will prefer a tetrahedral geometry (less steric repulsion between four ligands than between six).
3. CFSE is always larger for octahedral than for tetrahedral {except for d^0, d^5 (high spin) and d^{10}} because Δ is approximately twice as large in octahedral complexes.

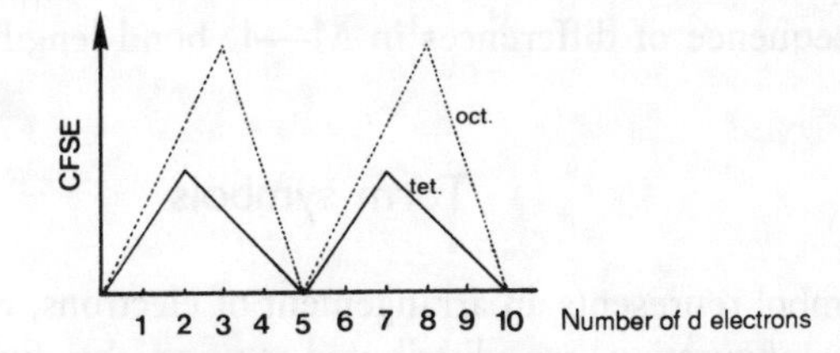

Thus, CFSE will always favour octahedral over tetrahedral {except for d^0, d^5 (high spin) and d^{10}} but the degree of favourability depends on the number of d electrons. The order of favourability for octahedral over tetrahedral is, for high spin complexes, d^3, d^8 (most favourable to have octahedral) > d^4, d^9 > d^2, d^7 > d^1, d^6 (where this order is calculated by subtracting the CFSE of an octahedral complex from that of a tetrahedral one). Thus, the tetrahedral geometry is most likely to occur, on the basis of CFSE, for d^1 (e.g. $[TiCl_4]^-$), d^5 (e.g. $[MnCl_4]^{2-}$) and d^6 (e.g. $[FeCl_4]^{2-}$), this is indeed found but, even for d^2 (e.g. $Cr(OR)_4$) and d^7 (e.g. $[CoCl_4]^{2-}$), the CFSE preference for octahedral is not large and tetrahedral complexes are quite common for these ions. For d^3 and d^8 there is a strong CFSE preference for the octahedral geometry and there are few tetrahedral complexes (e.g. $NiBr_2(PPh_3)_2$) with these numbers of d electrons. The CFSE preference for an octahedral geometry can be increased by having strong field ligands since these increase the value of Δ, and therefore magnify any differences in CFSE between octahedral and tetrahedral geometries. Also, strong field ligands can result in the formation of low spin octahedral complexes, which have large CFSEs (due to greater occupation of the lower energy d orbitals) and therefore a large preference for the octahedral geometry.

As stated above, CFSE is only one of a number of terms which are important in determining the geometry and very often it will not be the decisive factor.

Taking all things into account, *tetrahedral geometries will be most favoured over octahedral with large, negatively charged, weak field ligands*, such as Br^-.

LIMITATIONS OF CFSE

1. CFSE is generally only a small energy term, rarely more than about 10 per cent of the total energy involved in the formation of any complex and it can therefore be swamped by other energy terms such as bond energies, solvation energies, lattice energies, etc.
2. CFSEs are worked out in units of Δ and it is often forgotten that Δs vary between complexes. For example, if we want to compare the CFSEs of $[Cr(H_2O)_6]^{2+}$ and $[Fe(H_2O)_6]^{2+}$ we cannot simply say that it is $3/5\Delta$ for the former and $2/5\Delta$ for the latter since Δ is $14\,000\ cm^{-1}$ for the Cr complex and $10\,400\ cm^{-1}$ for the Fe one. The difference is mostly a consequence of differences in M—L bond lengths.

Term symbols

A term symbol represents an arrangement of electrons, e.g. how the electrons in a d^n configuration are distributed among the five d orbitals. These electronic arrangements will differ in energy depending on the extent to which the d electrons interact with each other, e.g. suffer coulombic repulsions.

FREE IONS We are principally concerned with the ground state. There is a simple method to predict the ground state term symbol of free (no ligands) transition metal ions. The following rules (Hund's rules) must be obeyed, *in this order*:

1. The ground state is that which has maximum S, i.e. the maximum number of unpaired electrons. For the ground state $S = M_s = \Sigma m_s$ where m_s

($= \pm 1/2$) is the spin of each electron.

2. The ground state has maximum L assuming that rule (1) is already obeyed. For the ground state $L = M_l = \Sigma m_l$ and m_l is related to the orbital angular momentum of each electron. m_l is the component of the orbital angular momentum in a specified direction, e.g. it tells us which particular p orbital (p_x, $m_l = +1$; p_y, $m_l = -1$; p_z, $m_l = 0$) an electron occupies.
3. The ground state has minimum J if the shell is less than half full and maximum J if it is more than half full. Where J, the total spin–orbit angular momentum quantum number, can take the values:

$$L+S, L+S-1, L+S-2, \ldots, L-S$$

EXAMPLES

d^2

Draw out a set of five boxes to represent the d orbitals, with $m_l = +2$ at the left-hand side and $m_l = -2$ at the right-hand side.

m_l	+2	+1	0	−1	−2

Put electrons in from the left-hand side, keeping the maximum number of unpaired spins (rule 1)

+2	+1			
1	1			

Now

$$S = M_s = \Sigma m_s = 1/2 + 1/2 = 1$$
$$L = M_l = \Sigma m_l$$

i.e.

$$L = 2 + 1 = 3$$

Different L values are given symbols,

$L =$ 0	1	2	3	4	5	...	cf.	$l =$ 0	1	2	3	...
S	P	D	F	G	H	...		orbital s	p	d	f	...

A term symbol is written as

$$^{2S+1}L_J$$

where $2S+1$ is the **multiplicity**, e.g. for d^2, the ground state term symbol is ^3F, i.e. a 'triplet F term'.

For states for which $S \neq 0$ and $L \neq 0$, spin–orbit coupling can be obtained. This is best considered as coupling between spin and orbital angular momentum, i.e. if there is a spin vector and an orbital angular momentum vector, they can couple together to give a resultant spin–orbit vector,

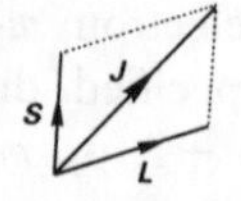

$\mathcal{J}$, the total spin–orbit angular momentum quantum number can take values,

$L+S$, $L+S-1$, $L+S-2$ ... $[L-S]$

thus, for d^2, $\mathcal{J}$ can equal 4, 3, 2.

Therefore, applying Rule 3, the full ground state term symbol for a d^2 (less than half-filled shell) electronic configuration is $^3F_2$

d^7

+2	+1	0	−1	−2
↑↓	↑↓	↑	↑	↑

$S = 3/2$ therefore $2S+1 = 4$
$L = 3$
$\mathcal{J}$ takes values 9/2, 7/2, 5/2, 3/2

Therefore the full term symbol for the ground state is $^4F_{9/2}$. Here $\mathcal{J}$ takes its maximum value because the shell is more than half full.

APPLICATION OF A CRYSTAL FIELD In the presence of a crystal field the ground state term symbol is split. There is no effect on the spin and hence the multiplicity remains the same. The term is split in the same way as atomic orbitals:

S → A_1
P → T_1
D → E and T_2
F → A_2, T_2 and T_1 – the T_2 term always come in the middle.

An A term is singly degenerate; an E term is doubly degenerate; a T term is triply degenerate.

Once again, *g* and *u* labels only apply to complexes with a centre of symmetry (see page 41).

Now, just as a set of d atomic orbitals is split by an octahedral crystal field into t_{2g} and e_g orbitals, a D term is split into an E_g term and a T_{2g} term. However, whereas in an octahedral complex the t_{2g} set of orbitals is always lower than the e_g set of orbitals, when a D term is split by an octahedral field, in some complexes the E_g term is the ground state term but in others the T_{2g} term is lower. *It must always be remembered that these term symbols refer to arrangements of electrons in the* t_{2g} *and* e_g *orbitals, where the* t_{2g} *orbitals are always lower in energy for octahedral complexes.*

A T label describes an electronic configuration with orbital angular momentum, i.e. $L = 1$ but for A and E terms there is no orbital angular momentum, i.e. $L = 0$.

We can predict the ground state term symbol by considering whether or not we have orbital angular momentum for a particular electronic configuration, e.g. for d^1 octahedral

e_g

t_{2g}

The d_{xy}, d_{xz} and d_{yz} orbitals (the t_{2g} set) are degenerate and have the same shape and hence it can be considered possible to rotate these orbitals into one another. If there is an electron in one of these orbitals it could be imagined that rotation of one orbital into another results in the electron rotating about the nucleus. Such an electron would have orbital angular momentum associated with it. For d^1, the free ion term symbol is ^2D. This will split into ^2E_g and $^2T_{2g}$ in a crystal field. A T term has some orbital angular momentum, an E term does not, and therefore for an octahedral case we have a $^2T_{2g}$ ground term.

d^3 octahedral

The free ion ground term is ^4F which, in a crystal field will split into $^4A_{2g}$, $^4T_{2g}$ and $^4T_{1g}$. The arrangement of electrons in the orbitals is

e_g

t_{2g}

Now it is not possible to rotate one t_{2g} orbital into another because each orbital contains an electron of the same spin and two electrons of the same spin cannot occupy the same region of space (Pauli principle). Therefore because the electrons are not rotating about the nucleus, there is no orbital angular momentum present and we have a $^4A_{2g}$ ground state (the $^4T_{2g}$ and $^4T_{1g}$ terms have orbital angular momentum).

d^4 octahedral (high spin)

The free ion ground term is ^5D.

In this case we have one electron in the e_g set. The e_g orbitals ($d_{x^2-y^2}$ and d_{z^2}) are not of the same shape and cannot be rotated into each other. Hence an electron in the e_g set cannot have orbital angular momentum.

e_g

t_{2g}

Therefore the ground term is ^5E_g.

Two important points:

1. This simple approach to working out term symbols *only applies to high spin complexes*. For low spin the crystal field splitting is larger than the inter-electronic repulsion and the situation is more complex.
2. We have used the **Russell-Saunders coupling scheme**. This considers spin–spin coupling between electrons (e.g. singlet versus triplet) to be greater than orbit–orbit coupling (e.g. P versus D) and then spin–orbit coupling is just regarded as a minor perturbation.

Electronic spectra of transition metal complexes

In this section we are concerned with electronic spectra. There are three types of transitions which give rise to spectra in transition metal complexes:

1. Charge transfer, either metal to ligand or ligand to metal.
2. Intra-ligand spectra – transitions between ligand orbitals.
3. d $\longleftrightarrow$ d transitions.

TYPE OF TRANSITION	REGION OF ELECTROMAGNETIC SPECTRUM
Charge transfer	Ultraviolet but can have discrete bands or tails in visible
Intra-ligand	Ultraviolet
d $\longleftrightarrow$ d	Usually visible region

Selection rules tell us whether a particular electronic transition is *allowed* or *forbidden*. The selection rules for electronic transitions are:

1. $\Delta S = 0$, i.e. there is no change in the total spin of the electrons
2. $\Delta l = \pm 1$
3. $g \longleftrightarrow u$

Rules 2 and 3 combine to give the **Laporte selection rule**, which states that a transition is forbidden if it simply involves a redistribution of electrons in the same type of orbital in a particular quantum shell: 3d $\longleftrightarrow$ 3d transitions are therefore forbidden.

CHARGE TRANSFER SPECTRA A charge transfer transition is one between two molecular orbitals, one of which has mostly ligand character and one which has mostly metal character. There are thus two types: metal to ligand or ligand to metal. For example:

- MnO_4^- and CrO_4^{2-}: the colours of these are due to ligand to metal charge transfer transitions as both metal ions are formally d^0 (Mn^{7+} and Cr^{6+} have no d electrons for d $\longleftrightarrow$ d transitions).
- M to L charge transfer can be seen in complexes such as $[Fe(phen)_3]^{2+}$, containing ligands with low lying π^* orbitals, which can accept electron density from the metal.

Ligand to metal charge transfer transitions are favoured by having a readily reducible metal ion (high oxidation state metal) and easily oxidizable ligands (i.e. more favoured for bromine than for chlorine). Metal to ligand charge transfer is favoured by an easily oxidized metal ion (low oxidation state) and a readily reduced ligand.

INTRA-LIGAND SPECTRA These involve transitions between two orbitals which have mostly ligand character. They are usually found in complexes containing aromatic ligands where π to π^* transitions occur, e.g. pyridine, bipy.

d $\longleftrightarrow$ d SPECTRA Electronic transitions between d orbitals of the same quantum shell are strictly forbidden by the Laporte rule, therefore they should never be obtained. However, partial relaxation of the selection rules can occur and *weak* bands are generally observed, principally in the visible region.

There are usually several d $\longleftrightarrow$ d absorption bands representing electronic rearrangements within the set of d orbitals, in the spectrum of a given complex. The number and position of the bands depends on:

1. *The metal and its oxidation state*, e.g. the spectra of $[Co(H_2O)_6]^{2+}$, $[Co(H_2O)_6]^{3+}$ and $[Ni(H_2O)_6]^{2+}$ are all very different: there are different numbers of bands, the bands have different shapes and different intensities.

   The factors that influence the spectra are:

   (a) the electronic configuration of the metal, i.e. the number of electrons and whether the metal complex is high or low spin – different electronic configurations give a completely different set of transitions;

   (b) the oxidation state of the metal. If the electronic configuration is the same but the oxidation state is different (e.g. Mn^{2+}, d^5 high spin and Fe^{3+}, d^5 high spin), the number and general appearance of the bands will be very similar but the increased Δ for the higher oxidation state will result in the bands coming at different energies.

2. *Changing the geometry*, e.g. the octahedral complex $[Co(H_2O)_6]^{2+}$ is pale pink but the tetrahedral $[CoCl_4]^{2-}$ is bright blue – see below.
3. *The ligands around the metal*, which is a manifestation of the spectrochemical series (see above), e.g. $[Ni(NH_3)_6]^{2+}$ is blue–violet but $[Ni(H_2O)_6]^{2+}$ is green. Here, the general form of the bands will be similar but the transitions will occur at different frequencies, due to different values of Δ.

INTENSITIES OF LINES: THE BREAKDOWN OF THE LAPORTE RULE Relaxation of the Laporte rule must occur for d ⟷ d transitions to be observed in transition metal spectra. There are a number of ways in which this can occur:

Octahedral complexes

1. Vibronic coupling, which is the main method for relaxation of the selection rules in octahedral complexes.

   For an octahedral complex there are allowed vibrations which produce an antisymmetric array of ligands about the metal at a particular moment in time. When no centre of symmetry is present the *g* and *u* labels are no longer applicable. *Electronic transitions occur much faster than vibrational motion* (the Franck–Condon principle)† and the electronic transitions may very crudely be thought of as occurring when the molecule has an antisymmetric distribution of ligands for which the centre of symmetry and, hence the *g* and *u* labels, have been destroyed. In the absence of a centre of symmetry mixing between d and p orbitals can occur because they are now of the same symmetry (in an octahedron the d orbitals are *g* symmetry but the p orbitals are *u* symmetry, therefore no mixing can occur) and so electronic transitions are no longer purely d ⟷ d but become partially d ⟷ p ($\Delta l = \pm 1$) and are therefore partially allowed.
2. π Acceptor and π donor ligands are able to interact with d orbitals on the metal to produce a bonding and an antibonding set of π orbitals (see spectrochemical series). In accordance with the LCAO method (page 18) the π orbitals will have a mixture of both ligand and metal orbital character and hence electronic transitions are no longer purely d ⟷ d, thereby relaxing Laporte rule 2.

† Electronic transitions are fast compared to the time taken for a molecule to vibrate, and therefore electronic transitions occur with all atoms 'frozen' in position.

How do these electronic spectra come about?
Only for the d^1 octahedral system can we talk of an electron being promoted from a t_{2g} to an e_g orbital. For the other electronic configurations movement of one electron influences all the other electrons and we must therefore use term symbols – *lines in a spectrum correspond to transitions between electronic terms* (the term symbols represent different arrangements of electrons within the set of d orbitals and so transitions between terms correspond to rearrangements of electrons within the d orbitals).

In the spectrum of $[Ti(H_2O)_6]^{3+}$ there is one band, corresponding to the transition $^2T_{2g} \rightarrow {}^2E_g$ (for d^1 this is the same as an electron being transferred from the t_{2g} to the e_g set of orbitals); the energy of this transition is Δ. However, in the spectrum of $[V(H_2O)_6]^{3+}$ there are three d $\longleftrightarrow$ d transitions (only two are actually observed, the third being masked by a charge-transfer band). The electronic ground term of the free V^{3+} ion is ^3F; in an octahedral field this is split into $^3T_{1g}$, $^3T_{2g}$ and $^3A_{2g}$ terms, with the $^3T_{1g}$ lowest. Thus, we get two transitions,

$$^3T_{1g} \rightarrow {}^3T_{2g} \text{ and } {}^3T_{1g} \rightarrow {}^3A_{2g}$$

The third transition arises from the next lowest term in the free ion being ^3P, which gives rise to $^3T_{1g}$ term in an octahedral crystal field. The third transition is

$$^3T_{1g}(F) \rightarrow {}^3T_{1g}(P)$$

These transitions can be summarized as follows:

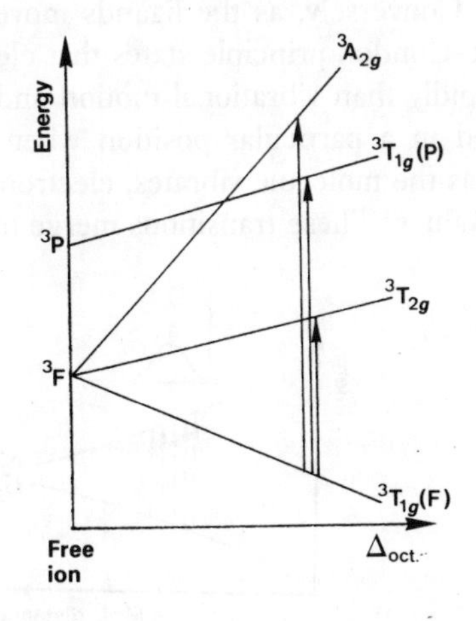

Tetrahedral complexes

The two main differences between the spectra of octahedral and tetrahedral complexes are:

1. For tetrahedral complexes the bands are further towards the lower energy end of the visible region than for octahedral: $\Delta_{tet.} \approx 1/2\ \Delta_{oct.}$.
2. The bands are usually much more intense (approx. 1000 times), e.g. $[CoCl_4]^{2-}$ is an intense blue colour.

The reason for the greater intensity is that tetrahedral complexes do not possess a centre of symmetry. The *g* and *u* labels no longer apply at all (the orbitals are t_2 and *e* and selection rule 3 is void) and much better mixing can occur between d and p orbitals than for octahedral complexes (where it only occurs for antisymmetric vibrational modes). The transitions are therefore partially d $\longleftrightarrow$ p and partially allowed. Although bands are more intense for tetrahedral than for octahedral, even for tetrahedral the transitions are not fully allowed because they are still only *partially* d $\longleftrightarrow$ p.

Band widths

Broad absorption bands are characteristic of the spectra of transition metal complexes. There are three main reasons for this:

1. *Vibrations*: if we consider the symmetric breathing mode of an octahedral complex then it is clear that the ligand field is not constant but continually varies between some maximum value, i.e. minimum M—L distance, and some minimum value, i.e. maximum M—L distance. Thus, as the ligands all move out together $\Delta_{oct.}$ decreases because there is less interaction between the ligands and the metal d orbitals, therefore less repulsion. Conversely, as the ligands move in $\Delta_{oct.}$ increases.

   The Franck–Condon principle states that electronic transitions occur much more rapidly than vibrational motion and therefore a molecule is essentially fixed in a particular position when an electronic transition occurs. Thus, as the molecule vibrates, electronic transitions occur for a range of $\Delta_{oct.}$ values. These transitions merge to give a broad absorption band.

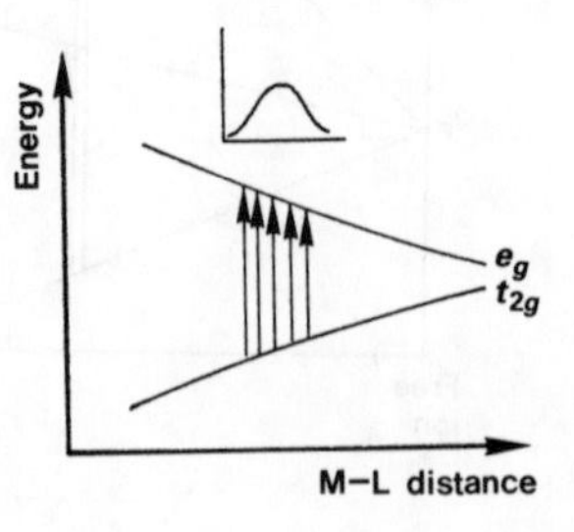

2. *Spin–orbit coupling*: this is small for the first row transition metals and so only a small splitting of upper and lower levels occurs. Transitions occur

between these levels but the energy differences are so small that instead of appearing as separate transitions they merge into a broad band.

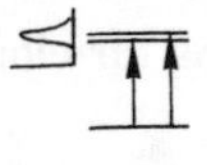

3. *The Jahn–Teller effect*: this can also cause small splitting and hence broadening of bands.

THE JAHN–TELLER EFFECT The Jahn–Teller theorem states that: **Any non-linear molecule in a degenerate electronic state will distort so as to remove that degeneracy**. Or, to put it another way, when a degenerate set of orbitals is occupied unsymmetrically then a distortion of the molecule will occur which will remove the degeneracy of the orbitals.

The best way to illustrate this is with an example: consider an octahedral Cu^{2+}, d^9, complex, the crystal field splitting diagram is as follows,

e_g

t_{2g}

The e_g set consists of the $d_{x^2-y^2}$ and d_{z^2} orbitals. There are three electrons to occupy these two degenerate orbitals and therefore an unsymmetrical electronic arrangement must result and a distortion will occur. If two electrons go into the d_{z^2} orbital and only one into the $d_{x^2-y^2}$, then ligands along the z axis will be repelled more than ligands in the xy plane, i.e. the ligands along the z axis will be further from the metal ion than those in the xy plane. A secondary effect is that ligands moving out along the z axis allow closer approach to the metal of those in the xy plane (less inter-ligand repulsion).

If the ligands along the z axis are further from the metal then the d_{z^2} orbital will be stabilized relative to the $d_{x^2-y^2}$ since the electrons in the d_{z^2} orbital will feel less repulsion from the ligand electrons.

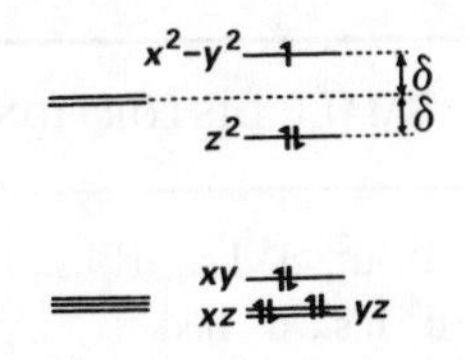

This distortion results in a stabilization energy of δ because two electrons in the d_{z^2} orbital are stabilized but only one in the $d_{x^2-y^2}$ is destabilized.

Although it seems perfectly reasonable that the two electrons should occupy the $d_{x^2-y^2}$ orbital, this does not occur and all complexes exhibit an elongation along the z axis. This is the result of mixing between the d_{z^2} and the s orbital, which can occur when the symmetry of the complex is reduced from octahedral by two ligands moving out along the z axis or four ligands moving out in the xy plane.

Consider the splitting diagrams for elongation in the z direction (**I**) and in the xy plane (**II**)

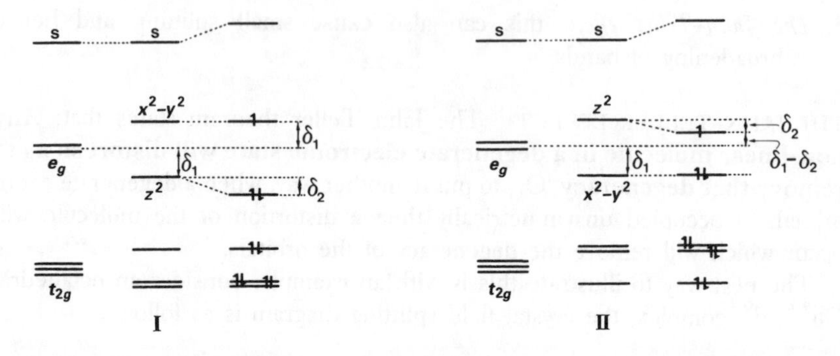

In **I**, s-d_{z^2} mixing results in the s orbital being raised in energy and the d_{z^2} lowered. Now the two electrons in the d_{z^2} orbital are stabilized by a total amount $2(\delta_1 + \delta_2)$, giving total stabilization $2(\delta_1 + \delta_2) - \delta_1 = \delta_1 + 2\delta_2$.

In **II**, only one electron is stabilized by an amount δ_2, therefore the overall stabilization is $2\delta_1 - (\delta_1 - \delta_2) = \delta_1 + \delta_2$. Therefore stabilization is not as large and elongation in the z direction is more favourable.

For elongation along the z axis the t_{2g} set will also be split; the d_{xz} and d_{yz} will be stabilized and the d_{xy} destabilized. The splitting of the t_{2g} set is much less than that of the e_g set since the t_{2g} orbitals point between the ligands and are therefore not so sensitive to changes in M–L distances. Unsymmetric occupation of the t_{2g} set (e.g. d^2) will result in similar distortions to those seen above but the magnitude of the distortion will be much smaller than for unsymmetrical occupation of the e_g set.

Predicted Jahn–Teller distortions in octahedral complexes:

LARGE DISTORTION	SMALL DISTORTION	NO DISTORTION
d^4 h.s., d^7 l.s., d^9	d^1, d^2, d^4 l.s., d^5 l.s., d^6 h.s., d^7 h.s.	d^0, d^3, d^5 h.s., d^6 l.s, d^8 h.s., d^{10}

Unequal occupation of metal d orbitals in tetrahedral complexes would also be expected to result in Jahn–Teller distortion but since none of the d orbitals points directly towards the ligands, the distortions are not so pronounced and far less well documented.

The Jahn–Teller effect can be detected by X-ray crystallography, i.e. differences in bond lengths, and is also encountered in electronic spectra. For example, the spectrum of $[Cu(H_2O)_6]^{2+}$ has a shoulder. This is due to two overlapping broad bands arising from transitions from the t_{2g} set to lower, d_{z^2}, and higher, $d_{x^2-y^2}$, orbitals.

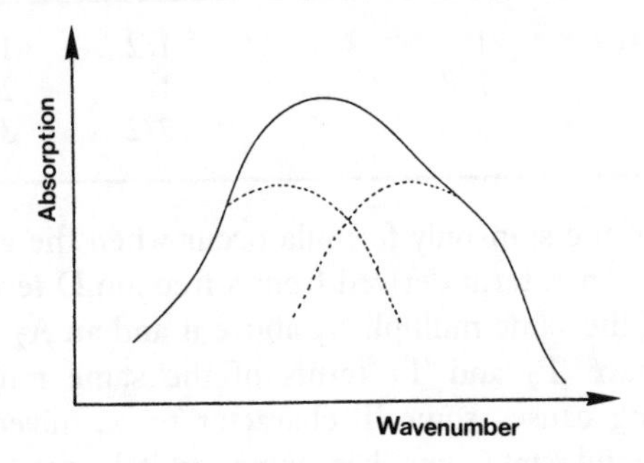

Magnetism

Here we shall be concerned with *magnetically dilute* substances, i.e. ones in which there is no coupling between magnetic moments on adjacent metal centres. Where there is coupling this leads to more complex forms of magnetism such as ferromagnetism.

There are two types of magnetism encountered in magnetically dilute substances:

1. Diamagnetism: this is possessed by all atoms that contain full shells of electrons and is due to the circulation of charge within these shells. Diamagnetism is therefore possessed by all substances except H atoms. Diamagnetic substances are repelled by a magnetic field. This effect is several orders of magnitude smaller than paramagnetism.
2. Paramagnetism: this is exhibited by substances that contain unpaired electrons. These substances are attracted by a magnetic field.

The first row transition metal complexes can be divided into two groups: those with A or E ground terms (no orbital angular momentum) and those with T ground terms (orbital angular momentum in ground state).

COMPLEXES WITH A AND E GROUND TERMS The magnetic moment for A and E ground terms is well approximated by the **spin-only** formula,

$$\mu_{s.o.} = \sqrt{4S(S+1)} \qquad S = \Sigma s$$

where $\mu_{s.o.}$ is the magnetic moment in Bohr magnetons (BM); a measure of the number of unpaired electrons.

NO. OF UNPAIRED ELECTRONS	S	$\mu_{s.o.}$ (BM)
1	1/2	1.73
2	1	2.83
3	3/2	3.87

Deviations from the spin-only formula occur when the ground term of the free ion is D or F. An E term derived from a free ion D term (D→E, T_2) will have a T_2 term of the same multiplicity above it and an A_2 term derived from an F term will have T_2 and T_1 terms of the same multiplicity above it. Spin–orbit coupling causes some T character to be mixed into the ground state and the ground state now has some orbital angular momentum, L, associated with it. The magnetic moment is now given by,

$$\mu_{eff.} = \mu_{s.o.}\left(1 - \frac{\alpha\lambda}{\Delta}\right)$$

$\alpha = 2$ for E ground terms and 4 for A.

λ is the spin–orbit coupling constant – a measure of the strength of spin–orbit coupling. Δ is the crystal field splitting energy.

This gives good agreement with observed values, e.g. the experimental magnetic moment for Ni^{2+} octahedral complexes is generally about 3.1 BM, as predicted by this equation but not by the spin-only formula which gives a value of 2.83 BM.

Mn^{2+} high spin octahedral complexes have a $^6A_{1g}$ (derived from an ^6S free ion term) ground state and obey the spin-only formula very closely since there are no T terms of the same multiplicity above the ground state.

The magnetic moments of complexes with A or E ground terms are independent of temperature. This is termed **temperature independent paramagnetism** (TIP).

COMPLEXES WITH T GROUND TERMS This is a much more complex situation and no simple formula exists for predicting the magnetic moments of complexes with T ground terms.

The full formula for the magnetic moment of an atom or ion with ground state $^{2S+1}L_J$ is

$$\mu = g\sqrt{J(J+1)}$$

where

$$g = 1 + \frac{J(J+1) - L(L+1) + S(S+1)}{2J(J+1)}$$

Even this formula does not give good agreement with experimental values. One reason for this is that the formula was constructed for a free ion or atom and therefore there is no real reason why it should apply to a complex where interaction with ligands can occur. Upon formation of a complex from a free ion some 'quenching' of orbital angular momentum occurs due to splitting of the d orbitals. (In the free ion, the d_{xy}, d_{xz}, d_{yz} and $d_{x^2-y^2}$ orbitals are all degenerate and the same shape. Therefore an electron can be imagined as rotating from one orbital to another, giving rise to orbital angular momentum. Splitting in a crystal field means that the $d_{x^2-y^2}$ is no longer degenerate with the other three orbitals, and the orbital angular momentum is consequently reduced, i.e. 'quenched'.) This can either be complete, which makes it appropriate to use the spin-only formula, or partial, as in the case of complexes with T ground terms. The partial quenching of orbital angular momentum means that the free ion expression for the magnetic moment is no longer applicable.

For ^3T the term is split into three levels, $J = L+S$, $L+S-1$, $L-S$ i.e. $J = 4, 3, 2$. The separation of the J states is of the order of kT and so the population of these states, and hence μ, varies with temperature, i.e. we have **temperature dependent paramagnetism.**

Sometimes several factors such as asymmetry of the ligand field and/or Jahn–Teller distortion, conspire to give a magnetic moment for complexes with T ground terms which is close to the spin-only formula.

EXAMPLE

Stereochemistry of Ni^{2+}

An interesting application of magnetic moments is to distinguish between the stereochemistries of Ni^{2+}.

The common stereochemistries of Ni^{2+} are octahedral, tetrahedral and square planar.

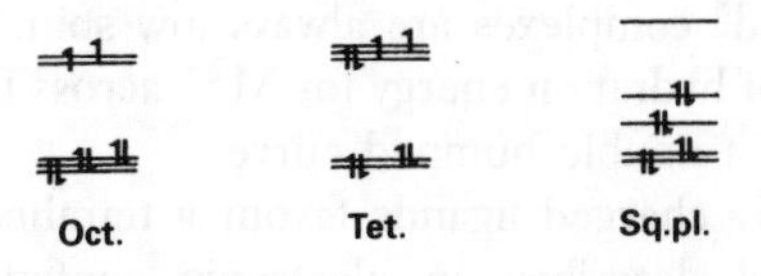

Square planar can immediately be distinguished from the other two because it is diamagnetic, the other two are paramagnetic with two unpaired electrons.

The ground term for the Ni^{2+}, d^8 free ion is ^3F. The octahedral complex does not have orbital angular momentum associated with the ground term whereas the tetrahedral one does. Therefore the octahedral complex has a

$^3A_{2g}$ ground term and the tetrahedral complex a $^3T_1$ ground term (see above). Hence the octahedral complex exhibits temperature independent paramagnetism and will give a magnetic moment value close to that value given by the spin-only formula for two unpaired electrons. However, the magnetic moment for the tetrahedral complex will be temperature dependent and should not be very close to the spin-only value.

Note: you can see whether the magnetic moment is temperature dependent or not simply by looking at the splitting diagram and working out whether or not there is any orbital angular momentum associated with it. This avoids the rigmarole of working out the term symbols.

Summary

1. When transition metals are oxidized they always lose their $(n+1)$s orbital electrons before their nd orbital electrons.
2. The crystal field splitting diagrams for octahedral, tetrahedral and square planar complexes are:

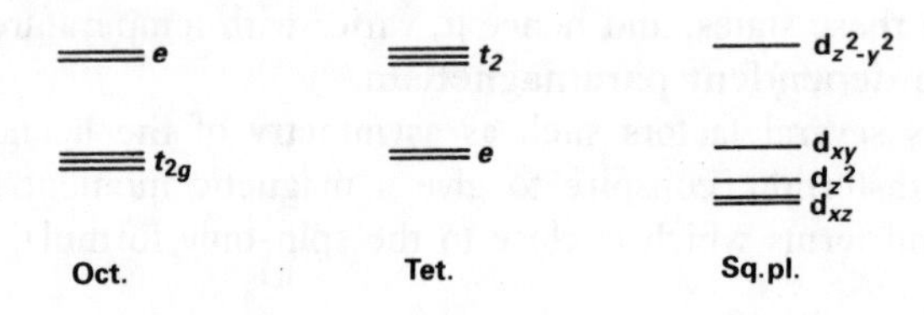

3. The crystal field splitting energy is largest in a complex of a high oxidation state third transition series metal with π acidic ligands.
4. The spectrochemical series is $I^- < Br^- < F^- < H_2O < PR_3 < CO$.
5. Octahedral complexes containing strong field ligands are usually low spin; weak field ligands generally give rise to high spin complexes.
6. Tetrahedral complexes are high spin.
7. Square planar d^8 complexes are always low spin.
8. The variation of hydration energy for M^{2+} across the first transition metal series follows a double humped curve.
9. Large negatively charged ligands favour a tetrahedral geometry.
10. A term symbol describes an electronic configuration and is written $^{2S+1}L_J$.
11. Term symbols are split by a crystal field.
12. The selection rules for observation of electronic spectra are:

$$\Delta S = 0,\ \Delta L = \pm 1,\ g \longleftrightarrow u$$

Thus d $\longleftrightarrow$ d electronic transitions are forbidden. However, relaxation of selection rules can occur, which results in transition metal complexes often being coloured.

13. The electronic spectra of transition metal complexes are affected by oxidation state of the metal, by geometry and by the surrounding ligands.
14. Absorption bands in electronic spectra are usually broad.
15. Cu^{2+} complexes show Jahn–Teller distortions.
16. Complexes with A and E ground terms obey the spin-only formula for magnetic moments and exhibit temperature independent paramagnetism.
17. The magnetic moments of complexes with T ground terms cannot be predicted by the spin-only formula and vary with temperature.

Section B: Some Aspects of the Chemistry of the First Row Transition Metals

Size, coordination number and stereochemistry

Atomic, and hence ionic (for a particular oxidation state), radius decreases from left to right. This is due to the nuclear charge increasing by one unit as we move through each element from Ti to Cu and the d electrons being very poor at shielding each other. There is thus an increase in effective nuclear charge felt by the outer electrons so that they are pulled in more closely to the nucleus.

This decrease in atomic/ionic radius has all the expected consequences, e.g. from left to right:

1. Electronegativity increases (see Chapter 1).
2. Heat of hydration ($\Delta H_{hyd.}$) shows a general increase, i.e. the energy change for the process,

   $$M^{2+}_{(g)} + aq. \rightarrow [M(H_2O)_6]^{2+}_{(aq)}$$

   (see CFSE above).

It also shows itself in the coordination numbers of the first row metals. In general, there is a decrease in the maximum coordination number from left to right and coordination numbers above six are much less common on the right-hand side. This is due to the smaller size of the elements at the right-hand side and the difficulty of packing more ligands around.

All the elements from Ti to Co show a maximum coordination number of 8, e.g. $Ti(NO_3)_4$, $[Cr(O_2)_4]^{3-}$, $[Co(NO_3)_4]^{2-}$.

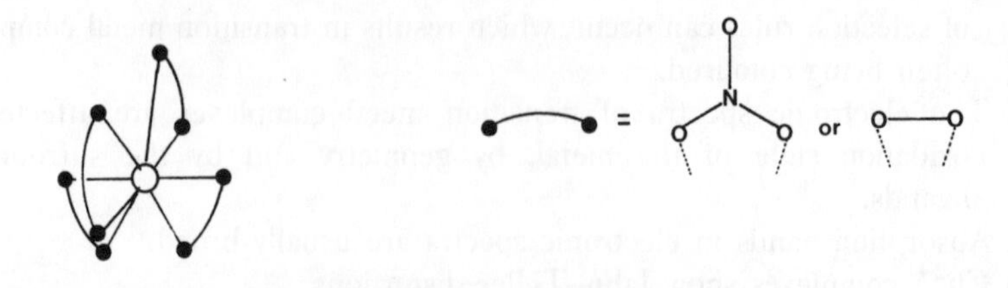

For Ti there are several compounds in which the Ti has a coordination number of 7 or 8 but for Co the nitrate complex is one of very few where Co has a coordination number above 6.

Ni has a maximum coordination number of 7 and this only occurs with macrocyclic ligands as in, e.g. $[Ni(DAPBH)(H_2O)_2]^{2+}$.

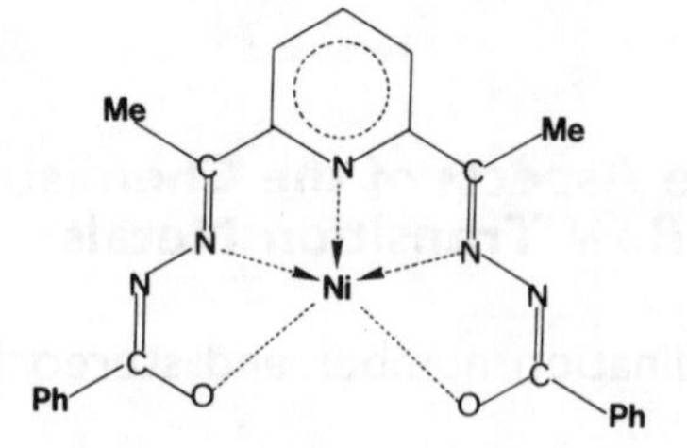

Cu forms no complexes in which its coordination number is greater than 6, e.g. $[Cu(H_2O)_6]^{2+}$.

Note: this discussion of coordination number assumes that a cyclopentadienyl ring (Cp, C_5H_5) occupies three and not five coordination sites on a metal, i.e. we would regard ferrocene as having a pseudo-octahedral structure,

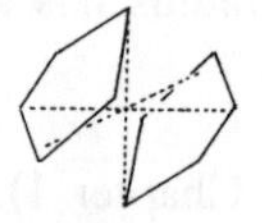

Some people do think of ferrocene ($FeCp_2$) as 10 coordinate Fe, therefore some books will contain references to, e.g. 10 coordinate Fe or 12 coordinate Ti, as in $[Ti(Cp)_2(CO)_2]$. This does not, however, affect the principles behind the above arguments.

Stereochemistry

Higher coordination numbers – coordination numbers of 7 and 8 are generally only shown with bidentate or multidentate ligands (these take up less room and there is a smaller decrease in entropy than occurs upon coordination of monodentate ligands – see chelate effect below). The possible ideal stereochemistries for eight coordination are dodecahedral, square antiprismatic and cubic.

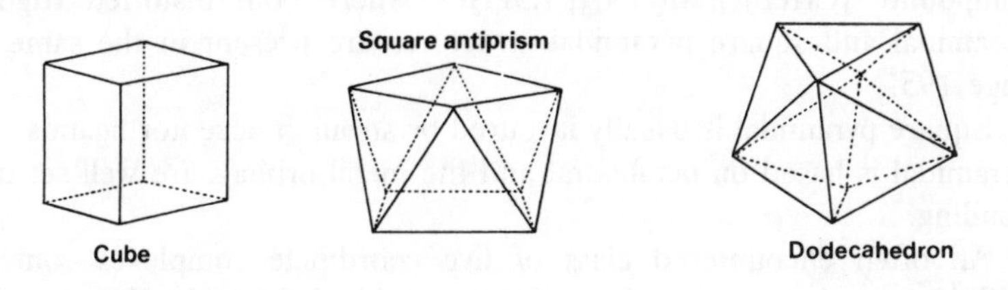

Of these, dodecahedral is slightly more stable than square antiprismatic, which, in turn, is generally considerably more favourable than cubic. Only dodecahedral/distorted dodecahedral stereochemistries are exhibited by the first row transition metals.

Both pentagonal bipyramidal, e.g. $TiCl(S_2CNMe_2)_3$, and capped trigonal prismatic, e.g. $Cr(CO)_2(diars)_2X$, can be obtained for coordination number 7. There is little energy difference between these two geometries.

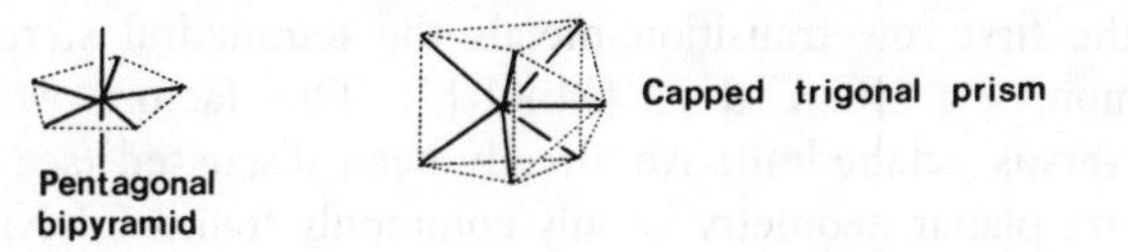

All the transition elements form six coordinate octahedral structures and this is by far the commonest stereochemistry for the transition metals. The favourability of the octahedral geometry comes from a combination of factors:

1. There is a good balance between forming the maximum number of strong bonds and not introducing too much steric repulsion between ligands.
2. CFSE is greatest for the octahedral structure.

The trigonal prismatic geometry is also possible for a six coordinate metal but this is not found in any discrete complexes of the first row transition metals.

Coordination number 5 is not as common as 4 or 6 but it is known for all the transition metals. The geometries adopted are trigonal bipyramidal, e.g. $CrCl_3(NMe_3)_2$ and square pyramidal, e.g. $[Co(CN)_5]^{3-}$. Square pyramidal structures are usually distorted, so that the metal has moved out of the square plane.

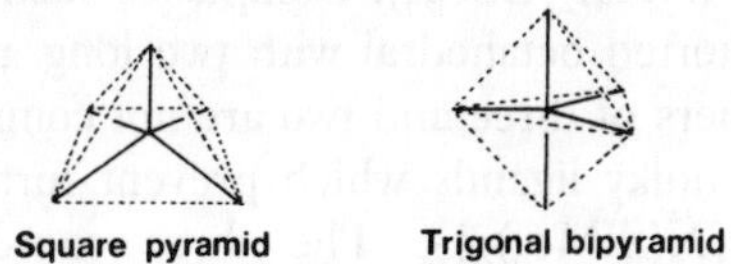

There is generally only a small energy difference between the two structures such that equilibria are often possible and there is even a compound, $[Cr(en)_3][Ni(CN)_5].1.5H_2O$, where both distorted trigonal bipyramidal and square pyramidal structures are present in the same crystal, page 195.

Square pyramidal is usually favoured by strong π acceptor ligands – square pyramidal is based on octahedral and the metal orbitals are well set up for π bonding.

An often encountered class of five coordinate complexes contains the $[VO]^{2+}$ 'pseudo-ion', e.g. the square pyramidal $VO(acac)_2$. The 'pseudo-ion' label comes from the fact that the unit contains a very strong V=O bond and tends to be retained in a number of compounds. The five coordinate VO complexes will often take up a weakly bound sixth ligand to give a distorted octahedral structure, e.g. $[VO(bipy)_2Cl]$. This is a fairly common phenomenon; five coordinate square pyramidal and lower coordination number complexes, with 'spare cordination sites' will often pick up solvent molecules. This comes from the drive to form the maximum number of bonds possible, allowing for steric factors.

For all the first row transition metals the tetrahedral stereochemistry is quite common, e.g. $[CoCl_4]^{2-}$, $[MnO_4]^-$. The factors which influence tetrahedral versus octahedral have already been discussed (see page 155).

The square planar geometry is only commonly found for Ni^{2+}, d^8, where its adoption allows the high lying $d_{x^2-y^2}$ orbital to remain unoccupied. Where it is found for other elements, and this is only for Mn, Co, Cu, it is with special 'planar' ligands, as in, e.g. $[Mn(phthalocyanine)]^{2-}$.

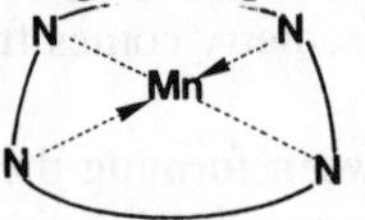

Just as five coordinate square pyramidal complexes can pick up a sixth ligand to generate octahedral complexes, so can square planar complexes pick up one or two ligands, e.g. the Lifschitz salts, $Ni(L—L)_2X_2$, where L—L is a substituted ethylenediamine (en) ligand and X can be a variety of anions. Some of these complexes are yellow, diamagnetic (see above) and planar, i.e. $[Ni(L—L)_2]^{2+}(X^-)_2$, whereas others are blue, paramagnetic and octahedral, i.e. $[Ni(L—L)_2X_2]$ or $[Ni(L—L)_2(solvent)_2]^{2+}(X^-)_2$. Which complex is obtained depends on L—L, X and the solvent.

In some cases the square planar stereochemistry has been wrongly assigned to Cu complexes, e.g. $(NH_4)_2[CuCl_4]$. Complexes such as this often turn out to be Jahn–Teller distorted octahedral with two long axial bonds.

Coordination numbers of three and two are not common and are generally only found with very bulky ligands which prevent further coordination, e.g. $Ti\{N(SiMe_3)_2\}_3$, $Co\{N(SiMe_3)_2\}_2$. The three coordinate complexes are found for all elements and all have a planar geometry which keeps the bulky ligands as far apart as possible. Cu^I (on the right-hand side, and therefore small) can form three coordinate complexes without the need for bulky ligands, e.g. the helical polymer $[Cu(CN)_2]^-$. It also has the greatest tendency towards two coordination, as in, e.g. $[CuCl_2]^-$ (linear).

The complex $[Cu(CN)_2]^-$,

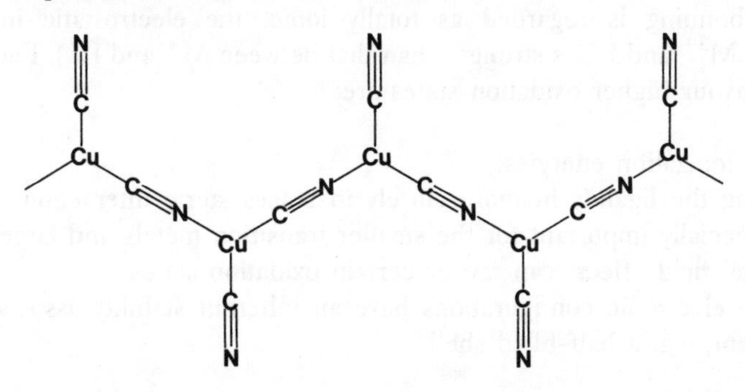

illustrates an important principle in that it is unwise to draw conclusions about stereochemistry and coordination number from the stoichiometric formula. When ligands adopt a bridging mode it is often associated with the metal's drive towards octahedral coordination, e.g. $NiCl_2py_2$

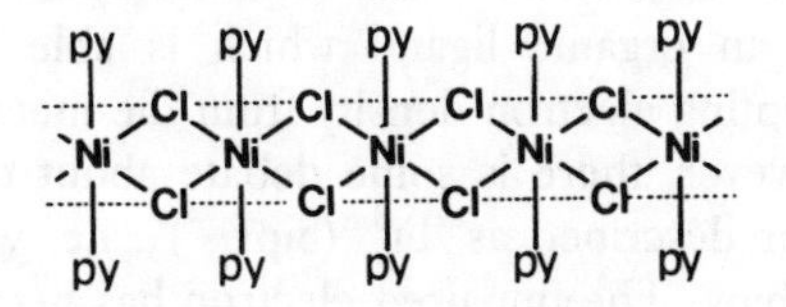

Halogens are the most commonly found bridging atoms.

Oxidation states

Ionization energy increases from left to right across the transition metal series – as the effective nuclear charge increases the outer electrons are pulled in more strongly, and are therefore more difficult to remove; this has important consequences for the oxidation states exhibited by the transition metals. The simplest view of oxidation states is to treat a compound as purely ionic, then the ease of attainment of a particular positive oxidation state, $n+$, depends on how easy it is to remove n electrons.

Positive oxidation states are the norm for transition metals, negative ones are only found with π accepting ligands, which are able to remove electron density from the metal and, hence, stabilize negative oxidation states, e.g. $[Fe(CO)_4]^{2-}$, with Fe in a −2 oxidation state (see Chapter 7).

In general, metals at the left-hand side of the transition metal series exhibit a wide range of positive oxidation states, up to and including the group valency (i.e. a $3d^0\,4s^0$ configuration). As we go from left to right the stability of the higher positive oxidation states decreases and that of the lower ones increases. A factor in favour of higher oxidation states is that the ligands are pulled in closer to the metal, so that stronger M–L interactions are obtained (if the bonding is regarded as totally ionic, the electrostatic interaction between M^{3+} and L^- is stronger than that between M^+ and L^-). Factors that can disfavour higher oxidation states are:

1. High ionization energies.
2. Pulling the ligands in more closely increases steric interactions – this is especially important for the smaller transition metals and large ligands.
3. Crystal field effects can favour certain oxidation states.
4. Some electronic configurations have an inherent stability associated with them, e.g. a half-filled shell.

The most common oxidation state for Ti is +4 and a wide range of compounds, such as TiO_2 and all the TiX_4 halides can be formed. Ti^{III} is also reasonably common although it is quite reducing and solutions are oxidized by air to give Ti^{IV}. The +2 oxidation state is very strongly reducing so that TiF_2 cannot be formed (see below). Lower oxidation states than this do occur, e.g. in $Ti^0(bipy)_3$ (bipy is an organic ligand which is able to stabilize lower oxidation states by accepting electron density from the metal into its π system – see Chapter 7). However, there is some debate about these systems, and they are probably better described as $Ti^{3+}(bipy\cdot^-)_3$, i.e. with an electron in the π* system of each bipy. The unpaired electron has been shown, by e.s.r., to spend at least part of its time on bipy.

For V, oxidation states up to +5 (d^0) are known (e.g. V_2O_5), although +5 is strongly oxidizing (easily reduced) and the +4 state is the most common (e.g. VCl_4), being neither strongly oxidizing nor reducing. There is also quite an extensive chemistry of V^{III} (e.g. $[V(H_2O)_6]^{3+}$) although this, and lower oxidation states, are unstable to oxidation by air.

All the elements up to and including Mn are capable of exhibiting the group valency, e.g. $[Cr_2O_7]^{2-}$ (oxidation state +6, d^0), $[MnO_4]^-$ (+7, d^0). The elements in the middle show a wide range of oxidation states – the ionization energy is not exorbitantly high, so that the higher positive oxidation states can be formed. Also, they have enough d electrons available for back donation and so can stabilize lower oxidation states by donating these electrons to acceptor ligands, i.e. electron density is removed from the metal centre {note that Ti does not form a binary carbonyl, $Ti(CO)_n$ – the stability of these depending very strongly upon π back donation from the metal (see Chapter 7) and Ti does not have enough electrons to do this efficiently}.

Mn has an extensive chemistry in the +2 (d^5) oxidation state (e.g. $[Mn(H_2O)_6]^{2+}$). This is by far the most stable oxidation state in aqueous solution, stability arising from the half-filled d shell, with all electrons unpaired.

Fe is the first element for which the group valency is not shown, the highest oxidation state being +6 (in $[FeO_4]^{2-}$) rather than +8 – once we reach Fe the effective nuclear charge felt by the d electrons is sufficiently high, so that the ionization energy to give +8 is too large to be compensated for by other factors. Oxidation states above +3 are rare and strongly oxidizing (want to be reduced), such that $[FeO_4]^{2-}$ will oxidize NH_3 to N_2 at room temperature. It must always be rembered that oxidation state is a formalism and does not represent the actual charge on an ion. However, the 'real charge' on Fe in oxidation state +8 would be higher than that in oxidation state +6 and the extra energy to produce the higher charge cannot be compensated for by other thermodynamic factors, such as increased lattice energies and bond energies associated with the more highly charged cation.

The chemistry of Fe is dominated by +2 and +3 oxidation states (e.g. $[FeCl_4]^{2-}$, Fe_2O_3), with most Fe^{II} salts being unstable to aerial oxidation to Fe^{III}.

Co is similar to Fe in being dominated by +2 and +3 oxidation states (e.g. $[CoCl_4]^{2-}$, $[CoF_6]^{3-}$) but when we get to Ni there are very few complexes in which the oxidation state is greater than +2. The maximum oxidation state for Ni is +4, in K_2NiF_6, but the chemistry of Ni is dominated by +2, e.g. $[Ni(H_2O)_6]^{2+}$.

The highest oxidation state for Cu is also +4 ($[CuF_6]^{2-}$). Again the chemistry is dominated by +2 but there is an extensive range of compounds containing Cu^I, e.g. CuI.

Halides and oxides

A good indication of the stability of oxidation states can be obtained from a study of the halides and oxides of the metals. In general, the highest oxidation states are only obtained in combination with the very electronegative O and F ligands. We do not tend to get the highest oxidation states for transition metals with I as the ligand, and we do not get the lowest oxidation states with F. This can be understood as follows:

IONIC BONDING The energy of an electrostatic interaction is proportional to the product of the charges on the anion and cation and inversely proportional to the distance between them. O^{2-} and F^- are small, giving rise to a strong electrostatic interaction. The size of this interaction increases as the charge on the metal ion increases. The I^- ion is large and, although the electrostatic interaction also increases as the charge on the metal ion increases, the larger inter-ionic distance makes the change proportionately less.

I^- is far more easily oxidized than F^-. F^- is sufficiently resistant to oxidation to be able to coexist with metals in high oxidation states; F^0, however, is usually a strong enough oxidizing agent to be able to oxidize metals in low oxidation states to higher ones. A metal in a higher oxidation state will be able to oxidize I^- to I_2, in the process being reduced to a lower oxidation state. This is the result of I^- being more polarizable than F^-, i.e. its outer electrons are not so strongly held. A metal in a higher oxidation state can thus pull an electron away from I^-, reducing the metal and oxidizing the I^- to I_2 (e.g. I^- will reduce Fe^{3+} to Fe^{2+}).

COVALENT BONDING The concentrated valence orbitals on the O and F ligands make for good overlap with metal orbitals and strong covalent bonds. If the bonds are strong there is a greater driving force to form more of them, something which is made easier by F and O being small, such that inter-ligand steric interactions are smaller. The I is large and its valence orbitals are diffuse, therefore M—I bonds are weak and forming lots of them is not going to release enough energy to offset large steric factors, therefore we get lower oxidation states.

The bonding in transition metal complexes is not simply ionic *or* covalent but involves contributions from both terms. The favourability of the higher oxidation states for O and F is therefore due to a combination of *both* ionic and covalent bonding.

For Ti, we get all the +4 halides but TiF_2 has not so far been made. For V, the only neutral +5 halide is VF_5 and the highest iodide is VI_3. For Mn, the highest oxidation state neutral fluoride is MnF_4 and the highest oxidation state iodide MnI_2. The group valency (+7) is only shown by the multiply-bonded oxide, $[MnO_4]^-$, probably due to steric factors – in MnF_7 the +7 oxidation state would result in the ligands being pulled in very close to the metal, thereby introducing a great deal of repulsion between the seven ligands. The highest oxidation state for Fe (+6) is also exhibited in a multiply-bonded oxide, $[FeO_4]^{2-}$. For Ni and Cu the highest oxidation state (+4) is shown only in fluoride complexes, $[MF_6]^{2-}$. The $M(Hal)_2$ compounds all exist for Ni but not for Cu; I can only bring out the +1 oxidation state for Cu, as in CuI.

High spin versus low spin

See above for general factors which affect the spin state that is exhibited by a particular transition metal.

As a broad generalization, +3 has a greater tendency towards low spin than +2 because of the much larger Δ for +3 (page 150). Most M^{2+} complexes are high spin, low spin usually being found only with strong field, π acceptor ligands. Thus, all $[M(H_2O)_6]^{2+}$ complexes are high spin and all $[M(CN)_6]^{4-}$ (still M^{2+}) low spin. Most $[M(H_2O)_6]^{3+}$ are low spin but there are exceptions and $[Fe(H_2O)_6]^{3+}$ is high spin.

Fe^{III} is unusual in having a strong tendency towards high spin, low spin only being obtained with strong field ligands such as bipy, CN^-. This is because Fe^{III}, d^5, high spin, has all its electrons unpaired and benefits from the 'magical stability of the half-filled shell' (exchange energy, page 8).

Co^{III} is also exceptional in having a very strong preference for low spin, such that there are only two known high spin Co^{III} complexes, $[CoF_6]^{3-}$ and $CoF_3(H_2O)_3$, both of which contain the weak field F ligand. This preference for low spin is a combination of the electronic configuration, $t_{2g}{}^6$, and the oxidation state. $t_{2g}{}^6$ has the maximum CFSE (−12/5Δ), an effect which is magnified by the large Δ for oxidation state +3. Fe^{II} also has the possibility of forming $t_{2g}{}^6$ but the smaller Δ makes CFSE a less important effect and Fe^{II} forms mostly high spin ($t_{2g}{}^4e_g{}^2$) complexes. The change in spin state for Fe^{II} comes between,

$$\underset{\text{high spin}}{[Fe(phen)_2(H_2O)_2]^{2+}} \rightarrow \underset{\text{low spin}}{[Fe(phen)_3]^{2+}}$$

Sometimes it is possible to get 'spin crossover' complexes, where we can induce interchange between low and high spin states by varying the temperature. An example of such a system is $Fe\{S_2CN(CH_3)_2\}_3$, which is low spin at low temperatures but as the temperature is raised we get population of the high spin state (followed by magnetic measurements). It is not possible to get total occupation of the high spin state and there is a Boltzmann distribution between high and low spin states; the magnetic moment is thus somewhere in between that for high and low spin.

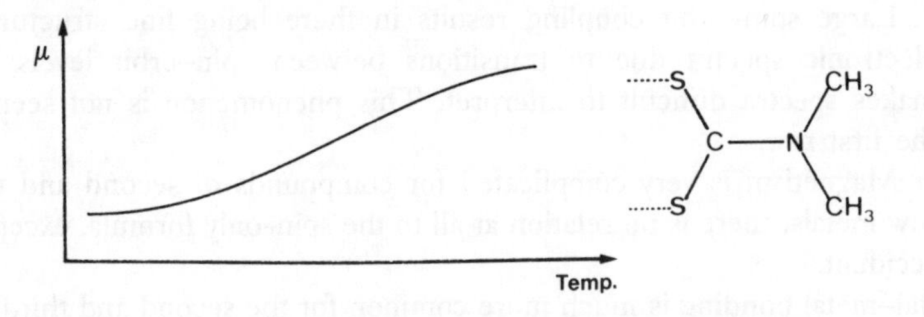

Differences between first row and second and third row transition metals

The following are the most important differences between first row and second/third row transition metals:

1. Going down a transition metal triad there is a large increase in size between the first and second row metals but between the second and third row there is very little size difference. This is a consequence of the **lanthanide contraction** (see below). Hence properties which depend on size, such as coordination number, hydration energies of ions, lattice energies, etc., are very similar for the second and third row but are very different from those of the first row.

   For the (larger) second and third row transition metals coordination numbers of 7, 8 and 9 are more common, e.g. OsF_7, $[Mo(CN)_8]^{4-}$, $[ReH_9]^{2-}$
2. The crystal field splitting energy, Δ, increases down a group. This is because the larger the d orbital, the more the electrons extend out towards the ligand electrons and, therefore, the greater is the repulsive interaction between metal and ligand electrons. From first to second row Δ approximately doubles because the 4d orbitals are a lot more diffuse than the 3d orbitals. Between the second and third rows there is only an approximately 25 per cent increase in Δ – the lanthanide contraction causes a smaller increase in size between 4d and 5d than between 3d and 4d orbitals.
3. The pairing energy of electrons decreases down a group. This is a consequence of the larger size of the d orbitals and hence the lower repulsion between the electrons within them.

   Points 2 and 3 together combine to make all second and third row transition metal complexes low spin, e.g. $[CoF_6]^{3-}$ is high spin but $[RhF_6]^{3-}$ is low spin.
4. Spin–orbit coupling increases down a group. This means that the Russell–Saunders coupling scheme is no longer applicable as it assumes that spin–orbit coupling is only a minor perturbation on spin–spin and orbit–orbit coupling. For second and third row metals the 'j–j coupling scheme' has to be employed.

   Large spin–orbit coupling results in there being fine structure in electronic spectra due to transitions between spin–orbit levels; this makes spectra difficult to interpret. This phenomenon is not seen for the first row.

   Magnetism is very complicated for compounds of second and third row metals; there is no relation at all to the spin-only formula, except by accident.
5. Metal–metal bonding is much more common for the second and third row than for the first row; metal–metal bond energies increase down a group. This is because the size of the d orbitals increases down the group and therefore better overlap is obtained at the longer bond lengths associated with metal–metal bonds (longer bond lengths reduce

repulsion between ligands on adjacent metal atoms). For the first row, metal–metal bonding is less favourable because of the shorter metal–metal distances (smaller atoms) and therefore greater repulsion between ligands on adjacent metal atoms.

6. There are proportionately fewer paramagnetic compounds for the second and third row transition metals. This is a consequence of the tendency towards low spin and the greater degree of metal–metal bonding (electrons paired up in metal–metal bonds), e.g. the complexes $[M_2Cl_9]^{3-}$, M = Cr, Mo, W, have the structure

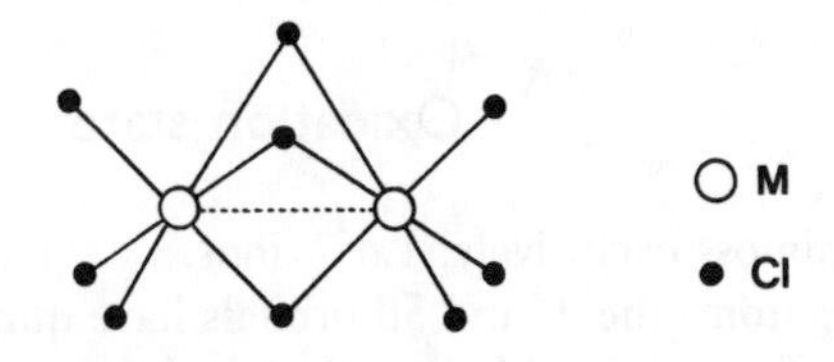

The metal–metal distances are:

Cr—Cr, 3.10 Å Mo—Mo, 2.67 Å W—W, 2.41 Å

This variation is due to increasing metal–metal bonding down the triad. For the Cr complex there is no M—M bonding and the magnetic moment is that expected for a d^3 configuration. For the W complex there is a triple bond between the metals so that the d^3 electrons are paired up and the complex is diamagnetic. The Mo complex exhibits intermediate behaviour.

7. The stability of higher oxidation states increases down the group. This is due to the lower ionization energies as the group is descended, e.g. $Os^{VIII}O_4$ exists but Fe is unstable in oxidation states above +3, and FeO_4 does not exist.

The lanthanide contraction

The lanthanides come between the second and third row transition elements. Across the lanthanide series the 4f orbitals are filled. Electrons in f orbitals are very bad at shielding each other from the nucleus, resulting in an increase in effective nuclear charge across the series and hence a decrease in size. The increase in effective nuclear charge is sufficient to offset the increase in size associated with filling of outer, 5d, orbitals for the third row as compared with the 4d orbitals of the second row and hence second and third row transition metals are very similar in size.

The Lanthanides (Ln)

The seven 4f orbitals are filled across the lanthanide series and hence there are 14 of them,

(La) Ce Pr Nd Pm Sm Eu Gd Tb Dy Ho Er Tm Yb Lu

Lanthanum is not strictly a lanthanide because it does not form any compounds in which it has a partially filled f shell (cf. Sc in the first transition metal series).

Oxidation state

They form, almost exclusively, Ln^{3+} ions.

In the free atoms the 4f and 5d orbitals have quite similar energies, e.g. the electronic configuration of Gd is $4f^7\ 5d^1\ 6s^2$, instead of $4f^8\ 6s^2$ (cf. Cr and Cu electronic configurations $3d^5\ 4s^1$ and $3d^{10}4s^1$). In order to obtain an ion we remove the outermost electron which requires energy. The 6s and the 5d orbitals are much better shielders than the 4f. As these orbitals are all of comparable energy in the free atom, it is much better to remove an electron from the 6s orbital because the net effect of this will be to greatly increase the effective nuclear charge felt by the outer electrons. The outer electrons are pulled closer towards the nucleus, which is favourable because we have increased attractive electrostatic interaction between the nucleus and the electron cloud. This effect would not be so large if electrons were removed from the 4f orbitals because there would still be electron density in the strongly shielding 6s orbitals. For positive ions, the increased effective nuclear charge results in the energy of the 4f orbitals falling much lower than that of the 5d and 6s. Therefore removal of electrons from 4f orbitals in the Ln^{3+} ion becomes difficult and electrons are only usually removed from the 6s and 5d orbitals. Removal of this electron density reduces the shielding of 4f electrons and they are pulled in even closer to the nucleus and lowered in energy. For Ln^{3+} the electronic configuration will be $4f^n$ and the orbitals are now too low in energy (core orbitals) for any further electrons to be removed.

Going from Ln to Ln^{3+} requires a large amount of energy but this is paid back by lattice or hydration energies. For Ln^{3+} to Ln^{4+} the large amount of energy required to remove an electron from a +3 ion cannot generally be offset by other thermodynamic terms, i.e. increased lattice energies and hydration energies associated with the higher cationic charge.

EXCEPTIONS Eu^{2+} (f^7), Ce^{4+} (f^0) and Yb^{2+} (f^{14}) can all exist in solution. This is due to the stability of the half-filled or filled shell. The electronic configuration is not, however, the only factor which has to be taken into account and Sm^{2+} (f^6) and Tm^{2+} (f^{13}) both exist. The existence of various oxidation states is best considered to be due to a combination of $\Delta H_{at.}$, $\Delta H_{latt.}$ and ionization energies (from a Born–Haber cycle).

The lanthanide contraction

Across the lanthanide series there is a substantial drop (by about 25%) in the ionic radius. This phenomenon is not unusual and contractions are seen across each row in the Periodic Table (e.g. Sc to Zn or Na to Ar) but the scale of the contraction for the lanthanides is larger. As we have discussed earlier, the reason for this is that the 4f electrons are very poor at shielding (order of shielding ability is $s > p > d > f$). The lanthanide contraction has three major consequences:

1. It almost exactly offsets the increasing size as we go from the second to third row transition metals, e.g. Ti (1.3 Å), Zr (1.45 Å), Hf (1.44 Å). Hence, for example, many sources of Zr are contaminated with Hf and vice versa.
2. For similar ligands, the coordination numbers of the lanthanides decrease from left to right across the series, e.g.

COMPOUND	COORD. NUMBER
$CeCl_3$	9
$TbCl_3$	8
$LuCl_3$	6

3. As most lanthanide ions exist in the +3 oxidation state their chemistry is very similar and they are very difficult to separate by chemical means. They can, however, be separated very effectively by ion-exchange chromatography, which basically exploits the lanthanide contraction.

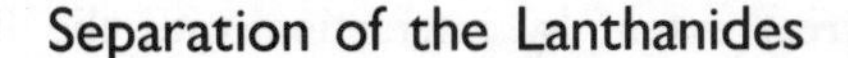

Separation of the Lanthanides

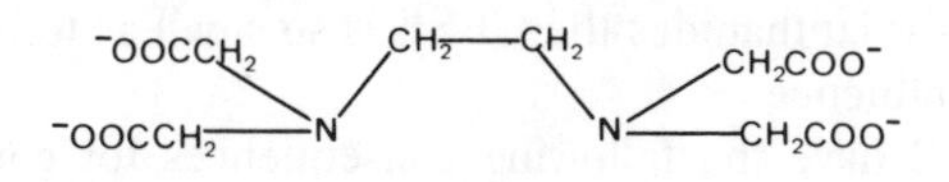

When a mixture of lanthanides is added to a polysulphonated (—SO_3H) ion exchange resin they displace the H^+ groups from the —SO_3H and are trapped on the resin ($-SO_3^-\ Ln^{3+}$). The equilibrium constant for this process is almost identical for all the lanthanides.

$$Ln_{resin} \rightleftharpoons Ln_{soln}$$

$$K_{resin} = \frac{[Ln_{resin}]}{[Ln_{soln}]}$$

A complexing agent, which exhibits preferential complexation between the lanthanides, is then added, e.g. EDTA (page 57). The interaction between EDTA and the lanthanides is primarily an ionic one, and as such depends on charge:radius ratio of Ln^{3+}. The EDTA removes the later lanthanides preferentially because their ionic potential is highest (ionic attraction for EDTA is going to be highest when the ions are smaller), i.e.

$$Ln_{resin} \rightleftharpoons Ln_{EDTA}$$

$$K_{dissoc.} = \frac{[Ln_{EDTA}]}{[Ln_{resin}]}\ (K_{dissoc.}\ Lu > K_{dissoc.}\ La)$$

Hence, the later lanthanides have a greater tendency to stay in solution longer whereas the early lanthanides have a greater tendency to stay on the resin, therefore the heavier lanthanides come out at the bottom of the column first.

Bonding and ligand field effects

For lanthanides in the +3 oxidation state the 4f orbitals are very low in energy so that they are well shielded from the effect of the ligands and unavailable for covalent bonding. This has two main consequences:

1. *The bonding in lanthanide compounds is almost totally ionic*; compounds can be regarded as containing Ln^{3+} ions surrounded by anions.
2. *Crystal field splitting of the 4f orbitals is minimal.*

Unlike, the d orbitals in transition metals, the f orbitals in lanthanides are not split to any great extent by interaction with the ligands. Lanthanide compounds have still smaller CFSEs than transition metal complexes. Thus, although the small CFSE in transition metal complexes can still have profound effects, for lanthanides the CFSE is so small as to be negligible and exerts very little influence.

Factors 1 and 2 have the following consequences for compounds of the lanthanides:

(a) For a particular coordination number, no particular geometry is favoured. The bonding is predominantly ionic, with the metal having little stereochemical requirements, so that the geometry is determined by the ligands and their ability to pack efficiently around the metal ion (maximizing anion–cation and minimizing cation–cation and anion–anion electrostatic interactions). For example, in the transition metal block, six coordinate complexes are almost exclusively octahedral due to the large crystal field stabilization energy associated with this geometry, whereas lanthanides often show trigonal prismatic, as well as octahedral, geometries.

(b) Lanthanides can have coordination numbers up to 12. This is due to a combination of their large size, no directional requirements of the 4f orbitals (ionic bonding) and minimal crystal field stabilization energy.

(c) Electronic spectra. Most Ln^{3+} ions are weakly coloured {examples of exceptions are La^{3+} (f^0), Gd^{3+} (f^7), Lu^{3+} (f^{14}), which are not coloured}.

The colours of transition metal complexes arise from transitions within the set of d orbitals whereas those of the lanthanides arise from transitions within the set of f orbitals. Both these types of transition are forbidden by the Laporte selection rule (page 160). In the transition metals the Laporte rule is relaxed by the following effects:

(i) Mixing between *n*d and (*n*+1)p orbitals (e.g. via vibronic coupling).
(ii) Covalent character in the bonding.

For lanthanides the equivalent of the first mechanism would involve mixing between 4f and 5d orbitals. Although oxidation state is a purely formal concept, the lanthanides really do have a fairly high positive charge on the ions, therefore the effective nuclear charge is very high, resulting in a large energy difference between the 4f and 5d orbitals because they are from different quantum shells. There is thus less mixing between the 4f and 5d orbitals and less effective relaxation of the selection rule.

Little interaction between the 4f orbitals and the ligands results in little covalency in the bonding and therefore very little relaxation of the selection rules via the second mechanism.

Transition metal spectra are characterized by broad lines, but in lanthanide spectra the lines are very narrow. The broadness of transition metal bands is due to three factors:

1. Vibration of the ligands causing variation in the crystal field felt by the metal and hence the splitting of the d orbitals.
2. Jahn–Teller effects.
3. Spin–orbit coupling.

The first two effects are less for the lanthanides due to reduced interaction of the f orbitals with the ligands. Spin–orbit coupling is much larger than for the first row transition metals and therefore distinct sharp transitions between the spin–orbit levels are obtained. However, the intensities of the lines are much lower than those seen for the transition metals because their predominantly ionic bonding does not relax the Laporte selection rules as effectively.

Magnetism

Magnetism of the Ln^{3+} complexes is very simple because small interaction between the f orbitals and the ligands means that in most cases the magnetic moment is the same as would be predicted for the free ion (no ligands) and is given by the formula,

$$\mu = g\sqrt{J(J+1)}$$

where

$$g = 1 + \frac{J(J+1) - L(L+1) + S(S+1)}{2J(J+1)}$$

To calculate the magnetic moment work out the free ion term symbol for the ground state using the rules discussed in the section on transition metals then substitute the appropriate values of S, L and J into the above equation. This relies on the fact that spin–orbit coupling is large and therefore only one J state is occupied at normal temperatures. This is true for all but two cases, i.e. Sm^{3+} and Eu^{3+}, where the separation between J states is small enough to give significant population of an upper state. If the ions are assumed to obey a Boltzmann distribution between these states then good agreement between measured and calculated magnetic moments is obtained.

The lanthanides are often regarded as being more similar to the alkaline earth metals (Group 2) than they are to the transition metals.

We have seen that ligand field effects, incorporating magnetism and spectra, are very different between the lanthanides and the transition metals. The lanthanides are in fact more similar to the alkaline earths than to the transition metals:

1. The alkaline earths and lanthanides have similar electronegativities (e.g. Ba 0.9, La 1.1 but Ti 1.6).
2. Lanthanides and alkaline earths tend to form predominantly ionic compounds/complexes, e.g. $PrCl_3$ and $BaCl_2$ are ionic salts, whereas those of the transition metals have much more covalent character, e.g. $TiCl_4$ is a low boiling point molecular liquid.
3. The lanthanide and alkaline earth elements all tarnish rapidly in air and

react with water to liberate hydrogen. The driving force for these reactions is the release of large amounts of lattice/hydration energy upon formation of oxides/hydrated ions. Transition metals are relatively unreactive to air and water under normal conditions.
4. Eu^{2+} occurs naturally in Ba^{2+} compounds.
5. Most alkaline earths and lanthanides dissolve in liquid ammonia giving an intense blue colour, characteristic of solvated electrons; transition metals do not do this.

The Actinides

(Ac) Th Pa U Np Pu Am Cm Bk Cf Es Fm Md No Lr

The electronic configurations of the actinide elements are uncertain because of the closeness in energy between 5f, 6d and 7s orbitals. This is due to the decreasing effect of the nucleus on the outer electrons, i.e the further we get from the nucleus, the less distinction there is between the various orbitals and the closer they become in energy (cf. the atomic spectrum of, e.g. H – as we get further from the nucleus the orbitals become closer in energy so that the lines in the spectrum get closer and closer and eventually merge into a continuum). Thus, for the actinides, the relative ordering of the outer orbitals is dependent upon the particular element and its oxidation state. This means that we are not quite sure how the available electrons distribute themselves among these orbitals.

The chemistry of the actinides is usually compared with that of the transition metals and the lanthanides. To which they are more similar depends on whether the 5f orbitals are valence or core orbitals. This can be considered in a series of points:

1. The early actinides (Th to Pu) resemble the transition metals in their chemistry:
   (a) They exhibit variable oxidation states, e.g. U can have oxidation states of 3, 4, 5, 6 (UCl_3, UBr_4, UF_6^-, UF_6). For the lanthanides the 4f orbitals were below the 5d and 6s orbitals, therefore not really available for bonding. However, in the early actinides, since the 5f, 6d and 7s are all of comparable energy the 5f electrons are now available as valence electrons for bonding, i.e. they are not buried as core electrons below the 6d and 7s electrons.

   For the lanthanides, on going to positive oxidation states, the 4f orbitals are pulled much lower in energy and effectively buried, becoming core orbitals. For the early actinides, increasing positive oxidation state does not pull the 5f orbitals down so much in energy (less affected by nuclear charge because they are further from the nucleus) and even when the +3 oxidation state is reached, the 5f orbitals are not buried and are still available as valence orbitals for chemical bonding, thus, higher oxidation states are possible for the actinides. With the 5f electrons available as valence electrons, the actinides, like the transition metals (where d electrons are available as valence electrons), can exhibit a range of oxidation states.

(b) Significant interaction with ligands results in:
   (i) Greater crystal field splitting than for the lanthanides.
   (ii) Bands in the absorption spectra of the actinides being about ten times as intense as for the lanthanides and considerably broader.
   (iii) The magnetic properties of the actinides being very complex, i.e. we cannot just treat the actinides as free ions.

(c) There is significant covalent character in the bonding. This is due to the 5f and 6d orbitals being better able to interact with the ligands than their counterparts in the lanthanides. e.g. $[UO_2]^{2+}$ $\{[O{=}U{=}O]^{2+}\}$. The U—O distances in $[UO_2]^{2+}$ are shorter than the sum of U^{6+} and O^{2-} ionic radii. The ion is linear and considerable π bonding, involving 6d and 5f orbitals on U, is postulated.

2. As we move across the actinide series, similarity with the lanthanides increases. This is because the 5f are poor at shielding and from left to right, across the series, the effective nuclear charge increases, so that the 5f orbitals are pulled in more. The change in character first becomes evident at Cm as the number of available oxidation states decreases. When we get to Cf, the elements are very similar to the lanthanides.

   The elements from Cf onwards resemble the lanthanides in the following ways:

   (a) The effective nuclear charge increases across the series, therefore the 5f–6d separation increases. When we get to Cf this separation is large enough to cause the +3 oxidation state to be shown almost exclusively, as for the lanthanides.
   (b) The compounds show increased ionic character. Study of the chemistry of these later actinides is, however, made extremely difficult by their being very radioactive. This radioactivity is often intense enough to break chemical bonds and it is therefore difficult to be certain about the extent of the similarity between the later

actinides and the lanthanides. The early actinides are not so radioactive and the chemistry of Th to U is quite extensive, therefore a comparison with the transition metals is quite valid.

The actinide contraction

Moving across the actinide series there is a decrease in size, resulting in an actinide contraction similar in magnitude to the lanthanide contraction.

This actinide contraction, together with the availabilty of various oxidation states, is used to separate the different actinides.

Magnetic and spectral properties

Magentism of the actinides is much more complicated than that of the lanthanides. This is the result of greater exposure, to the ligands, of the 5f compared to the 4f orbitals and spin–orbit coupling and crystal field splitting being of approximately the same magnitude. Magnetic moments are not given by Hund's formula (see above) and tend to be more temperature dependent than those of the lanthanides.

f ⟷ f Transitions in the electronic spectra are still Laporte forbidden but the greater influence of the crystal field on the 5f orbitals allows for better relaxation of the selection rules. Spectra are approximately ten times as intense as for lanthanides and about the same intensity as d ⟷ d transitions in octahedral transition metal complexes. Bands in actinide spectra are about twice as broad as those in the spectra of the lanthanides, which is due to the greater influence of ligand vibrations on the splitting of the f orbitals (see Franck–Condon principle, page 164).

Preparation of new actinides

The later actinides (Am to Lr) are produced by bombardment of earlier actinides with the elements He to Ne, e.g.

$$^{238}_{92}U + ^{14}_{7}N \rightarrow ^{248}_{99}Es + 4^{1}_{0}n$$

They can usually make about 50 atoms per experiment!
The experiments are extremely expensive.

Why bother?
It is believed that nuclear stability does not fall off linearly with increasing atomic number and it is thought that there are 'islands of nuclear stability'.

Just as certain electronic configurations are stable (i.e. 8, 18 . . . full shells) certain proton:neutron ratios are stable. It is predicted that the element $^{298}_{114}X$ will be stable and if this is prepared it could be the doorway to a new series of elements.

Summary

1. There is a general decrease in size, and hence coordination number, across the first row transition metal series.
2. There is a lower tendency to exhibit the group valency oxidation state at the right-hand side of the first transition metal series.
3. The sizes of second and third row transition metals are very similar.
4. All second and third row transition metal complexes are low spin.
5. Metal–metal bonding is more common for second and third row metals than for first row.
6. The stability of higher oxidation states increases down a particular triad.
7. Lanthanides exist almost exclusively in the +3 oxidation state in their compounds.
8. The bonding in lanthanide compounds is predominantly ionic.
9. The lanthanides can be successfully separated by ion exchange chromatography.
10. Crystal field effects are small for the lanthanides.
11. Lanthanides exhibit sharp atomic-like electronic absorption spectra.
12. Lanthanides are often considered to be more similar to the alkaline earths than to the transition metals.
13. The electronic configurations of the actinides are often difficult to predict.
14. The early actinides resemble transition metals in their chemistry, whereas the later ones are more similar to the lanthanides.

Section C: Thermodynamic stability of transition metal complexes

Consider replacement of water ligands in an aqua complex by some other ligand, L, i.e.

$$[M(H_2O)_n]^{x+} + L \rightleftharpoons [M(H_2O)_{n-1}L]^{x+} + H_2O$$

It is conventional to leave out the H_2O ligands, so that this equilibrium can be written as,

$M + L \rightleftharpoons ML \ldots 1$

and further equilibria as,

$ML + L \rightleftharpoons ML_2 \ldots 2$
$ML_2 + L \rightleftharpoons ML_3 \ldots 3$
$\vdots \quad \vdots \quad \vdots$
$ML_{n-1} + L \rightleftharpoons ML_n \ldots n$

Equilibrium constants for these equilibria are:

$$K_1 = \frac{[ML]}{[M][L]} \quad K_2 = \frac{[ML_2]}{[ML][L]} \quad K_3 = \frac{[ML_3]}{[ML_2][L]} \quad \ldots . K_n = \frac{[ML_n]}{[ML_{n-1}][L]}$$

The K_i values are called **stepwise formation (stability) constants.** **Overall formation (stability) constants,** β_i, can also be derived: for the equilibrium,

$$M + nL \rightleftharpoons ML_n \qquad \beta_n = \frac{[ML_n]}{[M][L]^n}$$

Trends in formation constants

For reactions in which ligand substitution involves no change in stereochemistry, $K_1 > K_2 > K_3 > \ldots > K_n$.

This is due to a number of factors:

1. *Statistical considerations*: the probability of replacing an H_2O ligand by L in a complex $[M(H_2O)_6]^{2+}$ is greater than that of replacing one in $[M(H_2O)_5L]^{2+}$. In the first complex every substitution involves replacement of H_2O by L but, in the second complex, only five out of every six substitution reactions result in H_2O being replaced by L, one involving L being replaced by L. The first equilibrium can be thought of as involving just H_2O in equilibrium with L, i.e.

   $H_2O \rightleftharpoons L$ (above M), $L \rightleftharpoons H_2O\text{–}M\text{–}H_2O \rightleftharpoons L$, $L \rightleftharpoons H_2O$ (below M) but, in the second, $H_2O \rightleftharpoons L$ (above M), $L \rightleftharpoons H_2O\text{–}M\text{–}H_2O \rightleftharpoons L$, $L \rightleftharpoons L$ (below M)

   i.e. part of the equilibrium involves $L \rightleftharpoons L$.
2. *Steric factors*: if L is larger than H_2O, then as more H_2O molecules are

replaced by L, it becomes successively more difficult to squeeze more L ligands into the coordination sphere of the M.

3. *Charge effects*: if L is negatively charged then as the neutral H_2O is replaced by L^{n-} it will become increasingly more difficult to force another L^{n-} in against a more negatively charged transition metal complex.

For a given hard metal ion, the overall stability constant for complexes with hard bases increases as the oxidation state of M increases. This is because a hard–hard interaction is primarily an electrostatic one and, as the charge on the metal ion increases, the electrostatic interaction (proportional to the product of the charges) increases (page 28).

For complexes with a given ligand the overall stability constants follow the order of ionic potentials (charge:radius ratio), i.e. for metal ions of the same charge the overall stability constants increase as the size of the metal ion decreases. This leads to the *Irving–Williams* series for stability of transition metal complexes with a *given ligand*: $Mn^{2+} < Fe^{2+} < Co^{2+} < Ni^{2+} < Cu^{2+} > Zn^{2+}$

There is a decrease in radius from Mn^{2+} to Ni^{2+} (increasing effective nuclear charge) but an increase in radius from Ni^{2+} to Cu^{2+} to Zn^{2+} {crystal field effects – for octahedral complexes, the electrons go into orbitals (d_{z^2} and $d_{x^2-y^2}$) which point directly at the ligands and so shield the ligands from the attractive force of the nucleus, therefore the ligands move out}. The 'anomalous position of copper' is due to the fact that we are not comparing like with like; this series refers to the stability of octahedral complexes, but, for Cu^{2+}, Jahn–Teller distortion leads to its forming complexes in which four ligands (in the *xy* plane) are bound very strongly and two others (along the *z* axis) are only weakly bound.

Crystal field stabilization energy also has some effect on the stability constants within the series; there is no CFSE for Mn^{2+} (d^5, high spin) and Zn^{2+} (d^{10}) complexes (lower stability constants).

The chelate effect

Complexes of multidentate ligands tend to be more stable than those with comparable monodentate ligands. For example, the following reaction is favourable:

$$[Ni(NH_3)_6]^{2+} + 3en \rightleftharpoons [Ni(en)_3]^{2+} + 6NH_3$$

(en is ethylenediamine, $H_2NCH_2CH_2NH_2$)

This is a purely **thermodynamic** phenomenon, i.e. it only affects the position of equilibrium and not how long it takes to reach it.

We must consider two factors:

1. In the above equilibrium the entropy of the system increases from left to right since more free molecules are produced. $\Delta G = \Delta H - T\Delta S$ and $\Delta G = -RT\ln K$, assuming that the enthalpy change is approximately equal for NH_3 and en (both bonding through the same atom in very similar environments, i.e. M—NH_3 versus M—NH_2R), ΔG is more favourable for formation of the en complex and therefore the equilibrium lies to the right.
2. Once one end of a chelating ligand is coordinated it is more favourable to coordinate the other end than to bring in an external ligand. This is because, with one end of the chelating ligand attached to the metal the other end is also relatively constrained, i.e. it does not have full freedom of movement, therefore not so much entropy is lost as would be the case if another ligand is brought in from the solution.

Three chelate rings in a complex represent a more stable situation than two chelate rings, and in turn this is a more stable situation than one ring.

Five membered chelate rings

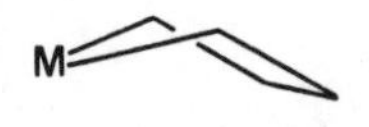

are more stable than six membered rings which, in turn, are more stable than seven membered rings. Rings smaller than five membered tend to be unfavourable. The relative stabilities of the various chelate rings is mainly dependent upon two factors:

1. Entropy effects: consider one end of a bidentate ligand already joined to a metal atom; the longer the chain, the more the unattached end waves about and therefore more entropy is lost when this becomes coordinated (more ordered).
2. Ring strain: the bond angles in a planar five membered ring are closest to the tetrahedral, sp^3, value of 109°. For rings with more than five members puckering occurs to give suitable bond angles. For rings with less than five members the bond angles are significantly smaller than the tetrahedral value and there is a large amount of strain in the ring.

Factors 1 and 2 make formation of five membered rings most favourable.

Isomerism

There are many types of isomerism encountered in transition metal complexes.

GEOMETRICAL ISOMERISM Square planar complexes, MA_2B_2 exhibit *cis/trans* isomerism, e.g. $Pt(NH_3)_2Cl_2$

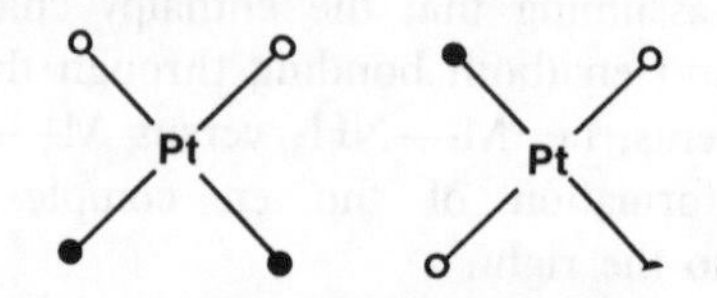

Tetrahedral complexes do not exhibit this type of isomerism.

Octahedral complexes MA_4B_2 can have *cis* and *trans* isomers, e.g. $[Co(NH_3)_4Cl_2]^+$.

Complexes MA_3B_3 can have *fac* and *mer* (facial and meridional) isomers, e.g. $Rhpy_3Cl_3$

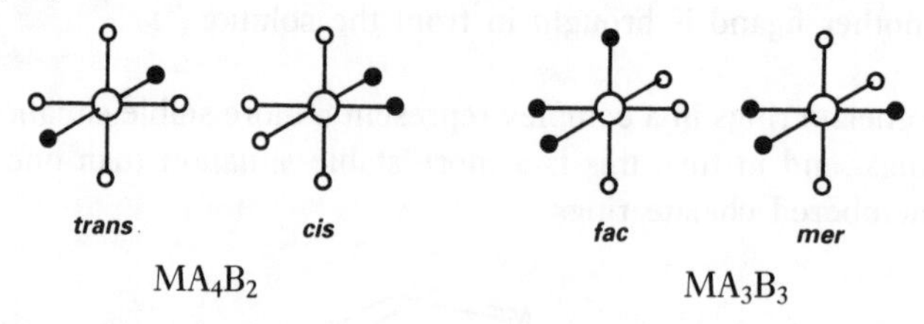

OPTICAL ISOMERISM To be chiral a complex must be asymmetric, i.e. there must be no plane or centre of symmetry. A chiral complex has a non-superimposable mirror image, e.g.

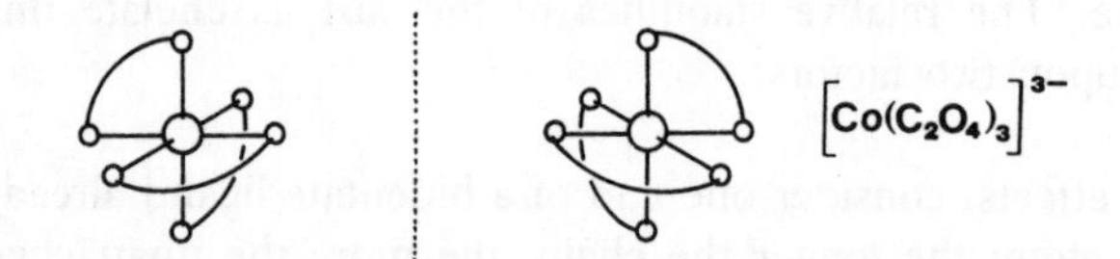

To generate optical activity it must be possible to resolve a chiral compound into enantiomers. Enantiomers rotate the plane of plane polarized light in opposite directions by equal amounts. Chiral complexes can only be resolved if the system is kinetically inert so that the process of ligand substitution is slow compared with the time taken to resolve the complex. Thus the inert (see later) chiral complex $[Co(C_2O_4)_3]^{3-}$ (d^6, low spin) can be resolved, but the labile complex $[Fe(C_2O_4)_3]^{3-}$ (d^5, high spin) cannot. Most complexes that have so far been resolved contain chelating ligands.

A convenient way of resolving a mixture of optically active cationic complexes is to react the mixture with an optically active anion to generate diastereoisomeric salts, which have different physical properties (melting points, solubilities, etc.).

STRUCTURAL ISOMERISM A metal in a transition metal complex can exhibit different stereochemistries. An elegant example of this is found in

$[Ni(CN)_5][Cr(en)_3].1.5H_2O$, where both trigonal bipyramidal and square based pyramidal $[Ni(CN)_5]^{3-}$ units exist in the same crystal.

Another example is $NiCl_2py_2$ which is a chain polymer in the solid state, with octahedral coordination at the Ni,

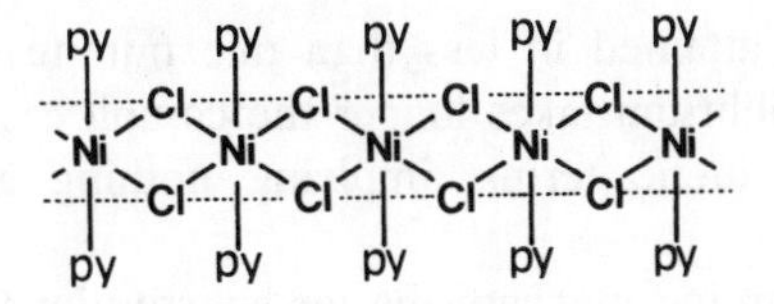

In solution either a tetrahedral or square pyramidal monomeric structure is adopted, depending on the solvent.

IONIZATION/HYDRATION ISOMERISM $CrCl_3.6H_2O$ has three isomers:

$[Cr(H_2O)_4Cl_2]^+.2H_2O$ (1)
$[Cr(H_2O)_5Cl]^{2+}(Cl^-)_2.H_2O$ (2)
$[Cr(H_2O)_6]^{3+}(Cl^-)_3$ (3)

These can be distinguished by the number of equivalents of silver nitrate solution with which they will react rapidly, i.e. according to how many uncoordinated Cl^- ions there are. Thus, **1** reacts with one equivalent of $AgNO_3$, **2** with two equivalents and **3** with three equivalents of $AgNO_3$.

LINKAGE ISOMERISM This occurs when it is possible for a ligand to coordinate through two different atoms, e.g. we can have $(Ph_3P)_2Pd(SCN)_2$ and $(Ph_3P)_2Pd(NCS)_2$. In the former, thiocyanate coordinates to the metal through the S and in the latter it coordinates through the N; thiocyanante is called an **ambidentate** ligand.

COORDINATION ISOMERISM The ligands are bound to different metal atoms in each isomer, e.g.

$[Co(NH_3)_6][Cr(CN)_6]$ and $[Cr(NH_3)_6][Co(CN)_6]$

The complex $Pt(NH_3)_2Cl_2$ exhibits several types of isomerism, i.e. we can have the following forms: *cis* $Pt(NH_3)_2Cl_2$, *trans* $Pt(NH_3)_2Cl_2$, $[PtCl_4]^{2-}[Pt(NH_3)_4]^{2+}$ and

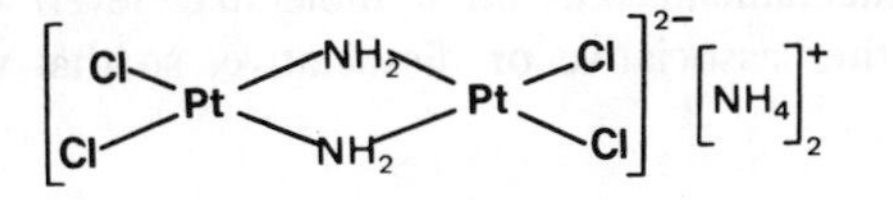

Transition metal reaction mechanisms

LIGAND SUBSTITUTION REACTIONS Consider a reaction,

$$L_nMX + Y \rightleftharpoons L_nMY + X$$

If equilibrium is attained in less than one minute the initial complex is called **labile**; if equilibrium takes longer the complex is called **inert**. Lability and inertness are kinetic terms, implying nothing about thermodynamic stability.

There are two limiting mechanisms for substitutional processes; these are A (Associative) and D (Dissociative). Both these mechanisms involve a detectable intermediate, i.e. a reaction profile **I**,

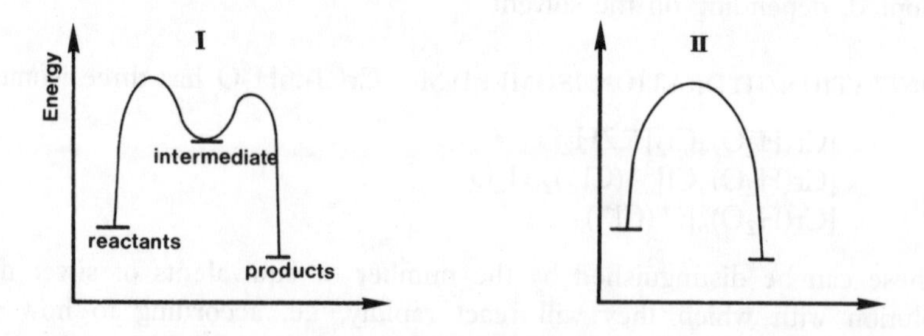

The rate determining step (RDS) is that which generates the intermediate.

In an A process the intermediate has higher coordination number than the initial complex, i.e.

$$L_nXM + Y \xrightarrow{\text{RDS}} L_nMXY \xrightarrow{\text{fast}} L_nMY + X$$

The rate of reaction depends on the concentration/nature of the incoming ligand (Y).

In a D process the intermediate has lower coordination number, i.e.

$$L_nMX + Y \xrightarrow{\text{RDS}} L_nM + X + Y \xrightarrow{\text{fast}} L_nMY + X$$

The rate of reaction is independent of the concentration/nature of Y.

If it is not possible to detect an intermediate then we have an I (Interchange) mechanism and a reaction profile, **II**.

The intimate mechanism (i.e. on a molecular level) of an interchange process can be either associative or dissociative, so that we have I_a and I_d mechanisms.

I_a

In the transition state there is substantial bonding to both the entering and leaving groups. The entering group plays an important part in determining the energy of the transition state and the reaction rate is dependent upon its concentration.

Transition state $[X \cdots M \cdots Y]^{\ddagger}$

I_d

There is only weak bonding to the entering and leaving groups in the transition state and the rate of reaction is largely independent of the concentration/nature of the entering group.

Transition state $[X \cdots M \cdots Y]^{\ddagger}$

In general, four coordinate planar transition metal complexes undergo nucleophilic substitution via an A mechanism because they can easily add another ligand. Six coordinate octahedral complexes react via D, I_d and, occasionally, I_a mechanisms; the A mechanism is not generally found because of the unfavourability of having a seven coordinate transition metal ion.

INERTNESS, LABILITY AND ELECTRONIC CONFIGURATION Remarkable success has been obtained in relating inertness/lability and relative reaction rates of substitution in octahedral complexes to crystal field stabilization energy (CFSE). For octahedral complexes undergoing substitution via a five coordinate square pyramidal transition state (D mechanism) crystal field activation energies (CFAEs) can be calculated by subtracting the CFSE of the initial octahedral complex from that of the square pyramidal transition state. Such calculations lead to the following predictions:

1. d^0, d^5 (high spin) and d^{10} complexes are labile since CFSE is zero in the ground and transition state.
2. For d^1, d^2, d^4(h.s.), d^6(h.s.), d^7(h.s.) and d^9 CFAEs are very small and these complexes are also labile.
3. CFAEs are large for d^3, d^4(l.s.), d^5(l.s.), d^6(l.s.), d^8 and these complexes are inert. The order predicted for decreasing reaction rates is $d^5 > d^4 > d^8 \approx d^3 > d^6$.

These predictions are born out by experimental observation.

No figures have been given here for the CFAE; this is because it is no simple matter to work them out. The problems with the application of CFAEs are:

1. It is assumed that the transition state has a regular square pyramidal geometry.
2. All bond lengths and the values of Δ are assumed to be the same as in the ground state. Five ligands will not, however, give the same value of Δ as six!

3. There are all the normal problems associated with CFSE, i.e. CFAE is only one of a number of terms involved in $E_{act.}$ and a totally ionic model is assumed (the basis of crystal field theory).

Despite these factors remarkably good agreement is obtained between calculated CFAEs and reaction rates.

THE WATER EXCHANGE REACTION

$$[M(H_2O)_6]^{n+} + H_2O \rightleftharpoons [M(H_2O)_6]^{n+} + H_2O$$

M	RATE (s^{-1})	M	RATE (s^{-1})
Cr^{2+}	7×10^9	Mn^{2+}	3×10^7
Fe^{2+}	3×10^6	Co^{2+}	1×10^6
Ni^{2+}	3×10^4	Cu^{2+}	8×10^9
Fe^{3+}	3×10^3	Cr^{3+}	3×10^{-6}
Rh^{3+}	4×10^{-8}		

The reaction follows an I_d mechanism, so that there is significant weakening of the M—OH_2 bond in formation of the transition state. The rate of reaction, thus depends strongly upon the M—OH_2 bond strength.

2+ ions

All the $[M(H_2O)_6]^{2+}$ complexes are high spin. The Cr^{2+} ($t_{2g}{}^3e_g{}^1$) and Cu^{2+} ($t_{2g}{}^6e_g{}^3$) complexes suffer large Jahn–Teller distortions, such that the axial bonds are lengthened and weakened, making the axial waters easier to replace, therefore giving the fastest rate for water substitution: d^4(h.s.) and d^9 are also predicted to be the most labile on CFAE grounds.

For d^5, d^6, d^7 the rates are quite similar but considerably slower than for the Cr^{2+} (d^4) and Cu^{2+} (d^9) complexes. The rates are slower because there is no significant Jahn–Teller weakening. Two factors affect the rate:

1. Decreasing size from d^5 to d^7: this means that the M—OH_2 bonds are shorter for d^7 and the H_2O more strongly held, which leads to the prediction of a slower rate of substitution for d^7.
2. CFAE: this predicts that all three complexes should be labile but that there should be a slight increase in lability from d^5 to d^7.

Factors 1 and 2 act against each other but the size effect dominates.

The relative inertness of the Ni^{2+} complex is as predicted from CFAE arguments and the smaller size of the d^8 ion.

3+ ions
The rates of substitution are all slower for the +3 complexes. This is consistent with stronger M—OH_2 bonds – the more highly charged metal pulls the ligands in more closely, making them difficult to replace.

Fe^{3+}, d^5, high spin – CFAE predicts its relative lability (CFSE is zero in both ground and transition states).

Cr^{3+}, $t_{2g}{}^3$ and Rh^{3+}, low spin, $t_{2g}{}^6$, are both predicted to be substitutionally inert on CFAE grounds. Rh^{3+} is more inert because of the larger Δ for the second row element, i.e. the magnification of crystal field effects.

THE *TRANS* EFFECT AND *TRANS* INFLUENCE The *trans* effect and *trans* influence are usually discussed with respect to square planar transition metal complexes.

The *trans* **effect** is a **kinetic** phenomenon and is the effect of a coordinated ligand on the rate of replacement of a ligand *trans* to itself. The *trans* effect series gives the ability of ligands to labilize the ligand *trans* to itself:

$$H_2O \approx OH \approx NH_3 \approx \text{amines} < Cl < Br < I < CH_3 < PR_3 \approx H^- < \text{olefins} \approx CO \approx CN$$

Thus, substitution of the Cl in *trans*-$[Pt(PEt_3)_2LCl]$,

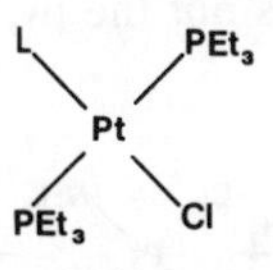

is about 10^5 times faster when L = H than when L = Cl.

The *trans* **influence** is a **thermodynamic**, ground state, phenomenon and is the extent to which a ligand weakens the bond *trans* to itself in the ground (equilibrium) state of a complex. Approximately the same series holds, such that the strongest *trans* effect ligand has the greatest *trans* influence, e.g. the Pt—Cl bond length in *trans*-$[Pt(PEt_3)_2Cl_2]$ is 0.12 Å shorter than that in *trans*-$[Pt(PPh_2Et)_2HCl]$.

The *Trans* Effect and Synthesis

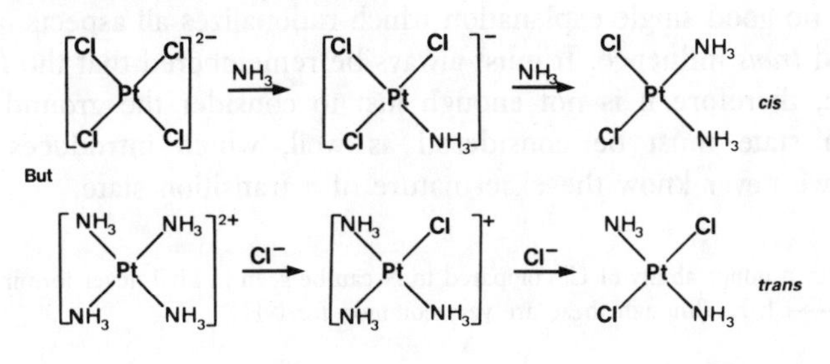

The products result as shown because Cl is a stronger *trans* directing ligand than NH_3, therefore ligands *trans* to a Cl are labilized in preference to those *trans* to NH_3.

Utilizing the *trans* effect and another general principle – all things being equal, an M—halogen bond is more labile than an M—N bond (the M—hal bond is weaker due to, e.g. Cl being a poorer σ donor than N).† We can make different isomers simply by varying the order of addition, i.e.

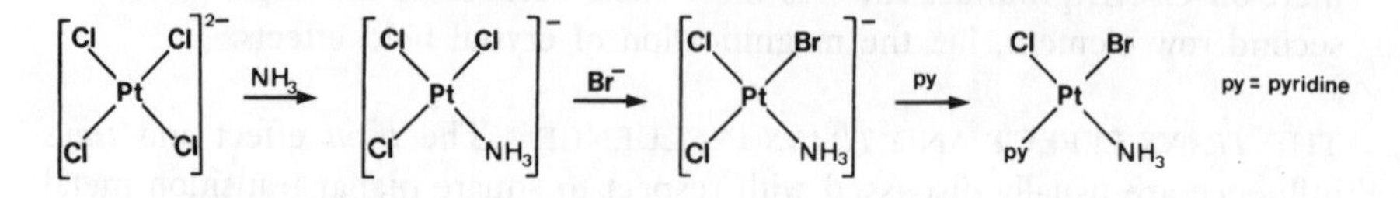

Trans effect order is Br > Cl > NH_3.

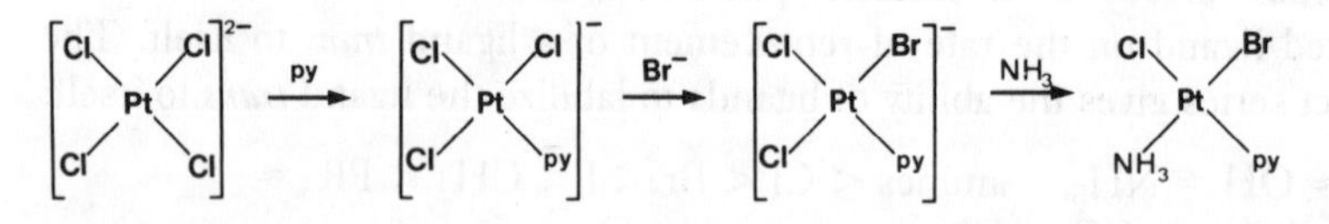

Trans effect series is Br > Cl > py.

The second step relies on the Pt—py (Pt—N) bond being less labile than the Pt—Cl bond, so that it is not the py, which is also *trans* to a Cl, which is replaced.

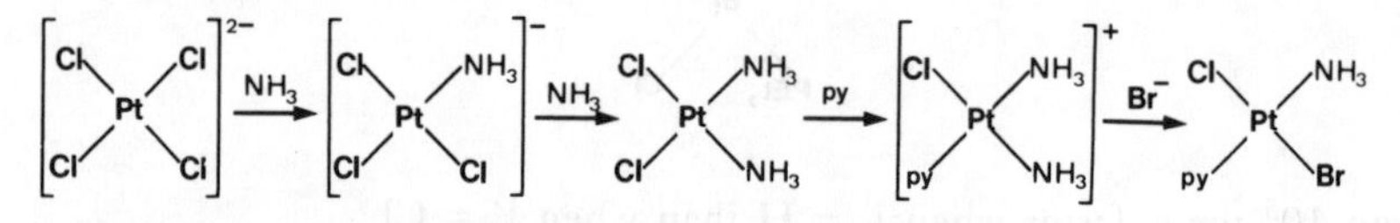

Trans effect series is Br > Cl > py > NH_3.

The third step is not what we would expect from the *trans* effect series, i.e. an NH_3 should be labilized in preference to the Cl. The actual product is the result of the inherent lability of the M—Cl bond.

Another problem of the *trans* effect series is that minor changes in order can occur depending on the incoming ligand.

Explanation of the *trans* effect and *trans* influence

There is no good single explanation which rationalizes all aspects of the *trans* effect and *trans* influence. It must always be remembered that the *trans* effect is kinetic, therefore it is not enough just to consider the ground state, the transition state must be considered as well, which introduces problems because we never know the exact nature of a transition state.

† The poorer σ donor ability of Cl compared to N can be seen in HCl never forming adducts of the type H—Cl: E, although these are very common for NH_3.

It is usual to divide explanations into σ and π components:

σ effects

1. *Electronegativity*: the length of the bond *trans* to a ligand, L, increases as the electronegativity of L decreases.

   For an X—M—L unit, a less electronegative L increases the electron density on M, therefore X to M σ donation is reduced and the X—M bond is weakened.
2. *Polarizability*

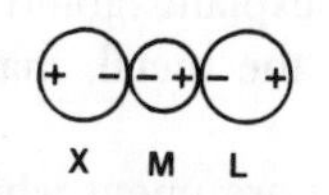

   If L is a polarizable ligand then the metal will induce a dipole as shown. This dipole on L will in turn induce a dipole on M which opposes the natural dipole of X and, therefore, weakens the X—M bond. This effect increases with the polarizability of the metal (a larger dipole will be induced on M), i.e. it is a larger effect for Pt^{II} than for Ni^{II} {Pt^{II} is larger, therefore the outer electrons are held less strongly by the nucleus}.

Both the above factors are concerned with a ground state bond weakening and refer to the *trans* influence. However, they can be seen to have some relevance to the *trans* effect once we realize that, in a square planar complex, a pair of *trans* ligands interacts with the same metal p orbital but in a trigonal bipyramidal transition state the interaction is not so direct.

Thus the electronegativity and polarization effects mentioned above are going to be smaller in the transition state if it is assumed that most interaction occurs through the same p orbital. Thus, a ligand which weakens the *trans* bond in the ground state does not weaken it as much in the transition state and is, in effect, destabilizing the ground state relative to the transition state, i.e. labilizing the *trans* ligand.

π effects

Strong *trans* directing ligands are very often those which are good π acceptors, e.g. CO (Chapter 7). If L and X are both π acids, then they will compete for the available d electron density on the metal.

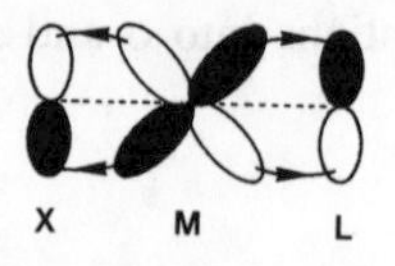

The greater the π acceptor ability of L, the more electron density X is deprived of and the weaker the M—X π bond. This is another ground state effect and explains bond length variations when both L and X have π orbitals but not if X has no π orbitals, e.g. if X = NH_3.

π Effects can be used to explain lability since a strong π acceptor will remove electron density from the metal, making it more willing to accept a fifth ligand.

These are the main lines of argument which have been put forward to try and explain the *trans* effect and *trans* influence. There are other theories and the truth is that nobody really knows exactly what is going on. The true explanation probably involves a combination of σ and π effects.

There is some evidence for the *trans* effect and *trans* influence in octahedral complexes. Not so much work has been done on these complexes but it has proved possible to derive an approximate series of *trans* labilizing ability, the order of which is essentially the same as for square planar. The situation is further complicated by the ligands showing varying degrees of *cis* labilization (many are able to labilize *cis* ligands far more effectively than *trans* ligands).

The ground state, *trans* influence, i.e. bond lengthening, is mostly seen in connection with complexes containing ligands *trans* to π acidic ligands and competing for π electron density, e.g.

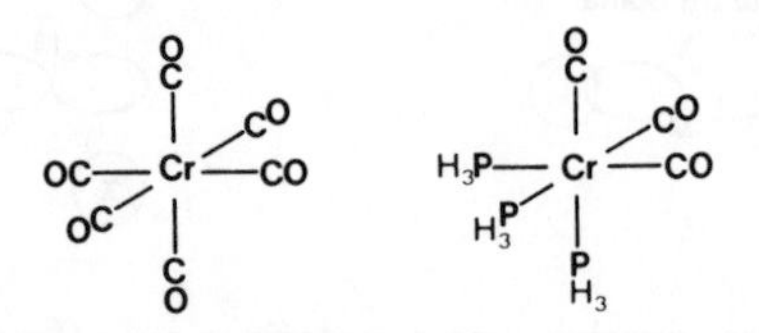

$Cr(CO)_6$, Cr—C = 1.91 Å $Cr(CO)_3(PH_3)_3$, Cr—C = 1.84 Å (page 232)

The mechanisms of the *trans* effect and *trans* influence in octahedral complexes are even less certain than for square planar. It is believed that similar factors are probably operating but much work has still to be done.

ELECTRON TRANSFER REACTIONS For example,

$$[^*Fe^{III}(CN)_6]^{3-} + [Fe^{II}(CN)_6]^{4-} \rightarrow [^*Fe^{II}(CN)_6]^{4-} + [Fe^{III}(CN)_6]^{3-} \quad (1)$$

where *Fe represents a radioactively labelled Fe atom,
or

$$[Co(NH_3)_5Cl]^{2+} + [Cr(H_2O)_6]^{2+} \xrightarrow{H^+} Co^{2+}{}_{(aq)} + 5NH_4{}^+ + [ClCr(H_2O)_5]^{2+} \quad (2)$$

There are two types of electron transfer processes:

1. *Outer sphere*: interaction between oxidant (the species which is reduced) and reductant (that which is oxidized) at the time of electron transfer is small. Each complex retains its own full coordination shell and the electron is transferred through both. Reaction 1, above, is an outer sphere process.
2. *Inner sphere*: a bridge is formed between the oxidant and reductant so that at least one ligand is common to the coordination spheres of both reactants; the electron is transferred through this ligand from one centre to the other. Reaction 2 is an inner sphere process.

Outer Sphere Reactions

In order to demonstrate unequivocally an outer sphere electron transfer it is necessary to have reactants in which ligand substitution is slower than the electron transfer process. Although this is necessary for proof it is not an essential condition for an outer sphere reaction to occur.

The simplest case for outer sphere reactions is when the reactants and products are chemically identical (Reaction 1). The rates of these reactions can be studied by, e.g. radio-isotope labelling and n.m.r. spectroscopy. ΔG for a process such as this is zero since there is no chemical change. The activation energy is not zero and is made up of three main components:

1. Repulsion between the interacting species as they are brought into close proximity. Thus, if both complexes are, e.g. positively charged, then energy is required to force the two positive charges together. There is also going to be repulsion between the closed shells of electrons of the ligands on the different complexes when they are close together.

   The rate of reaction is sensitive to the nature of other species, e.g. counterions, in solution. For example, the rate of the reaction,

   $$[{}^*MnO_4]^-{}_{(aq)} + [MnO_4]^{2-}{}_{(aq)} \rightarrow [{}^*MnO_4]^{2-}{}_{(aq)} + [MnO_4]^-{}_{(aq)}$$

   depends on which cations are present, i.e. the reaction rate decreases in the order $Cs^+ > K^+ > Na^+ > Li^+$; this is due to ion pairing. Close association between anion and cation means that we are now considering interaction between essentially neutral oxidant and reductant species, which are easier to push together than negatively charged anions with no ion pairing. The ion pairing is greatest with Cs^+ since this has the smallest hydrated radius (see page 6) and therefore experiences the strongest electrostatic interaction with the anion.

2. Each complex is solvated and so energy is needed to distort the solvation shells if the complexes are to get close together.
3. Franck–Condon restrictions – the act of electron transfer is fast compared to the time taken for a molecule to vibrate and therefore electron transfer occurs with all atoms effectively 'frozen' in a particular position, i.e. bond lengths are fixed.

Consider,

$$[{}^*Fe^{III}(H_2O)_6]^{3+} + [Fe^{II}(H_2O)_6]^{2+} \rightarrow [{}^*Fe^{II}(H_2O)_6]^{2+} + [Fe^{III}(H_2O)_6]^{3+}$$

The Fe—O bond lengths in the Fe^{III} complex are significantly shorter than in the Fe^{II} complex (with the more positively charged metal the ligands are pulled in closer). Thus, when an electron is transferred from Fe^{II} to Fe^{III}, an Fe^{III} complex with long bond lengths and an Fe^{II} complex with short bond lengths are produced, i.e. both molecules are generated in excited vibrational states and will release energy to their surroundings as they relax back to the ground state. The overall reaction involves no net chemical change, therefore by just having a solution of Fe^{II} and Fe^{III} we are generating energy without anything else happening, i.e. we have a limitless energy source, which violates the first and second laws of thermodynamics.

To avoid this some distortion of the complex has to occur before reaction can take place. This requires energy and forms a part of the activation energy for the process. One activated complex that could be imagined is where the Fe^{II} bond lengths have been compressed to the equilibrium Fe^{III} distances (energy = A) and the Fe^{III} bond lengths have been elongated to the ground state Fe^{II} lengths (energy = B and the total activation energy is $A + B$) but the lowest activation energy pathway actually turns out to be when the Fe^{II}—L and Fe^{III}—L distances are equal in the transition state (activation energy = $C + D$).

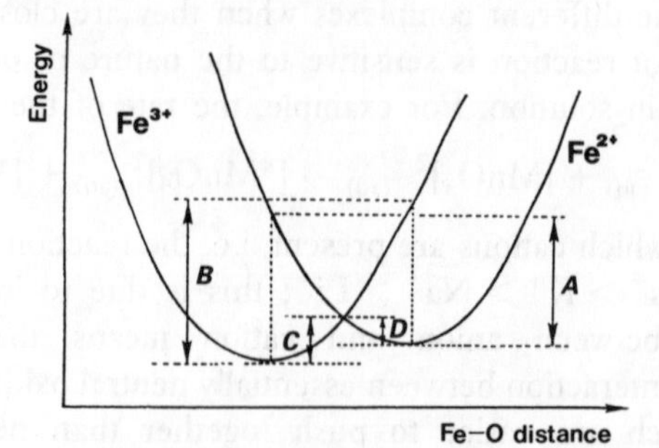

Reactions thus tend to be slower if there is a large difference in bond lengths between the interacting complexes since a greater amount of energy has to be supplied for distortion (rate constant, $k \propto e^{-E_{act.}}$).

Related to this is that reactions involving an e_g electron on one metal going to an e_g orbital on another metal, or $t_{2g} \longleftrightarrow t_{2g}$, tend to be much faster than $t_{2g} \longleftrightarrow e_g$ electron transfers. This is because the t_{2g} orbitals are σ non-bonding but the e_g orbitals are σ* antibonding and transfer of an electron from a non-bonding to an antibonding orbital, or vice versa, is going to result in a large change in bond lengths, e.g. the reaction between V^{II} and $[Co(NH_3)_6]^{3+}$ $\{V^{II}\ (t_{2g}{}^3) \rightarrow V^{III}\ (t_{2g}{}^2)\}$ is faster than that between Cr^{II} and $[Co(NH_3)_6]^{3+}$ $\{Cr^{II}\ (t_{2g}{}^3 e_g{}^1) \rightarrow Cr^{III}\ (t_{2g}{}^3)\}$.

When a reaction involves some chemical change and is exothermic the Franck–Condon restrictions are not so severe, i.e. there is no need for equilibration of bond lengths, since energy released by relaxation from excited vibrational states can form part of the overall energy released during the reaction. For endothermic reactions the restrictions of the Franck–Condon principle are more severe. Thus, exothermic reactions tend to be faster than those in which there is no change in free energy, and these, in turn, are faster than endothermic reactions.

The intimate mechanism of an outer sphere redox process would seem to involve close interaction between the two complexes (like two billiard balls sitting touching) and then electron transfer via a quantum mechanical tunnelling process through the barrier provided by the ligand shells.

Inner sphere reactions

An elegant example of an electron transfer which must involve a bridge between the interacting metal centres is:

$$\underset{\text{inert}}{\underset{Co^3\ d^6\ (t_{2g}{}^6)}{[Co(NH_3)_5Cl]^{2+}}} + \underset{\text{labile}}{\underset{Cr^{2+}\ d^4\ (t_{2g}{}^3 e_g{}^1)}{[Cr(H_2O)_6]^{2+}}} \xrightarrow{H^+}$$

$$\underset{\text{labile}}{\underset{Co^{2+}\ d^7\ (t_{2g}{}^5 e_g{}^2)}{[Co(H_2O)_6]^{2+}}} + 5NH_4{}^+ + \underset{\text{inert}}{\underset{Cr^{3+}\ d^3\ (t_{2g}{}^3)}{[ClCr(H_2O)_5]^{2+}}}$$

The Cl atom in the initial complex can be radioactively labelled and if the reaction is done in the presence of free unlabelled Cl^- only labelled Cl ends up in the final Cr^{3+} complex.

If the reaction went by an outer sphere mechanism then electron transfer from $[Cr(H_2O)_6]^{2+}$ would result in generation of an inert $[Cr(H_2O)_6]^{3+}$ (d^3) complex which would be unwilling to pick up a Cl^- ion from solution (generated by decomposition of the labile Co^{2+} complex); even if it did so, there should be no specificity for labelled Cl. The only way in which we can get complete transfer of labelled Cl to the final complex is if the labelled Cl bridged between the initial Co^{3+} and Cr^{2+} complexes and then remained attached to the inert Cr^{3+} when the bridge broke.

The intimate mechanism of the inner sphere electron transfer is:

$$[Co^{III}(NH_3)_5Cl]^{2+} + [Cr^{II}(H_2O)_6]^{2+} \rightarrow [(NH_3)_5Co^{III}\text{-}Cl\text{-}Cr^{II}(H_2O)_6]^{4+}$$

Precursor complex ↓

$$[(NH_3)_5Co^{II}\text{-}Cl\text{-}Cr^{III}(H_2O)_6]^{4+}$$

Successor ↓ complex

$$[Co(H_2O)_6]^{2+} + 5NH_4^{+} \underset{\leftarrow}{H^+} [ClCr(H_2O)_5]^{2+} + [Co(NH_3)_5Cl]^{2+}$$

The formation of the initial bridged (precursor) complex requires loss of an H_2O from the Cr^{2+} (d^4) complex, which is easy because of the labilty of the complex (d^4 is labile). When the successor complex breaks up the Cl remains with the Cr^{III} since this is substitutionally inert (d^6 low spin) and the Co^{II} (d^7) is labile.

In general:

1. The bridging ligand usually belongs to the species to be reduced (the oxidant).
2. It is not necessary for the bridging ligand to dissociate from the metal centre to which it was originally joined – there are three possibilities,

$$\begin{aligned} R + X{-}O &\rightarrow O + X{-}R \quad && R = \text{Reductant} \\ &\text{or } O{-}X + R && O = \text{Oxidant} \\ &\text{or } O + X + R && \end{aligned}$$

3. It is possible for any of the following steps to be rate determining:

Bridge formation

This will tend to occur when ligand substitution in the initial complex is slower than electron transfer (a ligand must dissociate, to produce a vacant coordination site, before a bridge can be formed). For this type of reaction the rate is strongly dependent on the nature of the bridge, i.e. (assuming the bridging group comes from the oxidant) the rate will depend on the affinity of the reductant for the bridging group.

Electron transfer

The rate here is also very sensitive to the nature of the bridge. This is partly due to the initial pre-equilibrium involved in formation of the bridged complex (as in bridge formation above) and partly due to the the actual act of conducting an electron through the bridge. Because there are two factors involved it is difficult to really give any generalizations about the ability of simple ligands to act as bridges. For example, for halogen bridges the order of reaction rate depends on the hardness/softness of the reductant (i.e. its affinity for the hard F or the soft Br) and the polarizability of the bridge (i.e. Br is more polarizable than F, which should make electron transfer easier). For reduction by hard cationic complexes the order of reaction rates is F > Cl > Br but when soft cation complexes comprise the reductant the order is reversed (see HSAB theory, page 28).

With polyatomic bridges conduction is often made easier by having a conjugated system between the metal centres, e.g.

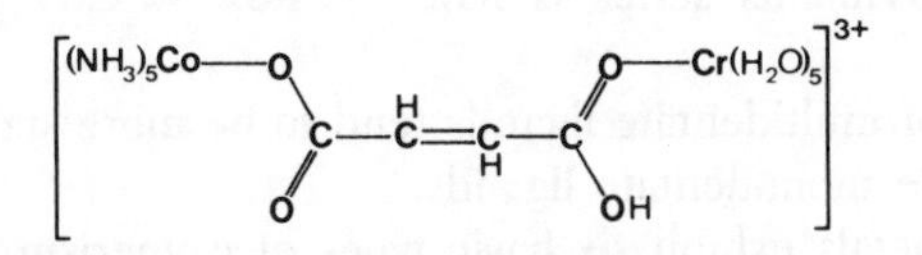

The intimate mechanism for electron transfer can be of two types:

1. A chemical mechanism involving transfer of an electron from the reductant to the bridging ligand and from thence to the oxidant. Rates of electron transfer in this case are going to be faster when the donor orbital on the reductant, the ligand and oxidant acceptor orbitals are all of the same symmetry (i.e. σ or π). This is because if all orbitals are of the same symmetry we can imagine there being a direct path through from reductant to oxidant.

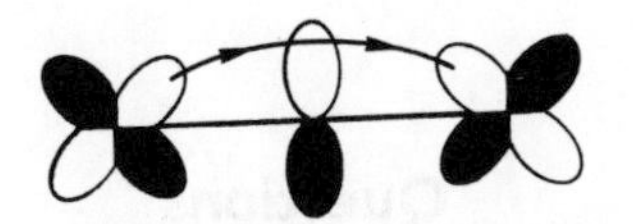

When there is a mismatch of symmetry then electron transfer is slower and it is sometimes possible to detect the presence of an intermediate in which the ligand has been reduced, e.g.

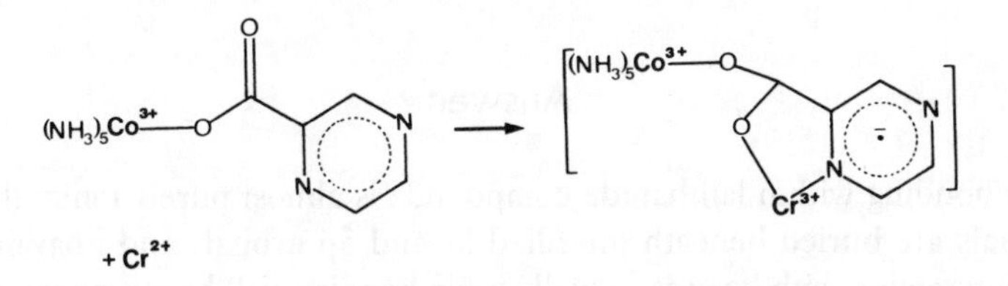

2. Quantum mechanical tunnelling through the potential energy barrier constituted by the bridging ligand.

Bridge breaking

Bridge breaking is the rate determining step. In this case it is sometimes possible to isolate the successor complex, e.g. in the reaction between

$[Fe(CN)_6]^{3-}$ and $[Co(CN)_5]^{3-}$ the complex $[(NC)_5Fe—CN—Co(CN)_5]^{6-}$ can be isolated. This provides further evidence for the inner sphere mechanism.

Summary

1. The Irving–Williams series is $Mn^{2+} < Fe^{2+} < Co^{2+} < Ni^{2+} < Cu^{2+} > Zn^{2+}$.
2. Complexes of multidentate ligands tend to be more stable than those with comparable monodentate ligands.
3. Transition metals exhibit six basic types of isomerism.
4. There are four basic mechanisms for substitution in transition metal complexes: D, A, I_a and I_d.
5. With regard to ligand substitution processes, d^0, d^5(h.s.), d^{10}, d^1, d^2, d^4(h.s.), d^6(h.s.), d^7(h.s.), d^9 are labile, whereas d^3, d^4(l.s.), d^5(l.s.), d^6(l.s.) and d^8 are inert.
6. The *trans* effect series is $H_2O \approx OH^- \approx NH_3 \approx$ amines $< Cl^- < Br^- < I^- < CH_3^- < PR_3 \approx H^- <$ olefins $\approx CO \approx CN^-$.
7. There are two types of electron transfer reactions, inner sphere and outer sphere.

Questions

1 Why is it that lanthanide ions form the strongest bonds with the most electronegative ligands but M—L bond strengths for the first row transition metal series follow an order, $CO > NH_3 > H_2O > F^-$?

Answer

The bonding within lanthanide compounds is almost purely ionic; the 4f orbitals are buried beneath the filled 5s and 5p orbitals and unavailable for interaction with ligands. Totally ionic bonding will be strongest when the ligands are small and highly charged (electrostatic energy is proportional to charge and inversely proportional to inter-ionic distance), i.e. strongest interaction with F^-. Note that it is generally the most electronegative ligands which have the greatest tendency to form anions.

Transition metal complexes have a considerable amount of covalency in their bonding due to the d orbitals acting as true valence orbitals and being available for overlap with ligand orbitals. A good illustration of the availability of the 3d orbitals compared to the 4f comes from the different magnitude of crystal field effects – for the lanthanides the splitting of 4f orbital energies is rarely more than a few hundred cm^{-1}, whereas, for the first row transition metals the splitting is generally of the order of 10 000 cm^{-1}.

The bonding within transition metal complexes is made up of a combination of σ and π effects:

(a) CO is a poor σ donor but readily accepts electron density into its C—O π* orbitals to give very strong π bonding. The 4f orbitals in lanthanides are unable to participate in π bonding and lanthanide carbonyl complexes tend to be extremely unstable (see page 256).

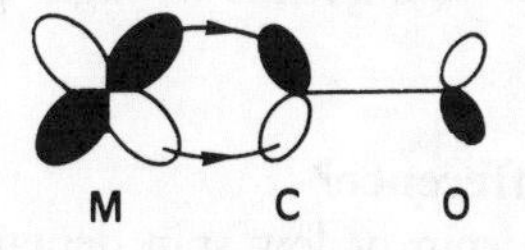

(b) NH_3 and H_2O do not participate in π interactions and the bonding is mostly the result of σ donation from the ligand into a vacant orbital on the metal. There is also an ionic contribution between the positively charged metal and the negative end of the dipole on the ligand. Although this electrostatic effect is larger in H_2O compounds (O more δ− than N) the lower electronegativity of N makes it more willing to let go of its lone pair of electrons, i.e. NH_3 is a better σ donor and the M—N bond is stronger.

(c) F^- is not a good σ donor and although it can participate in L → M π donation, this effect is also not large. The inability of F^- to be a good σ or π donor can be related to the low polarizability of F^- (i.e. in simple compounds MX, F has the greatest tendency of all the halogens towards ionic bonding). The bonding in F compounds will have quite a large ionic contribution but this is generally not enough to compensate for the small σ and π bonding.

2

Compound	Magnetic moment (BM)
$[CoF_6]^{3-}$	4.9
$[RhF_6]^{3-}$	0.0
$NiCl_2(PPh_3)_2$	3.3
$NiCl_2(PEt_3)_2$	0.0

Comment on these magnetic moments.

Answer

Co^{3+} and Rh^{3+} are both d^6 octahedral – two possible electronic configurations,

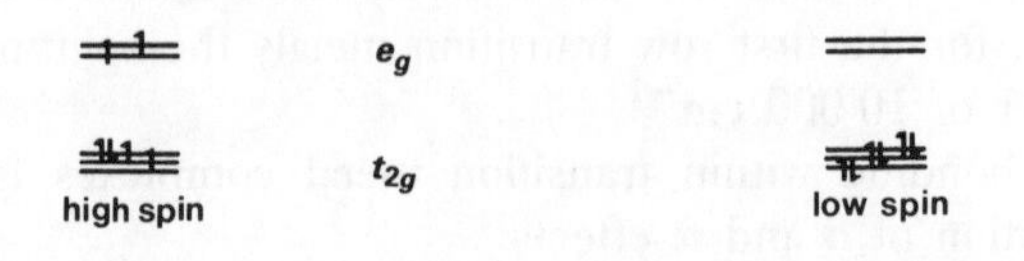

total electron spin $= \Sigma s = S = 4 \times 1/2 = 2$ total spin $= 0$

Substitute these values into the spin-only formula, $\mu = \sqrt{4S(S+1)}$. For $S = 2$, $\mu = 4.9\,\text{BM}$ and for $S = 0$, $\mu = 0$, therefore we can conclude that the Co^{3+} complex is d^6 high spin and the Rh^{3+} complex is d^6 low spin.

Why is there a difference?

Whether we get high spin or low spin depends on the relative sizes of crystal field splitting energy, Δ, and the pairing energy, P (energy to pair two electrons in the same orbital) – if $\Delta > P$ we get low spin and if $P > \Delta$ we get high spin. There is a large increase in Δ as we go down a group, due to the outer d orbitals being more diffuse and interacting more strongly with ligand electrons; this leads to all complexes of the second and third row transition metals being low spin. (Note that there should also be a decrease in P as we go to the second and third rows – orbitals bigger, less repulsion between electrons in them – but this is more than cancelled out by the increase in Δ. It should be pointed out that this is one of the few high spin Co^{3+} complexes. Also, the ground state for Co^{3+} high spin octahedral is $^4T_{2g}$ and the magnetic moment therefore would not be expected to obey the spin-only formula. Although the spin-only formula might not be strictly applicable for a T ground state, it will usually provide a guide to the number of unpaired electrons.

Both Ni^{2+} complexes are d^8 four coordinate. The two possible stereochemistries are square planar and tetrahedral,

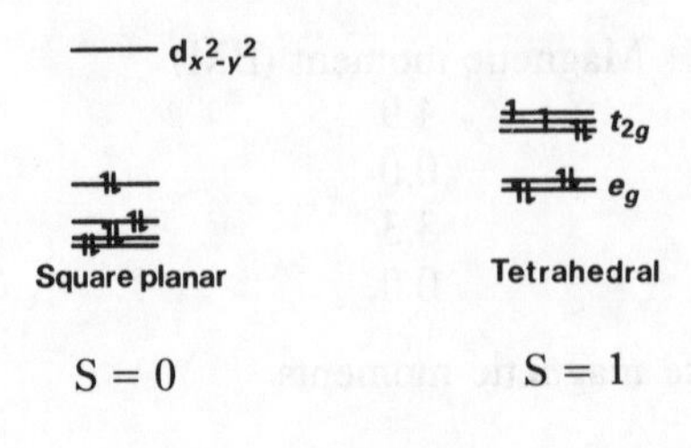

S = 0 S = 1

A low spin complex, with no unpaired electrons ($\mu = 0$, diamagnetic), is preferred for d^8 square planar as this avoids occupation of the high energy $d_{x^2-y^2}$ orbital. $NiCl_2(PEt_3)_2$ is, thus, square planar.

The tetrahedral complex should have a magnetic moment corresponding to two unpaired electrons, i.e. $S = 1$ and, if we substitute this into the spin-only formula we get $\mu = \sqrt{4 \times 1(1+1)} = 2.83$ BM, which corresponds reasonably closely to the value given in the question, therefore $NiCl_2(PPh_3)_2$ is tetrahedral. The difference between the calculated magnetic moment and that given in the question is the result of the ground state being $^3T_1$, and the spin-only formula being strictly not applicable. {Note that another possible formulation for $NiCl_2(PPh_3)_2$ is that it contains octahedrally coordinated Ni in a Cl bridged chain structure (see $NiCl_2py_2$). However, the ground state for Ni^{2+} octahedral is $^3A_{2g}$, which would be expected to give rise to a magnetic moment much closer to the spin-only value of 2.83 BM.}

The difference between these two complexes is due to a combination of steric and CFSE terms:

(a) The tetrahedral geometry keeps the four ligands as far apart as possible and is going to be favoured on steric grounds.

(b) The crystal field stabilization energy is larger in the square planar complex (remember, this is based on an octahedron).

Therefore, if steric effects are not large the square planar geometry (ligands 90° apart) is preferred but with the very bulky PPh_3 ligands steric repulsion is so large that the tetrahedral structure (ligands 109° apart) has to be adopted to minimize the interaction.

3 Account for the following reactions:

$$[CoCl(NH_3)_5]^{2+} + [Cr(OH_2)_6]^{2+} \rightarrow [CrCl(OH_2)_5]^{2+} + [Co(H_2O)]^{2+} + 5NH_{3(aq)}$$

$$[Co(NH_3)_6]^{3+} + [Cr(OH_2)_6]^{2+} \rightarrow [Cr(OH_2)_6]^{3+} + [Co(H_2O)]^{2+} + 6NH_{3(aq)}$$

Answer

Both reactions are electron transfer processes, i.e.

$$Co^{3+} + Cr^{2+} \rightarrow Co^{2+} + Cr^{3+}$$

The first reaction goes via an *inner sphere* mechanism, with the Cl acting as a bridging ligand, i.e.

$$[CoCl(NH_3)_5]^{2+} + [Cr(OH_2)_6]^{2+} \rightarrow$$
$$[(NH_3)_5Co^{III}\text{---}Cl\text{---}Cr^{II}(OH_2)_5]^{4+}$$
$$\downarrow \text{electron transfer}$$
$$[(NH_3)_5Co^{II}\text{---}Cl\text{---}Cr^{III}(OH_2)_5]^{4+}$$
$$\downarrow$$
$$[Co(OH_2)(NH_3)_5]^{2+} + [CrCl(OH_2)_5]^{2+}$$

The starting materials are an inert (d^6 low spin) Co^{3+} complex and a labile Cr^{2+} (d^4 high spin complex). Dissociation of a ligand from the labile complex allows the Cl to bridge between the metals (note that the bridging group comes from the inert complex). Electron transfer generates a labile Co^{2+} (d^7 high spin) and an inert Cr^{3+} (d^3) centre and when the bridging complex breaks up the Cl remains attached to the inert, Cr^{3+}. The $[Co(OH_2)(NH_3)_5]^{2+}$ then exchanges ligands with the solvent to generate $[Co(OH_2)_6]^{2+}$, i.e. $Co^{2+}{}_{(aq)}$.

In the second reaction the electron transfer occurs by an *outer sphere* mechanism due to the inability of the NH_3 to act as a bridging group – coordination to one metal makes the nitrogen coordinatively saturated. The mechanism involves close approach of the two complexes and then electron transfer through the coordination spheres of both. Thus, no NH_3 enters the coordination sphere of the Cr^{2+} during the progress of the reaction and once the inert Cr^{3+} complex has been generated there is no possibility of exchange with NH_3 present in solution.

4 $[Ni(CN)_4]^{2-}$ is square planar but $[Ni(CN)_4]^{4-}$ and $[NiCl_4]^{2-}$ are tetrahedral; explain.

Answer

Both $[Ni(CN)_4]^{2-}$ and $[NiCl_4]^{2-}$ contain Ni^{2+} d^8. The difference between them is due to

- π bonding
- the strength of the ligand field
- steric effects

(a) CN^- with vacant, fairly low energy π^* orbitals (it is isoelectronic with CO) is able to participate in π bonding (acting as a π acceptor) with suitable orbitals on the metal. The orbitals are much better set up for π bonding in the square planar structure than in the tetrahedral one, therefore CN^- gives rise to a square planar geometry

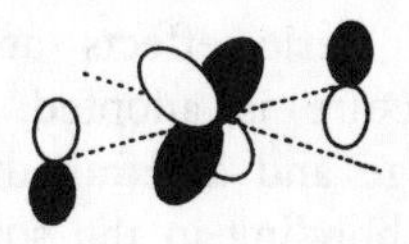

Cl^- can also participate in some π bonding (acting as a π donor) but this effect is not so large that it cannot be more than counteracted by steric factors (less inter-ligand repulsion in tetrahedral).

(b) CN^- is a very strong field ligand (π acceptor). Cl^- is a weak field ligand (π donor).

CFSE is much larger in the square planar structure than in the tetrahedral one. Square planar is based upon an octahedron with two vertices removed,

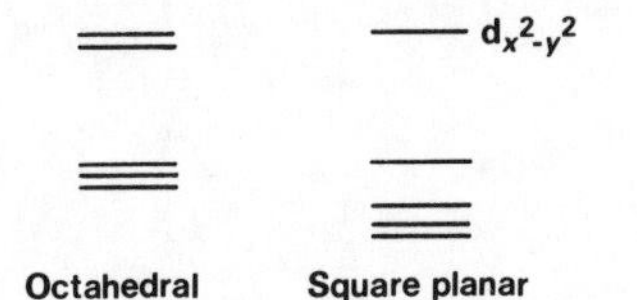

The orbitals in the square planar complex are lower in energy than in the octahedron thus, all things being equal, the CFSE should be larger for square planar than for an octahedral structure (all things will never be equal as six ligands will always produce a larger crystal field than four). We have already seen that CFSE is larger for an octahedron than for a tetrahedron, therefore it is readily appreciated that the CFSE in a square planar complex will be substantially larger than that in a tetrahedron, especially with the $d_{x^2-y^2}$ orbital unoccupied in d^8 square planar.

CFSE is proportional to Δ, therefore any effects due to it are magnified by having a strong field ligand (CN^-) instead of a weak field one (Cl^-). Thus, CN^- has a much greater tendency towards a square planar structure than does Cl^-.

(c) Cl^- occupies a much larger volume in the coordination sphere of the Ni than does CN^-, therefore Cl^- will favour the less sterically crowded structure, i.e. a tetrahedron, where the ligands are as far apart in space as possible. The volume occupied in the coordination sphere of the metal is in this case just related to the size of the donor atom,

Thus, for $[Ni(CN)_4]^{2-}$ steric effects are small but CFSE is large and a square planar structure is adopted, but for $[NiCl_4]^{2-}$ CFSE is small, steric effects are large and a tetrahedral complex is the result. For $[Ni(CN)_4]^{2-}$ the greater π bonding in the square planar complex, although favourable, is not nearly as important as the other two effects. This can be seen from the fact that $[Ni(CN)_4]^{4-}$ is tetrahedral and not square planar although π bonding should be larger with Ni in zero oxidation state, i.e. more willing to donate electrons to a π accepting ligand.

In $[Ni(CN)_4]^{4-}$ there is no CFSE as the metal is d^{10} and therefore no CFSE preference for any geometry. The structure is dictated by steric effects and a tetrahedral geometry is adopted. Thus, as would be expected, Ni^0, without a partially filled d shell, acts as a main group element and obeys VSEPR rules.

CHAPTER 7

Transition Metal Organometallic Chemistry

μ and η Notation

This is a commonly encountered notation used to describe the bonding modes of ligands in organometallic complexes.

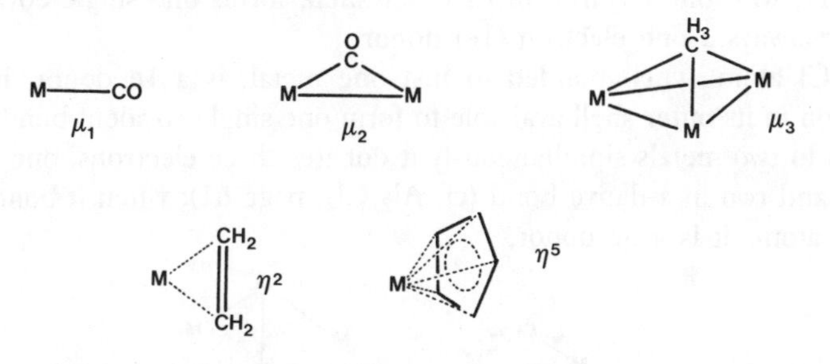

If a ligand adopts a μ_n bonding mode it is bonded to *n metal atoms*. If a ligand is exhibiting an η^n bonding mode then *n ligand atoms* are being used to bond to one or more metal atoms.

Counting Electrons

A great deal of organometallic chemistry can be rationalized from a knowledge of the number of electrons in the valence shell of the metal. Therefore, it is very important to be able to count electrons in organometallic complexes; to do this we employ the following set of rules:

1. Consider the metal to be in a *zero* oxidation state.
2. Add or subtract electrons according to the overall charge on the complex (add for negative charge and subtract for positive).
3. Ligands are considered as neutral atoms or groups. How to work out the number of electrons that a particular ligand donates is explained below.
4. A single metal–metal bond donates one electron to each metal atom or two electrons to the overall complex.

Note: this is just a formal way of counting electrons and we are not, for instance, implying that a metal is necessarily in a zero oxidation state.

How many electrons does a ligand donate?

This can be worked out by adopting a simple valence bond approach to the bonding within a complex, i.e. regarding each M—L bond as either a normal single bond in which the ligand contributes one electron to the valence shell of the metal or a dative covalent bond in which the ligand donates two electrons to the metal.

EXAMPLES

H, with one electron in its outer shell, forms one single covalent bond and is always a one electron (1e) donor.

A Cl atom, when bonded to just one metal, is a 1e donor, having one electron in its outer shell available to form one single covalent bond. When Cl bonds to two metals simultaneously it donates three electrons, one in a single bond and two in a dative bond (cf. $Al_2 Cl_2$, page 61); when it bonds to three metal atoms it is a 5e donor,

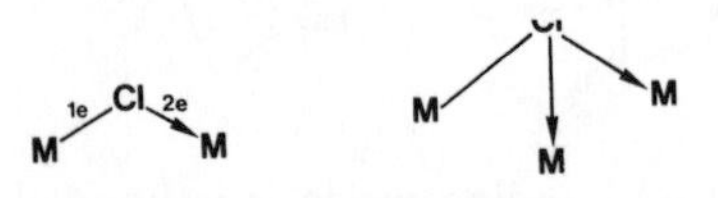

cf. Al_2Cl_6

These representations are only a formalism, a convenience which allows us to count electrons more easily, and in the two cases all M—Cl bonds are equivalent.

CN and CH_3 are 1e donors,

NO can act as a 1e or 3e donor. When the M—N—O unit is bent NO is a 1e donor; when the unit is linear NO is a 3e donor, i.e.

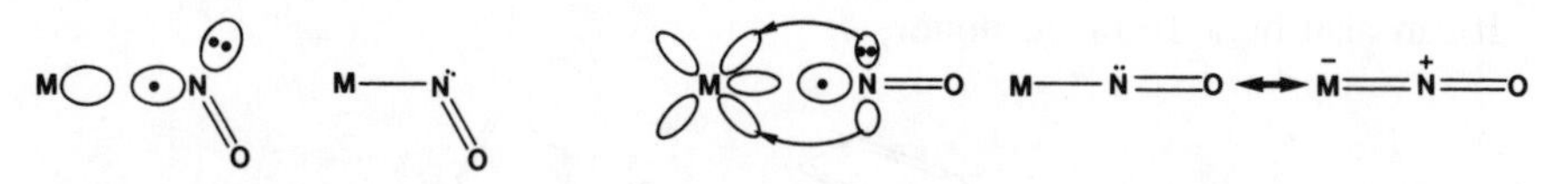

CO is virtually always a 2e donor, bonding through a lone pair of electrons on the C, i.e.

The number of electrons CO donates does not depend on the number of metal atoms to which it is bonded (there are about two or three exceptions to this rule – see later). This is a rare example where the valence bond approach is not really adequate and the bonding has to be regarded as involving multi-centre 2e bonds, i.e.

An alkene donates two electrons, the pair which makes up its C=C π bond,

Allyls (C_3H_5) are either 1e or 3e donors, i.e.

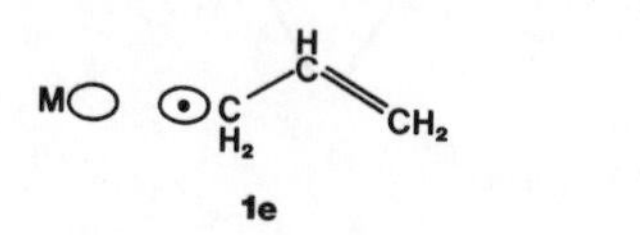

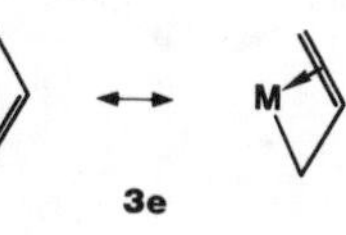

This can also be drawn as

i.e. like half a benzene ring, with one electron in each p orbital on the carbons.

Cyclobutadiene (C_4H_4) is a 4e donor, i.e.

Cyclopentadienyl (Cp, C_5H_5) is usually a 5e donor, i.e.

It can also be a 1e or 3e donor,

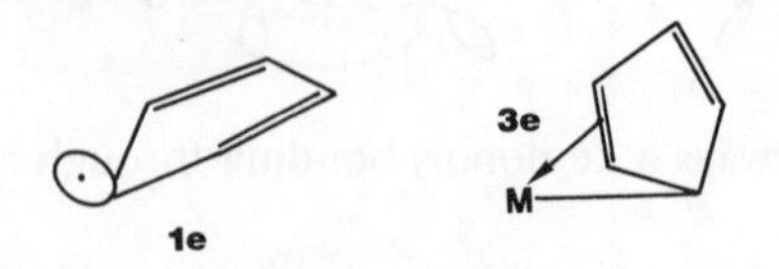

Note: you will sometimes see Cp written as a 6e donor, i.e. Cp^-. In this case the metal is regarded as being in a positive oxidation state (+1), and the overall electron count will be unchanged – see below.

Benzene is usually a 6e donor,

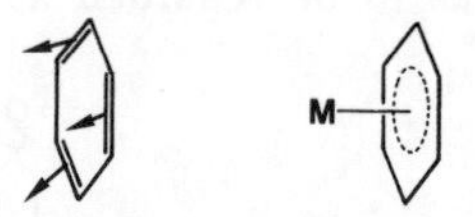

It can also be a 2e or 4e donor (by using one or two of the C=C double bonds, respectively, in σ bonding to the metal).

Multidentate ligands

These coordinate to the metal through more than one donor atom, e.g. ethylenediamine (en) and bipyridyl (bipy) are bidentate 4e donors.

EXAMPLES OF HOW TO COUNT ELECTRONS IN ORGANOMETALLIC COMPLEXES

$Fe(CO)_5$

	No. of e^-
Fe in oxidation state 0 has eight d electrons (see note)	8
5 × CO contribute 5 × 2e to the valence shell of the Fe	10
Total number of electrons in the valence shell of the Fe	18

(*Note*: we normally think of most transition metals as having an electronic configuration d^ns^2 but in complexes it is more convenient to regard all the electrons as being in the d orbitals, i.e d^{n+2}. This is not as unreasonable as it might sound since the electronic distribution is going to be different in a complex from that in a free atom.)

$Fe(C_5H_5)_2$, i.e. $FeCp_2$ (ferrocene)

	No. of e^-
Fe^0 d^8	8
2 × Cp	10
Total number of electrons in the valence shell of the Fe	18

An alternative approach to the electron count in ferrocene considers the Fe to be in a +2 oxidation state and Cp^- as a 6e donor:

	No. of e^-
Fe^{2+} d^6	6
2 × Cp^-	12
Total number of electrons in the valence shell of the Fe	18

$Co_2(CO)_8$

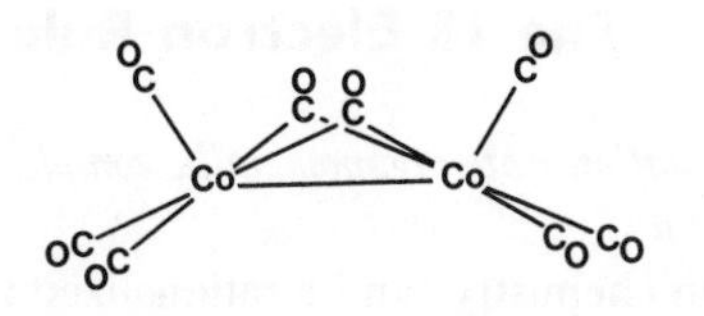

Consider each Co separately

	No. of e^-
Co^0 d^9	9
3 × μ_1-CO (terminal)	6
2 × μ_2-CO (bridging)	2
M—M bond	1
Total per Co	18

Note that each bridging CO can be conveniently regarded as donating 1e to each Co, i.e. a total of 2e per CO.

There is also an alternative $Co_2(CO)_8$ structure, which also has a Co—Co bond but no bridging carbon monoxides; the electron count is the same.

$[Fe_2(CO)_8]^{2-}$

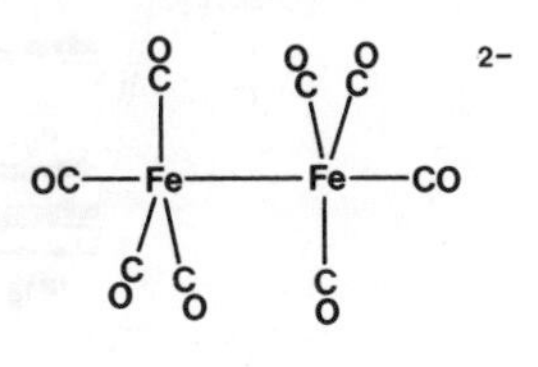

Again, we consider each Fe separately

	No. of e^-
Fe^0 d^8	8
1− charge per Fe unit	1
4 × CO	8
M—M bond	1
Total per Fe	18

Note that the charge is shared equally between the two Fe atoms.

All the organometallic complexes here have 18 electrons in the valence shell of the metal. This is a fairly general phenomenon and is known as the **18 electron rule**.

The 18 Electron Rule

The majority of low oxidation state organometallic complexes have 18 electrons in the valence shell of the metal.

A lot of main group chemistry can be rationalized from the desire to have a full valence shell of electrons – the octet rule. In organo-transition metal complexes there are nine valence atomic orbitals available on the metal (5 × d, 1 × s and 3 × p) and the formation of 18 electron (18e) complexes can most simply be regarded as arising from a similar desire to fill up all these orbitals. The actual reasoning behind the 18e rule is slightly more involved and we must consider MO theory.

Let us start with an octahedral geometry:

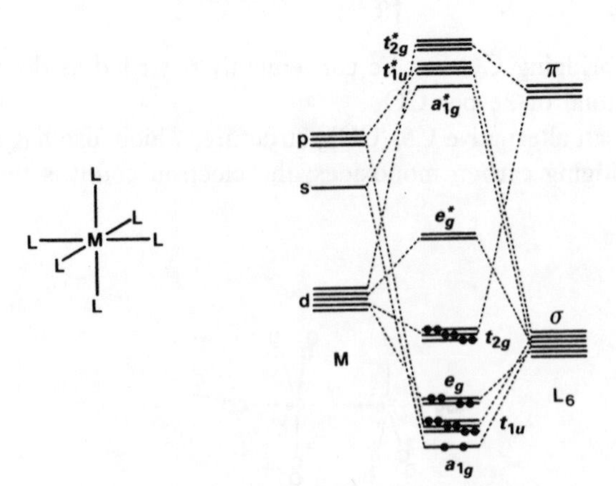

The MOs produced by interaction of the six organic ligands with the metal can be classified in four groups:

1. Six MOs involved in M—L σ bonding. These MOs are the most stable and lowest in energy.
2. The metal t_{2g} set which, although σ non-bonding, can participate in π bonding. Most organic ligands possess some sort of π symmetry acceptor orbitals (e.g. C—O π* in CO) which can interact with the metal t_{2g} set.

B shows the interaction between a ligand group orbital and a metal d orbital.

The set of three molecular orbitals labelled t_{2g} are M—L π bonding MOs (see page 152, where we discussed CO as a strong field ligand). (*Note*: it is possible to construct 12 ligand group orbitals using two π* orbitals on each CO but only three of these are of the correct symmetry to interact with metal orbitals, the others remain non-bonding.)
3. The metal e_g orbitals which are σ* antibonding and π non-bonding.
4. Higher lying antibonding orbitals, with which we need not concern ourselves.

There are thus nine molecular orbitals which have either σ or π bonding character and exactly 18 electrons are required to fill them.

The 18 electron rule thus arises from a desire to fill up all σ and π bonding orbitals.

Now let us look at a tetrahedral ML_4 complex.

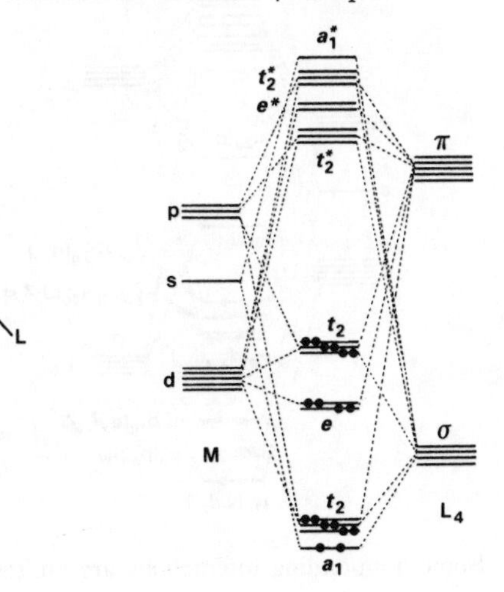

Again we can classify the orbitals:

- There are four M—L σ bonding MOs.
- The five orbitals which make up the *e* and t_2 sets are all able to participate in π bonding.

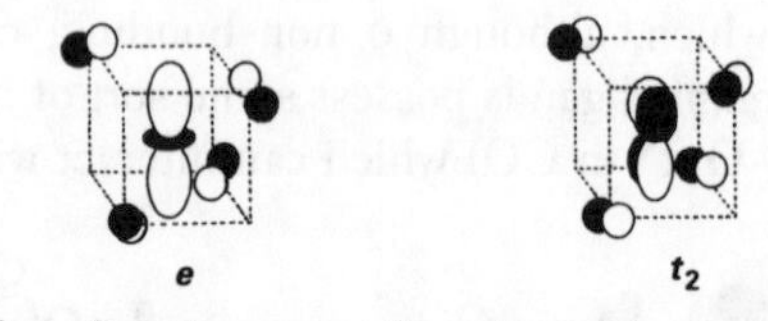

Although this π bonding is not as directional (and therefore as strong) as in the octahedral case, the *e* and t_2 sets are both stabilized by π interaction with vacant orbitals on the organometallic ligands.

Thus, there are nine MOs which are stabilized by either σ or π bonding character and 18 electrons are required for them to be completely filled.

The t_2 set is actually σ* antibonding, but not strongly so {mixing between d and p orbitals (see electronic spectra of tetrahedral complexes, page 164) results in these orbitals being mostly σ non-bonding} and the π bonding is able to counteract this effect. It will be seen later, that in many organometallic complexes, the desire to participate in strong π bonding often dominates any σ bonding effects.

The same approach can also be used to rationalize the 18e counts which are observed for five coordinate trigonal bipyramidal and square pyramidal organometallic complexes.

Square planar complexes

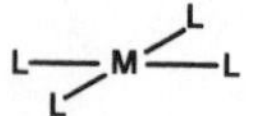

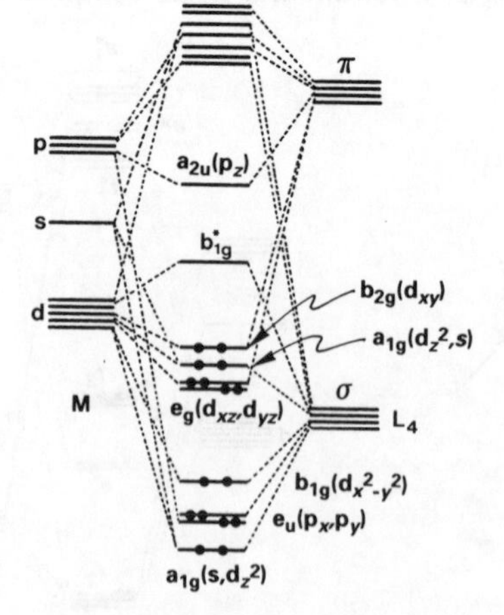

Some antibonding interactions are omitted.

Three of the d orbitals (d_{xy}, d_{xz}, d_{yz}) are of the correct symmetry to participate in π bonding. To fill up all the bonding MOs, we occupy the four M—L bonding MOs and the d orbitals up to d_{xy}. Filling all the π bonding orbitals means that the d_{z^2}, which has no π bonding character but comes below the highest of the π bonding orbitals (d_{xy}), must also be filled. The $d_{x^2-y^2}$ is quite strongly σ* antibonding and has no π bonding character; therefore it remains unoccupied.

There are thus only eight MOs to be occupied and 16 rather than 18 electrons are characteristic of square planar complexes, e.g. $[HPt(PMe_3)_3]^+$,

	No. of e
Pt^0 d^{10}	10
1+ charge	−1
$3PMe_3$	6
H	1
Total no. of e in valence shell of Pt	16

Where the 18 electron rule does not work

The electronic configuration is not the only factor that governs the stability of transition metal complexes.

The only complexes for which the 18e rule works are low oxidation state organometallic complexes. This is because the organic ligands generally have π acceptor orbitals and stabilize the t_{2g} set of the metal by π bonding. The formation of an 18e complex fills up the σ and π bonding orbitals and avoids occupation of σ* orbitals. The stabilization of the t_{2g} set in an octahedral complex results in a large value of Δ (π accepting organic ligands are strong field), making occupation of the e_g* set (σ* antibonding) for an octahedral complex particularly unlikely (Δ > P, therefore low spin complexes are obtained). However, this is not the only factor – both $Fe(CO)_5$ and $[Fe(CO)_6]^{2+}$ would be 18e complexes but only the former exists. In this case reduced π back bonding for Fe^{2+} (i.e. weak M → L π back bonding) makes the cationic complex unstable (see later).

The stability of transition metal complexes depends mainly on:

1. *The number and strength of M—L bonds*: stable complexes are favoured by strong bonds, the more the merrier, allowing for steric factors.
2. *Steric factors (inter-ligand repulsion)*: the difficulty in crowding ligands around a metal atom.
3. *The overall charge on the complex*: complexes cannot support large negative or positive charges – it is rather difficult to keep taking electrons out of a

complex to produce a large positive charge (cf. ionization energies, page 7). Similarly, putting electrons in, to give large negative charges, results in exorbitantly high electron–electron repulsions (cf. endothermic second electron affinities, page 9).

4. *The electronic configuration/electron count.*
5. *Solvent and entropy effects.*

We have discussed the significance of electron count in organometallic complexes. Consider now a different type of complex, e.g. $[Ti(H_2O)]_6{}^{3+}$, a stable 13e complex. There are six M—L σ bonding MOs, which are filled, giving 12 electrons. The thirteenth electron goes into the t_{2g} set of the metal. Formation of an 18e complex would require addition of a further five electrons, which would go into either the non-bonding t_{2g} set (no π bonding with water, since it does not have any correctly orientated donor or acceptor orbitals) or the $e_g{}^*$ set which is σ* antibonding. However, there is no energy advantage in occupation of non-bonding or antibonding orbitals; in fact there is a disadvantage due to increased electron–electron repulsions. Also, addition of five electrons, to generate an 18e complex, would result in the Ti having a formal oxidation state of −2 and, whereas carbon monoxides can stabilize a negative charge by accepting electron density, from the metal, into the C—O π* orbitals, H_2O is not able to do this. Thus, not only would we localize a large negative charge on the Ti but we would introduce a large repulsive term between the lone pairs on the H_2O and the Ti^{2-} which would weaken the M—L σ bonding.

Following this argument it would be expected that $[Ti(H_2O)_6]^{4+}$, a 12e complex in which electrons just occupy M—L bonding orbitals, would be more stable than $[Ti(H_2O)_6]^{3+}$ since there will be less inter-electron repulsion and stronger M—L σ bonding (ligands pulled in more closely due to the higher charge on the metal). However $[Ti(H_2O)_6]^{4+}$ does not exist. The reason for this is that the Ti^{4+} ion is small and highly polarizing, and so will polarize the water molecules so much that hydrolysis occurs,

$$[Ti(H_2O)_6]^{4+} + H_2O \rightarrow [Ti(H_2O)_5(OH)]^{3+} + H_3O^+.$$

Halide complexes do not generally conform to the 18e rule, e.g. $[CrF_6]^{3-}$ is a 15e complex. The bonding in these halide complexes has much more ionic character than in organometallic complexes (electronegativities: Cr 1.7, C 2.6, F 4.0), so that an MO description of the bonding is no longer valid. The stability of these complexes depends on thermodynamic factors such as ionization energies (charge on the metal), hydration energies and various other factors in a Born–Haber cycle. Some compounds, e.g. $[CoF_6]^{3-}$, do appear to conform to the 18e rule, but this is more a matter of chance and the 18e complex just happens to be the thermodynamically favoured one.

In conclusion, *electron count is only one of many factors which influence the stability of transition metal complexes.*

The 18 electron rule and M—M bonds

In an organometallic complex containing n metal atoms we can predict the number of M—M bonds by considering by how much the electron count falls short of $18n$.

e.g. $Mn_2(CO)_{10}$

The total number of electrons:	$2 \times Mn$	14e
	$10 \times CO$	20e
	Total	34e

If we divide the electrons up between the two Mn then each Mn will only have 17e, i.e. one short of the 'magical 18'. The electron count can be made up to 18 by formation of one M—M bond, i.e. Mn_a contributes 1e to the valence shell of Mn_b and vice versa (cf. Cl_2, where formation of a Cl—Cl bond results in both Cl atoms having full outer shells),

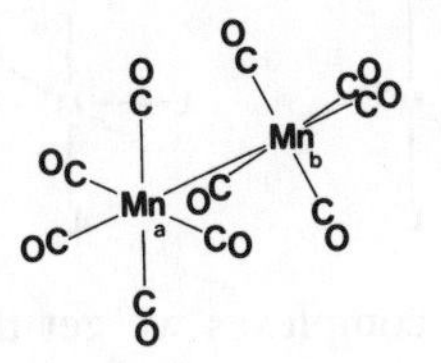

There is a simple formula for working out the number of M—M bonds, m.

$$m = \frac{18n - (\text{total number of valence electrons})}{2}$$ where n is the number of metal atoms

So, for $Fe_3(CO)_{12}$ we have,

total number of valence electrons $= 3 \times 8 + 12 \times 2 = 48e$

$$m = \frac{(18 \times 3) - 48}{2}$$

i.e. $m = 3$, which agrees with the structure obtained by X-ray crystallography,

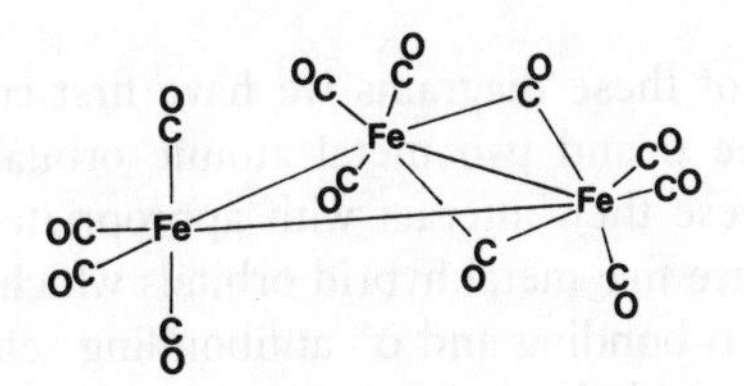

The Isolobal Relationship

This provides a bridge between inorganic and organic chemistry in that a realization of the similarity between the frontier orbitals of inorganic and organic moieties allows the rationalization of the existence of a great number of organometallic compounds and a prediction of their reactivity.

The **frontier orbitals** are the orbitals of a molecular fragment, e.g. $Fe(CO)_3$, which can participate in bonding, i.e. they are the molecular equivalent of valence atomic orbitals. These orbitals are available for donation or acceptance of electron density and are the HOMO (Highest Occupied Molecular Orbital), LUMO (Lowest Unoccupied Molecular Orbital) and perhaps a few other orbitals of comparable energy in a molecular fragment.

Before looking at the use of the isolobal principle it is important to understand the nature of the frontier orbitals of ML_n complexes.

Consider ML_n fragments derived from an octahedron,

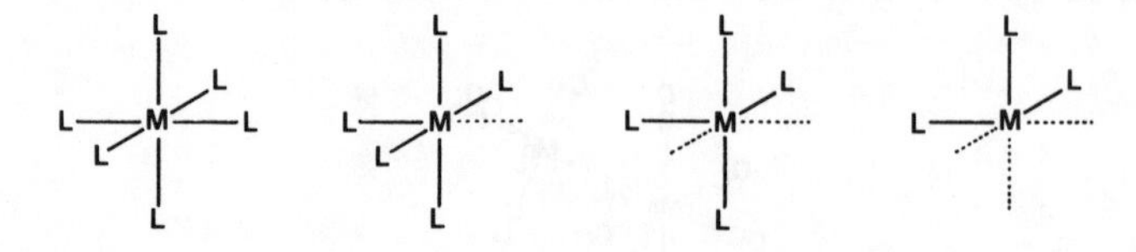

For the ML_6 and ML_5 complexes we get the following MO diagrams,

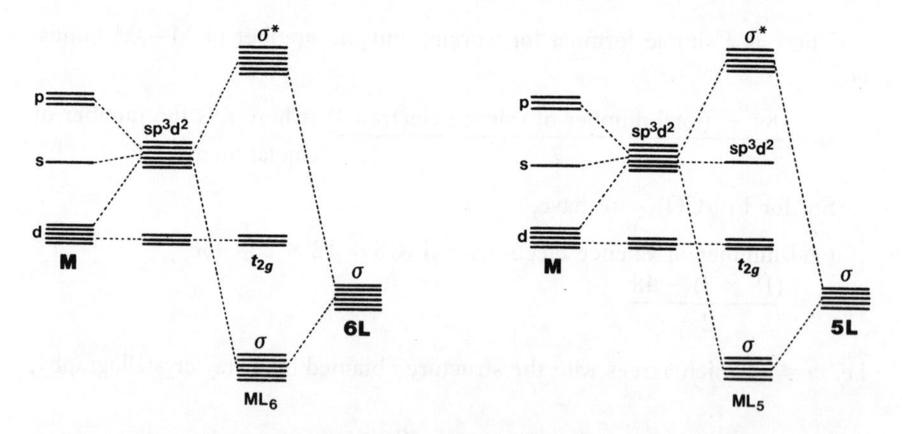

In construction of these diagrams we have first considered the hybridization of one s, three p and two metal atomic orbitals to generate six sp^3d^2 hybrid orbitals. These then interact with appropriate ligand orbitals.

For ML_5 there are five metal hybrid orbitals which interact with five ligand orbitals to produce σ bonding and σ* antibonding sets containing five orbitals each. One hybrid orbital, the one pointing towards the vacant site, is unchanged in energy and becomes a frontier orbital.

If there are four or three ligands, i.e. ML_4 or ML_3 fragments, then two or three orbitals, respectively, are left unchanged in energy and we have the following MO diagrams,

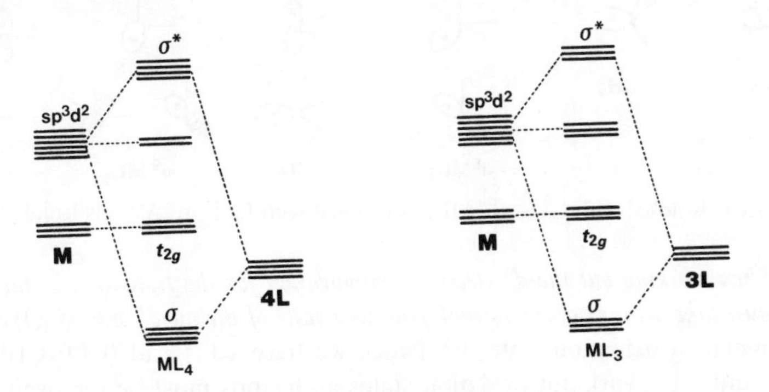

For ML_4 and ML_3 these non-bonding sp^3d^2 orbitals are equivalent to

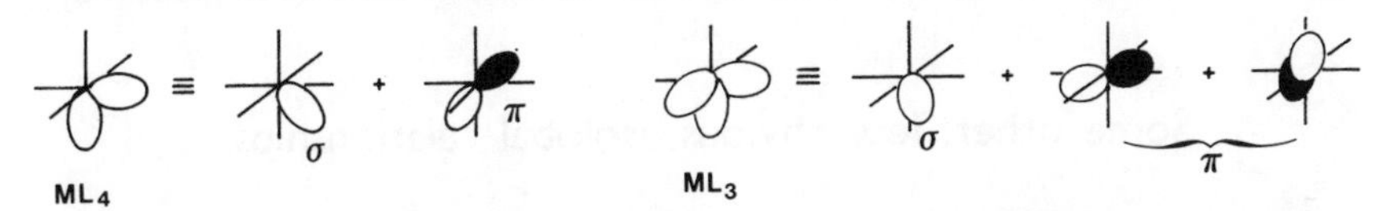

These are obtained by just altering the hybridization scheme slightly. (hybridization is just a way of looking at orbitals that makes it easier for us to visualize the bonding – it is not an absolute quantity and the two schemes given above are entirely equivalent pictures of the same thing).

This gives rise to the following MO diagrams,

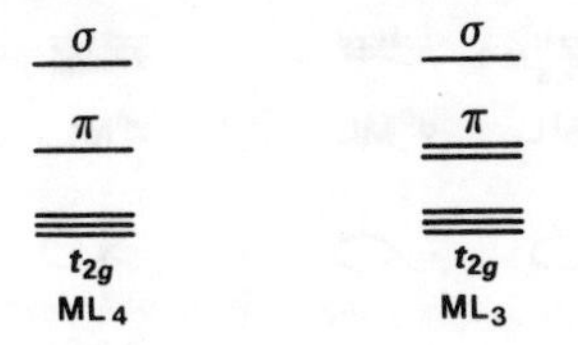

There are three basic principles on which we will base the rest of the discussion:

1. *Fragments are isolobal if the number, symmetry, extent in space and energy of their frontier orbitals are similar.*
2. *For fragments to bond together they must be isolobal.*
3. *If two fragments are isolobal and have the same number of electrons they are going to bond in the same way and would be expected to form similar complexes.*

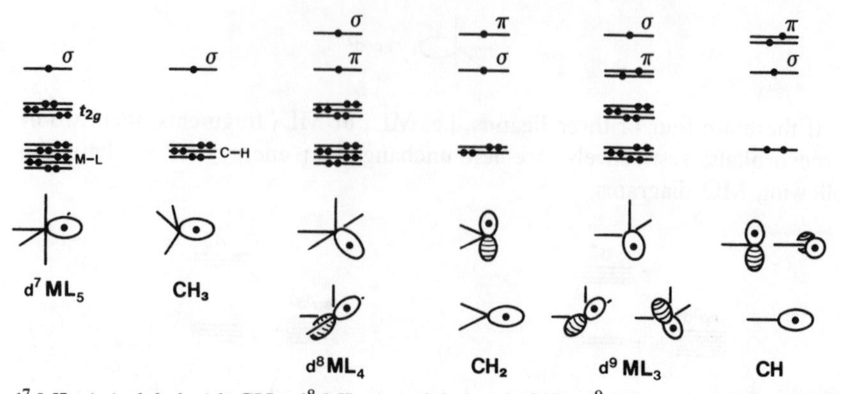

d^7 ML_5 is isolobal with CH_3, d^8 ML_4 is isolobal with CH_2, d^9 ML_3 is isolobal with CH.

When working out the d^n *electron configuration for the transition metal fragment we must take account of the formal oxidation state of the metal*, e.g. $(CO)_4ClW$ has the metal in oxidation state +1 (since we have Cl^-) and $(CO)_4ClW$ is a d^5 ML_5 unit. To work out oxidation states all ligands must be removed with full outer shells, e.g. Cl as Cl^-, CH_3 as CH_3^-, CO as CO, then the oxidation state is the charge left on the metal after all the ligands have been removed.

Some other, less obvious, isolobal relationships

If we remove two ligands from ML_5 or ML_4 fragments a useful extension of the isolobal analogy emerges,

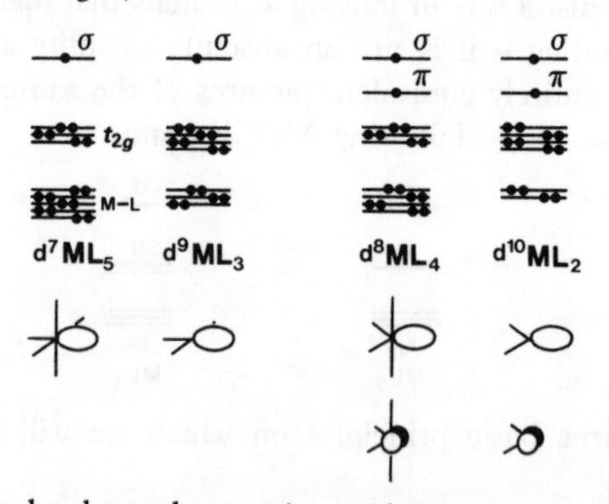

Removal of two ligands along the z axis results in the d_{z^2} orbital becoming non-bonding and joining the t_{2g} set. Hence *ML_5 d^n is isolobal with d^{n+2} ML_3 (T-shaped) and d^n ML_4 is isolobal with d^{n+2} ML_2 (V-shaped)* since the extra two electrons go into the d_{z^2} orbital which does not participate in bonding, therefore, overall, the frontier orbitals are the same. Similarly, *d^n ML_3 (T-shaped) and d^{n+2} ML are isolobal.*

Thus, d^9 ML_3 (T-shaped) is isolobal with d^7 ML_5, which is isolobal with CH_3 and d^{10} ML_2 is isolobal with d^8 ML_4, which is isolobal with CH_2.

INVOLVEMENT OF t_{2g} ORBITALS IN ISOLOBAL FRAGMENTS Not only is d^7 ML_5 isolobal with CH_3 but d^6 ML_5 is isolobal with CH_2 and d^5 ML_5 is isolobal with CH. This can be understood by considering involvement of the t_{2g} orbitals. For an ML_5 fragment, two of the t_{2g} orbitals are of π symmetry and the other of δ symmetry with respect to the incoming ligand.

For a d^6 configuration the orbitals can be filled as follows,

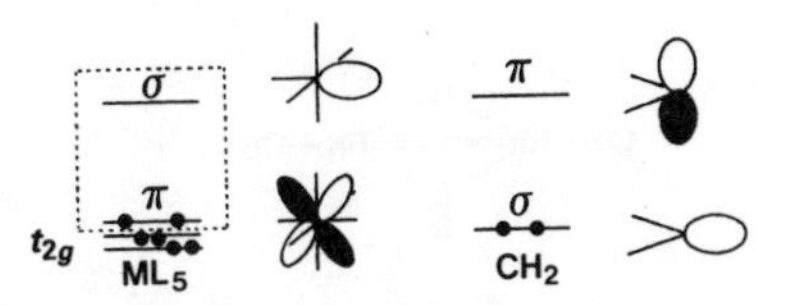

This is isolobal with CH_2 (except that here there is a filled π orbital and a vacant σ one, whereas for CH_2 it is the other way around).

Similarly, d^5 ML_5, using the two π symmetry members of the t_{2g} set, is isolobal with CH.

Another set of isolobal relationships involving use of the t_{2g} orbitals is:

- d^8 ML_3 is isolobal with CH_2 where the ML_3 unit is T-shaped.
- d^7 ML_3 is isolobal with CH.

Consider the MO diagram for the d^8 ML_3 (T-shaped) fragment derived from an ML_5 unit by removal of two ligands along the z axis. The removal of these ligands causes the d_{z^2} orbital to become part of the t_{2g} set (above). There is a vacant orbital of σ symmetry and a filled orbital of π symmetry (part of the t_{2g} set).

Removal of one electron from the t_{2g} to give d^7 ML_3 results in a CH analogue.

GOLD COMPLEXES $[R_3PAu]^+$, H^+, $Cr(CO)_5$ (d^6 ML_5), $CH_3{}^+$ are all isolobal.

$[R_3PAu]^+$ has just one σ symmetry frontier orbital available for bonding,

$AuPR_3$

The most important isolobal relationship is between $[R_3PAu]^+$ and H^+, so that it often proves possible to replace H in organometallic complexes by R_3PAu (see below).

APPLICATIONS OF THE ISOLOBAL PRINCIPLE d^7 ML_5 is isolobal with CH_3, e.g. $Mn(CO)_5$ is isolobal with CH_3, and just as CH_3 is not found as a discrete molecule, but rather dimerizes to give ethane ($H_3C—CH_3$), $Mn(CO)_5$ dimerizes to give $(CO)_5Mn—Mn(CO)_5$ (i.e. $Mn_2(CO)_{10}$). Other analogues of ethane are $(CO)_5Re—Re(CO)_5$ (i.e. $Re_2(CO)_{10}$) and $Cp(CO)_2Fe—Fe(CO)_2Cp$ (i.e. $Fe_2(CO)_4(Cp)_2$). Cp^- is equivalent to three CO ligands – it donates 6e and occupies three coordination sites. If we regard Cp as −1 then the oxidation state of Fe in $Fe(CO)_2(Cp)$ is +1 to give overall electrical neutrality, i.e. we have d^7 Fe^I ML_5.

d^8 ML_4 is isolobal with CH_2, e.g. Cp(CO)Rh, is isolobal with CH_2, and $Rh_2(Cp)_2(CO)_2$,

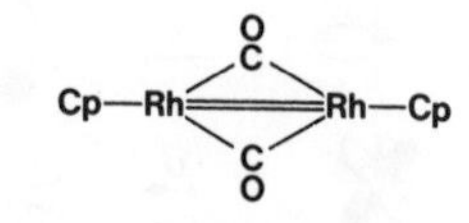

can be regarded as an analogue of ethene. The presence of bridging CO does not affect the basic isolobal analogy.

The same isolobal relationship leads to the following structures all being equivalent,

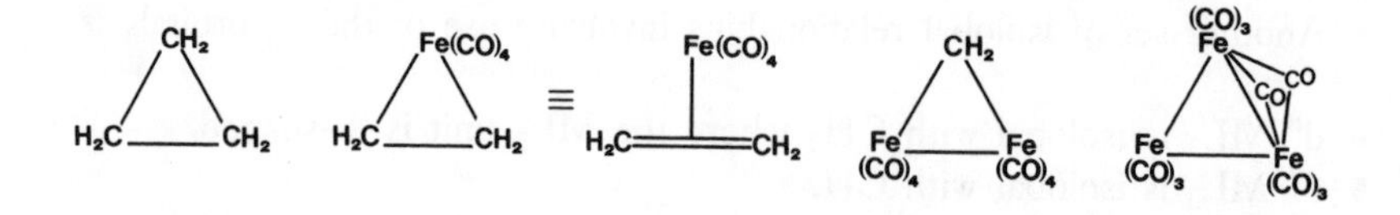

i.e. $Fe(CO)_4(C_2H_4)$, $Fe_2(CO)_8(CH_2)$ and $Fe_3(CO)_{12}$ are all cyclopropane analogues.

d^{10} ML_2 is isolobal with d^8 ML_4 and CH_2, and another cyclopropane analogue is

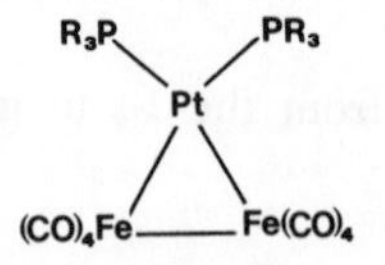

$Pt(PPh_3)_2(RC_2R)$ and $Pt(PPh_3)_2(RC{\equiv}W(CO)Cp)$,

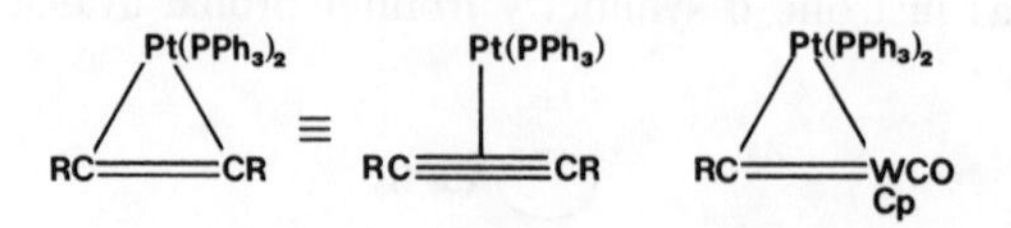

are equivalent to cyclopropene.

d^5 ML_5 is isolobal with CH, e.g. $(CO)_2(Cp)W{\equiv}CR$ and $W_2Cp_2(CO)_4$ are ethyne analogues (where W is W^I),

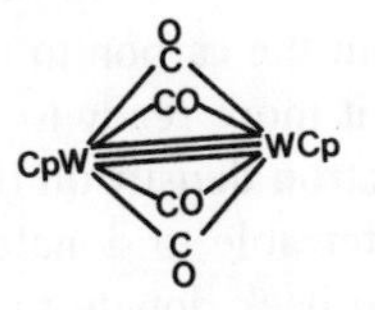

d^9 ML_3 is also isolobal with CH, therefore we predict ethyne analogues, $(CO)_3Co{\equiv}CR$. This, however always dimerizes to

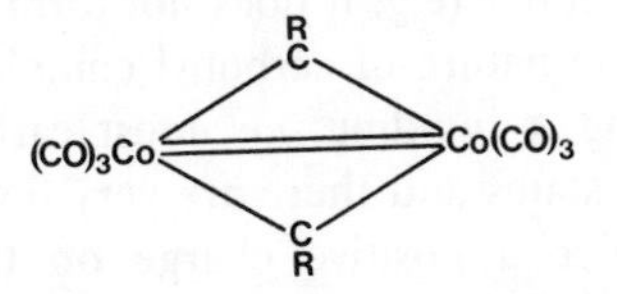

and it is important to realize that *the isolobal principle does not say absolutely that a compound will exist, it only rationalizes the bonding in existing organometallic complexes by analogy with main group organic compounds.*

$[R_3PAu]^+$ is isolobal with H^+, $H_2Os_3(CO)_{10}$ and $(AuPEt_3)_2Os_3(CO)_{10}$ can both be made,

Virtually all of organometallic chemistry can be rationalized by application of the 18e rule and the isolobal principle.

Synergic Bonding

Metal carbonyls

The bonding in an M—CO unit involves two components:

1. σ Donation of a lone pair on the carbon to a vacant metal σ symmetry orbital.
2. π Back donation from a filled metal d orbital to a π* C—O antibonding orbital.

Donation of a lone pair from the carbon to the metal increases the electron density on the metal, making it more ready to donate electrons to the CO π^* orbital. An increase in the electron density on the CO by *back donation* into the π^* orbital makes the CO better able to donate electrons to the metal, which makes the metal better able to back donate to the CO, which makes the CO better able to donate electron density to the metal. . . The σ and π bonding components thus help each other, and the bonding is termed **synergic**.

CO is a very poor σ donor (e.g. it does not form an adduct with the Lewis acid, BF_3, page 29). The nature of carbonyl complexes is usually dictated by the drive to have strong π bonding, i.e. most carbonyl complexes are with metals in low oxidation states and there are very few cationic transition metal carbonyl complexes since a positive charge on the metal will reduce its tendency to back donate. Positive charge on the metal would be expected to increase σ bonding (by pulling the ligands in closer). The absence of cationic complexes is an illustration of the dominance of π bonding effects in the coordination of CO to transition metals.

The π bonding involves donation of electron density into the C—O π^* MO; this has two effects:

1. An increase in M—C bonding, i.e.

   This results in partial double bond character in the M—C bond. This can be seen by X-ray crystallography and infrared (IR) spectroscopy in that the M—C bond lengths are shorter and the M—C stretching frequencies higher for carbonyls than for alkyls. For example, in the complex, $CpMo(CO)_3(C_2H_5)$, the average $Mo—C_{CO}$ bond length is 1.97 Å, whereas the $Mo—C_{Et}$ bond length is 2.38 Å.
2. A decrease in C—O bonding as electron density is pushed into the C—O π^* antibonding orbitals. Thus, X-ray crystallography shows longer C—O bond lengths and IR spectroscopy shows lower C—O stretching frequencies ($\upsilon_{C—O}$) for coordinated CO than for free $CO_{(g)}$. The degree of $M \rightarrow C$ π back donation is dependent upon four factors:
   (a) *Charge on the complex*: the greater the positive charge on the metal, the lower its tendency to donate electron density into the CO π^* antibonding orbitals, e.g.

	$[V(CO)_6]^-$	$Cr(CO)_6$	$[Mn(CO)_6]^+$
$\upsilon_{C—O}$ (cm^{-1})	1859	2000	2096
$\upsilon_{M—C}$ (cm^{-1})	460	441	416

In all these complexes the metals have the same number of d electrons and so differences are solely due to the different charges.

(b) *The effect of other ligands*: as we substitute carbonyls for stronger σ donating ligands then there is more electron density on the M available for π back donation to the CO, and the C—O stretching frequency decreases, e.g.

	$Ni(CO)_4$	$Ni(CO)_3PMe_3$	$Ni(CO)_2(PMe_3)_2$
$\upsilon_{C—O}$ (cm^{-1})	2128	2063, 1943	1994, 1934

Although it is a stronger σ donor a phosphine is not as good a π acceptor (using 3d orbitals, which are too high in energy to participate in good π bonding with the metal) as CO, therefore it does not compete so strongly for available π electron density on the metal; another factor which increases the π donation to CO.

(c) *The position of the metal in the Periodic Table*: M—C π bonding increases down the group. The 4d and 5d orbitals are more diffuse than the 3d orbitals and reach out further towards the CO π*, giving better overlap and, hence, bonding.

(d) *The bonding mode of the CO*: there are three commonly encountered bonding modes for CO, i.e.

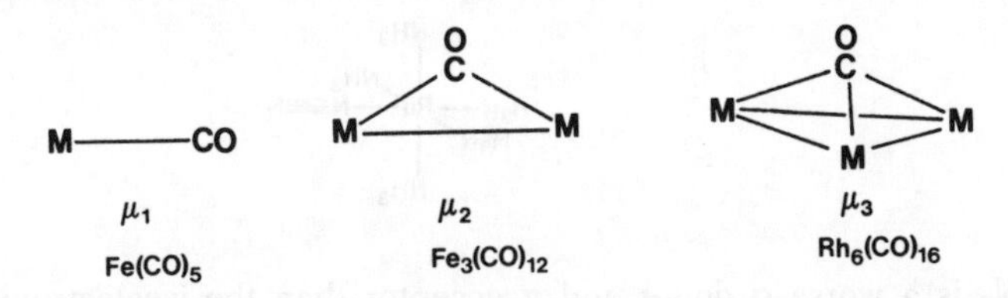

The CO stretching frequency decreases with the number of metal centres to which the CO is coordinated – three metals donate more electron density to the CO π* than do two metals, which donate more than one metal. The approximate ranges for C—O stretching frequencies in the different bonding modes are:

Bonding mode	μ_1	μ_2	μ_3
$\upsilon_{C—O}$ (cm^{-1})	2150–1900	1900–1750	1750–1600

There are two other known bonding modes for CO, i.e.

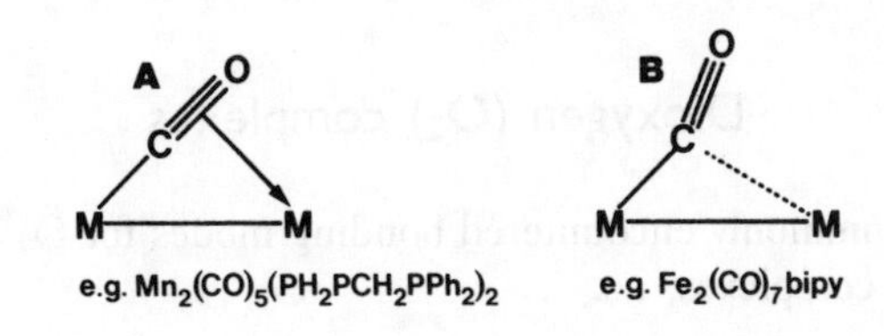

In A the CO is coordinated to one of the metals by using the CO π bonding electrons to form a σ bond (cf. alkenes, below), it is, thus, a 4e donor.

TERMINAL VERSUS BRIDGING CO Bridging CO is less commonly found as we descend a transition metal triad. This is, at least in part, due to the increase in M—M bond length down the group and the CO having greater physical difficulty interacting efficiently with two metals simultaneously.

Bridging CO becomes generally more common as the negative charge on a complex is increased. This is due to bridging CO being better able to remove electron density from the metals. The converse is true for positively charged complexes.

Dinitrogen (N_2) complexes

N_2 only exhibits an end-on bonding mode, similar to a terminal CO. The M—N≡N unit is linear, e.g. $[Ru(NH_3)_5(N_2)]^{2+}$

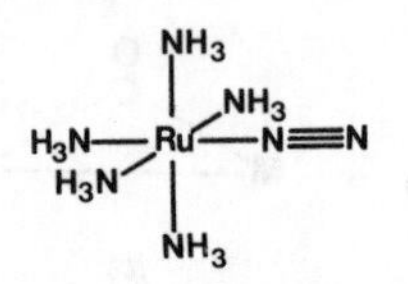

N_2 is a worse σ donor and π acceptor than the isoelectronic CO. It is a worse σ donor because N is more electronegative than C and therefore less ready to let go of its electrons. N_2 is a worse π acceptor because the π* antibonding molecular orbitals are very high in energy and energetically incompatible with the metal orbitals. The high energy of the π* orbitals is a result of the N—N π bonding being very strong (because the p orbitals are of the same energy), which causes the π bonding molecular orbitals to be very low in energy and the π* antibonding orbitals to be very high in energy. N_2 complexes are relatively rare and generally quite unstable because of the weakness of the bonding to a transition metal.

Dioxygen (O_2) complexes

There are two commonly encountered bonding modes for O_2 in mononuclear transition metal complexes, i.e.

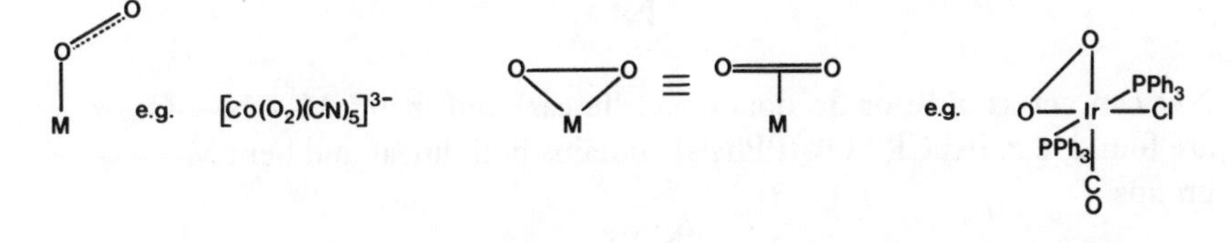

O_2 complexes are more common than N_2 ones:

1. For an end-on bonding mode, O_2 is going to be a worse σ donor than N_2 (O is more electronegative and will not have tendency to donate its lone pair) but a better π acceptor – the O—O π bonding is weaker in O_2, therefore the π* orbitals are lower in energy and better able to interact with the metal orbitals.
2. In the side-on bonding mode, the O_2 will be a better σ donor (the O—O π bonding molecular orbital is higher in energy than the equivalent orbital in N_2, and therefore more compatible in energy with the metal orbitals) and a better π acceptor than N_2.

In O_2 complexes the end-on bonding mode is favoured for metals in high positive oxidation states whereas the side-on mode is more commonly encountered in low oxidation state complexes. This is a combination of σ and π bonding effects:

- Significant L → M σ donation is required to stabilize metals in high oxidation states. For end-on bonding we are donating a lone pair from the O, which is higher in energy than the O—O π bonding molecular orbitals, therefore it is more readily donated to the metal to stabilize the higher positive charge (i.e. an end-on O_2 is a better σ donor than a side-on O_2).
- Significant M → L π back donation is required to stabilize metals in low oxidation states. In the side-on mode the π* orbital is better orientated to accept electron density from the metal, i.e. better able to stabilize the lower oxidation state.

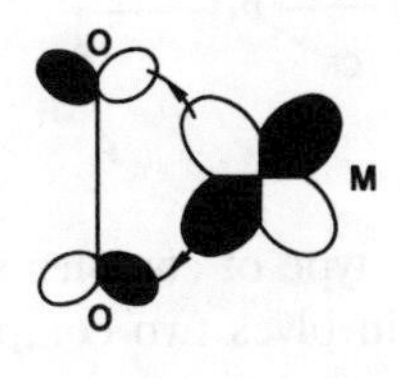

NO

NO can act as a 1e or 3e donor, i.e. 'linear' and 'bent' M—N—O groups are found, e.g. $[RuCl(NO)_2(PPh_3)_2]$ contains both linear and bent M—N—O groups.

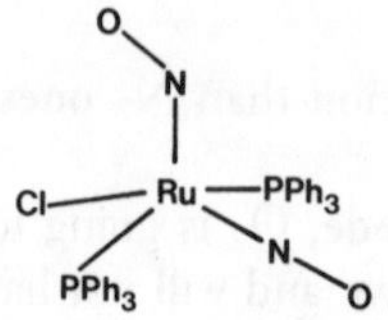

When the M—N—O unit is bent the NO acts as a 1e donor since the lone pair is not in the correct position to interact with the metal. In the linear molecule the NO donates its single electron and the lone pair on the N (see page 270).

Because the NO can easily switch bonding modes (linear ⟷ bent) transition metal–NO complexes can change quite readily between 16e and 18e counts, and can often interact without ligand dissociation or some other major rearrangement of the structure.

Bonding in alkene (olefin) complexes

The bonding of alkenes to transition metals provides another very important example of synergic bonding, which is a role model for bonding in many complexes containing unsaturated organic ligands.

The example most commonly encountered is the anion of Zeise's salt, i.e $[Cl_3Pt(C_2H_4)]^-$ (16e square planar, d^8 ML_3, T-shaped, is isolobal with CH_2 and so this is a cyclopropane analogue). This is a square planar complex with the C_2H_4 group perpendicular to the square plane.

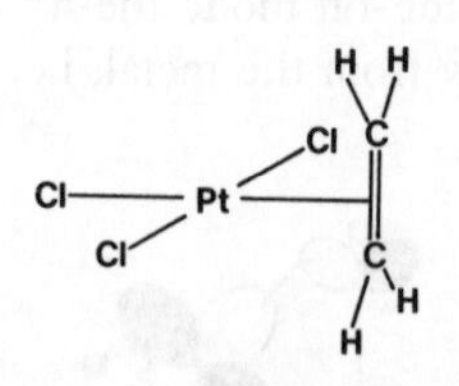

A bonding scheme for this type of complex was first put forward by Chatt, Dewar and Duncanson and involves two components:

- σ Donation from the C=C π orbital to a vacant orbital on the metal.
- π Back donation from a suitable filled d orbital on the metal into the C—C π* antibonding orbital of the alkene.

Both these components result in a weakening of the C—C π bond, therefore C—C bond lengths are longer and C—C stretching frequencies lower than in the uncoordinated alkene.

If we look closely at many alkene complexes we see that the hydrogens are bent back away from the metal and the bonding can be regarded as somewhere in between the two extremes,

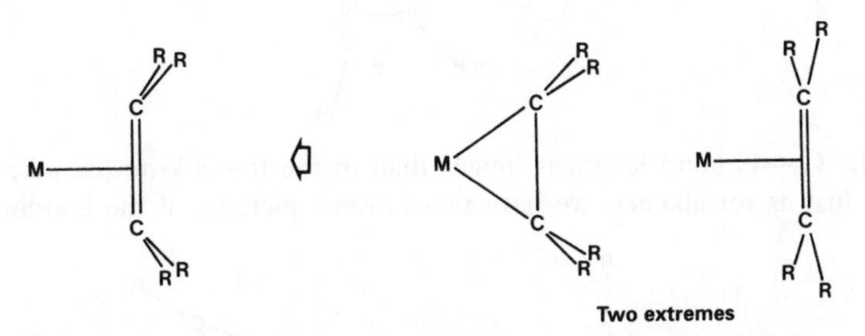

so that the C is not just simply sp^2 hybridized, as in an uncoordinated alkene, but has partial sp^3 (C—C single bond) character.

Alkenes are better σ donors than CO and tend to stabilize higher positive oxidation states on the metal. Thus, many alkene complexes exist with metals in high (+2) oxidation states, whereas such complexes are non-existent for CO (e.g. $[Fe(CO)_6]^{2+}$ does not exist, see page 223), and the C_2H_4 in $[Cl_3Pt(C_2H_4)]^-$ cannot be replaced by a CO.

Why should alkenes prefer the perpendicular bonding mode in square planar complexes?

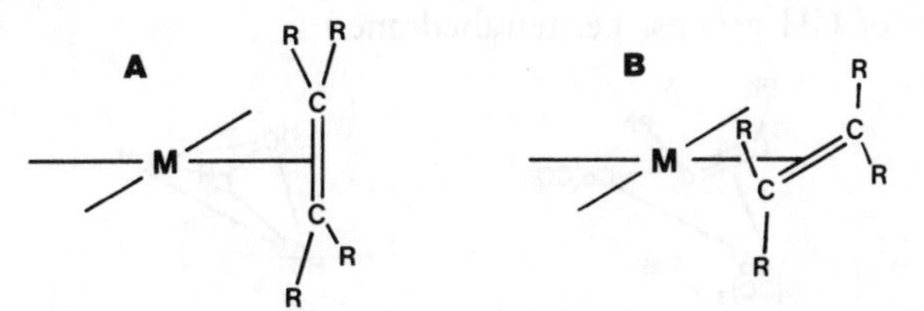

A is much more common than **B**.

The orientation of the alkene is due to steric factors – there is less steric interaction with other ligands when the alkene is perpendicular to the square plane.

Rotation of the alkene, about the mid-point of C=C (like a propeller), is possible. The activation energy for this process is not high, suggesting that the energy difference between the two configurations is small.

Alkyne complexes

Alkynes can be 2e or 4e donors. The bonding is very similar to that in alkene complexes, with σ donation from the C—C π bonding orbitals and back donation into C—C π* antibonding orbitals.

An alkyne acts as a 2e donor in monometal complexes, when only one π bond is being used for bonding to the metal, e.g. $(PPh_3)_2Pt(C_2Ph_2)$ (16e cyclopropene analogue, since Pt(0) d^{10} ML_2 is isolobal with CH_2).

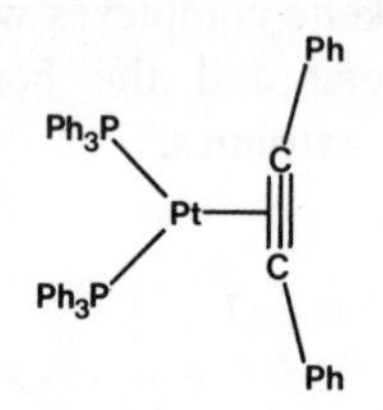

The C—C bond length is longer than in the free alkyne (cf. alkenes).

Just as for alkenes, we have two extreme pictures of the bonding,

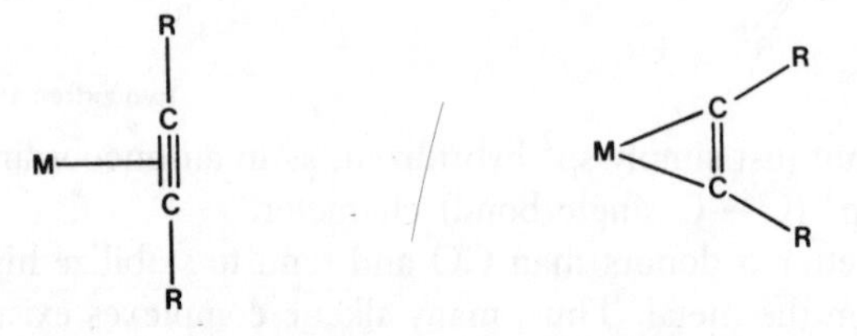

The bonding in these complexes lies somewhere between the extremes, with the R groups bending back by as much as 40°.

Four electrons can be donated when the alkyne bonds to more than one metal, using its two π bonds, e.g. $Co_2(CO)_6(C_2Ph_2)$. (18e. Each $Co(CO)_3$ group, d^9 ML_3, is isolobal with CH and the structure is equivalent to a tetrahedron of CH groups, i.e. tetrahedrane.)

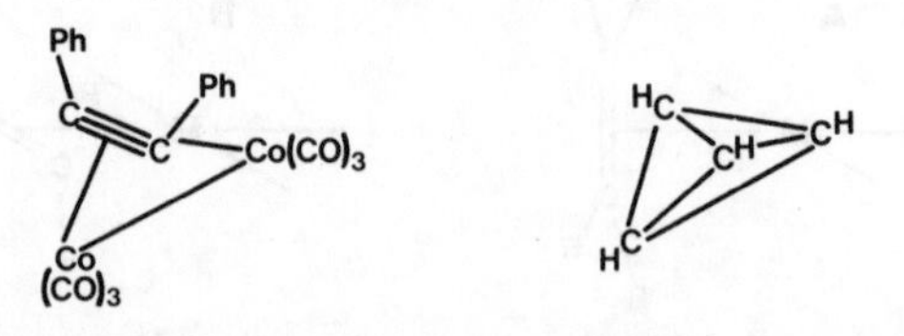

The C≡C group is perpendicular to the M—M bond. The C—C bond length is longer than in free alkynes and for alkynes coordinated to just one metal; this is consistent with increased π back donation from *two* metal atoms into alkyne π* orbitals and σ donation from *two* π bonds on the alkyne.

Carbene complexes

A carbene group is $=CR_2$, as in, e.g. $(CO)_4Fe{=}C(OEt)Ph$ (18e alkene analogue – Fe(0) d^8 ML_4 is isolobal with CR_2).

Bonding of a carbene to a transition metal can be considered as involving σ donation of a lone pair (in an sp^2 orbital) from the C to a vacant orbital on the metal and then π back donation from a filled orbital on the metal to the vacant p orbital on the C, i.e. similar to bonding in M—CO complexes.

The bonding is synergic. The planarity of the MCR_2 group and restricted rotation about the M—C bond (as is shown in n.m.r. experiments) are consistent with this view of the bonding.

The most stable carbene complexes occur with atoms other than C joined to the carbene C, e.g. $(CO)_5W{=}C(OMe)Ph$ (18e alkene analogue – W(0) d^6 ML_5 isolobal with CH_2).

The lone pair on the heteroatom can be donated into the vacant p orbital on the C, giving more extensive delocalization, i.e.

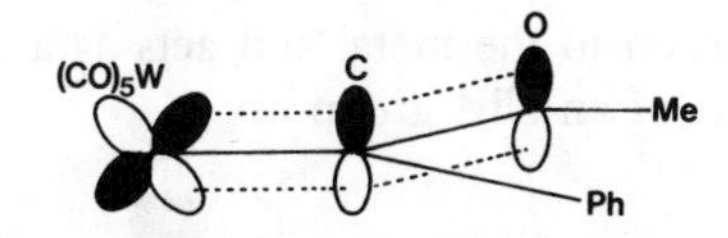

which stabilizes the system (cf. BF_3, page 62). This can also be represented in terms of resonance structures,

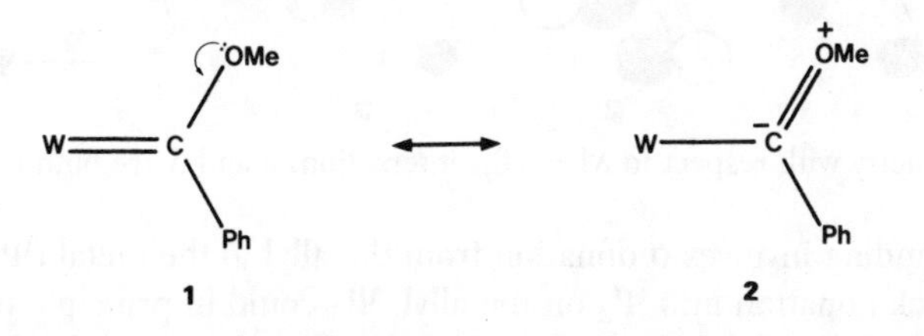

X and M compete with each other for the vacant p orbital on the C. The presence of the second resonance form reduces M=C double bond character. The more electronegative is X, the less likely it is to donate its lone pair to the C (or the less stable is a resonance form in which it is positively charged), therefore resonance form **2** is less important and we get more M=C double bond character. Rotation about M—C is thus easier (lower activation energy) with an NMe_2, rather than an OMe group, attached to the C, since there is less M=C double bond character in the NMe_2 complex.

Carbyne (CR) complexes

A carbyne group is $\equiv$CR, as in, e.g. $(CO)_4ClW{\equiv}(CPh)$ (18e alkyne analogue – W^I d^5 ML_5 is isolobal with CH).

The bonding in carbyne complexes is very similar to that in carbonyl complexes. There is a lone pair in an sp orbital on the C which is used in σ donation to the metal; π back donation from metal π symmetry orbitals into two perpendicular p orbitals (one empty and the other containing one electron) on the C results in a metal–carbon triple bond and a linear M—C$\equiv$R group.

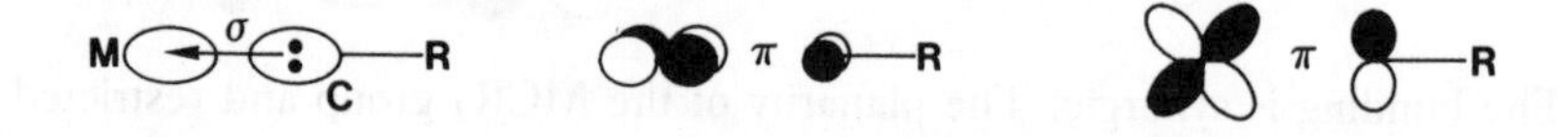

Allyl complexes

An allyl ligand is H_2CCHCH_2, as in, e.g. $Ni(C_3H_5)_2$ (16e),

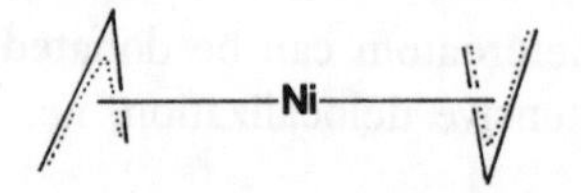

The allyl here sits side-on to the metal and acts as a 3e donor.

The frontier orbitals of an allyl group are

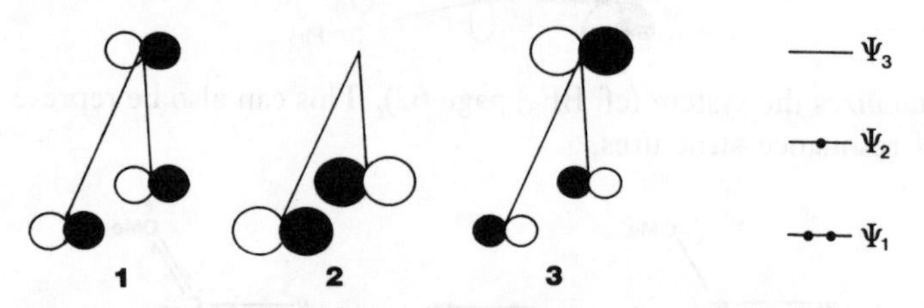

1 is σ symmetry with respect to M − allyl interaction. 2 and 3 are both of π symmetry.

The bonding involves σ donation from the allyl to the metal ($\Psi_1 \rightarrow$ M) and then π back donation into Ψ_2 on the allyl. Ψ_3 could in principle participate in bonding but it is too high in energy (compared to the metal d orbitals) to play a major role in bonding.

The system can be represented as

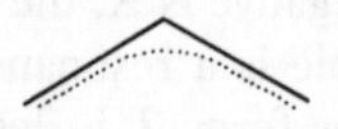

with the electron density delocalized over the three carbon atoms and both C—C bond lengths equal, or the bonding can be approached from a valence bond point of view,

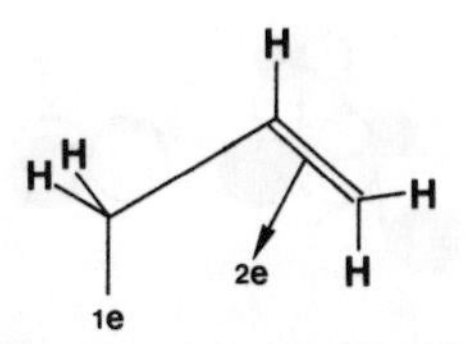

In some complexes, with unequal C—C bond lengths, this second picture actually seems to provide a better representation of the bonding.

When the allyl acts as a 1e donor the bonding just involves a σ component – an M—C σ bond, e.g. $Cp(CO)_3Mo—CH_2—CH{=}CH_2$.

Butadiene complexes

Butadiene is a 4e donor ligand which coordinates side-on to a metal as in, e.g. $Fe(CO)_3(C_4H_6)$ (18e).

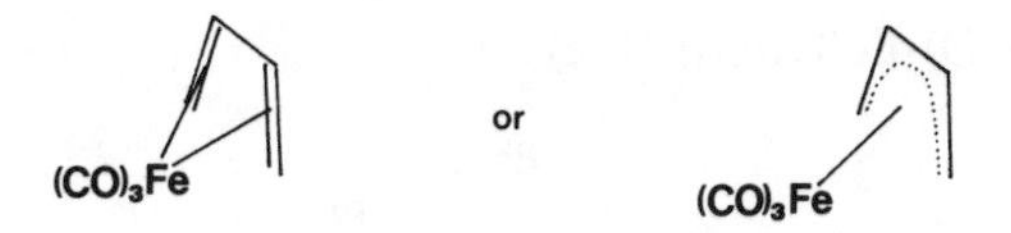

Note that butadiene adopts a *trans* conformation when free but a *cis* conformation when coordinated to a transition metal; this is so that both double bonds can interact well with the transition metal.

The C_4H_6 unit is planar with the carbons best regarded as sp^2 hybridized. Each one then has a p orbital perpendicular to the C_4 plane. The frontier orbitals which are important for bonding the C_4H_6 to a metal atom are

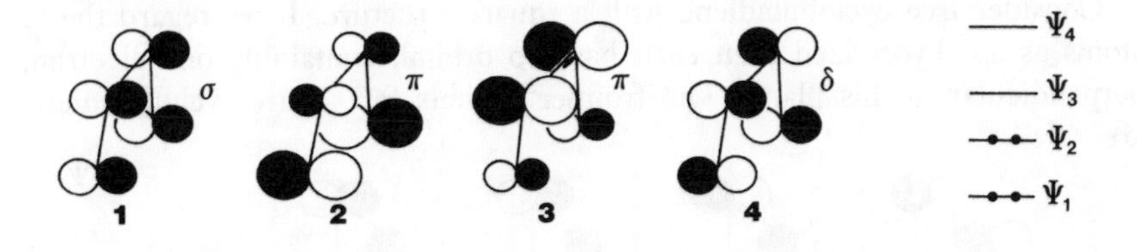

Ψ_2 is bonding between $C_{(1)}$ and $C_{(2)}$ and $C_{(3)}$ and $C_{(4)}$ but antibonding between $C_{(2)}$ and $C_{(3)}$. Therefore, occupation of just Ψ_1 and Ψ_2 gives us the picture of butadiene as containing one single and two double bonds. Bonding to a metal involves σ donation (L → M) *from* Ψ_1, π donation (L → M) *from* Ψ_2 and π back donation (M → L) *into* Ψ_3.

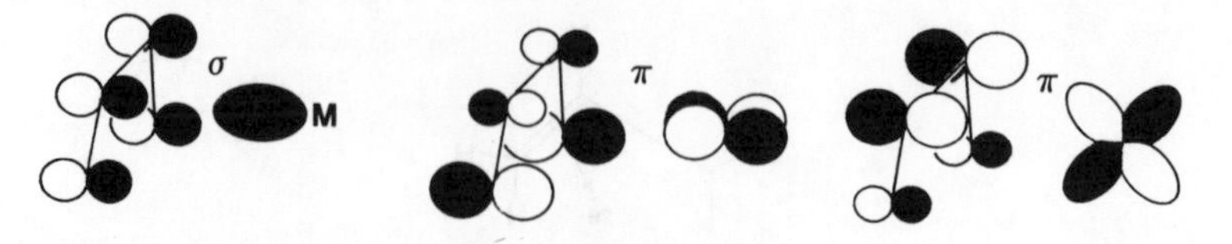

Ψ_4 is too high in energy to interact well with metal d orbitals and of δ symmetry (not easy to get good interaction).

σ Donation from Ψ_1 will weaken all the C—C bonds. π Donation from Ψ_2 weakens $C_{(1)}$—$C_{(2)}$ and $C_{(3)}$—$C_{(4)}$ but strengthens $C_{(2)}$—$C_{(3)}$ since antibonding electron density is being removed from between these two carbon atoms. π back donation into Ψ_3 weakens $C_{(1)}$—$C_{(2)}$ and $C_{(3)}$—$C_{(4)}$, since this MO is antibonding between these carbon atoms, but strengthens $C_{(2)}$—$C_{(3)}$. The overall effect of this is to make all C—C bonds in coordinated butadiene approximately equal, e.g. in $Fe(CO)_3(C_4H_6)$, C—C = 1.46 Å (cf. free butadiene where bond lengths are 1.34 and 1.47 Å).

Cyclobutadiene complexes

For example $(CO)_3Fe(C_4Ph_4)$ (18e)

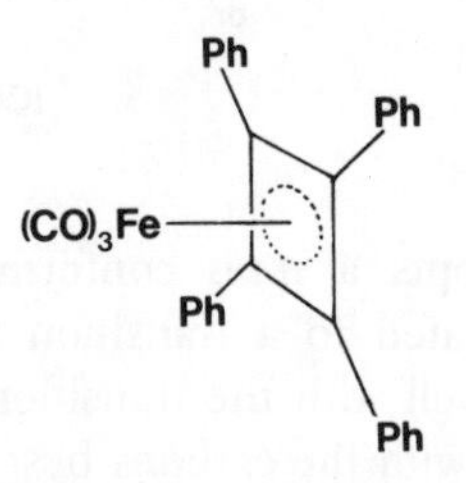

Free cyclobutadiene (C_4H_4) is a very short lived, unstable molecule, however its stability is greatly increased when it becomes coordinated to a metal atom.

Consider free cyclobutadiene with a square structure. If we regard the C atoms as sp^2 hybridized then each has a p orbital, containing one electron, perpendicular to this plane. The frontier orbitals for square cyclobutadiene are

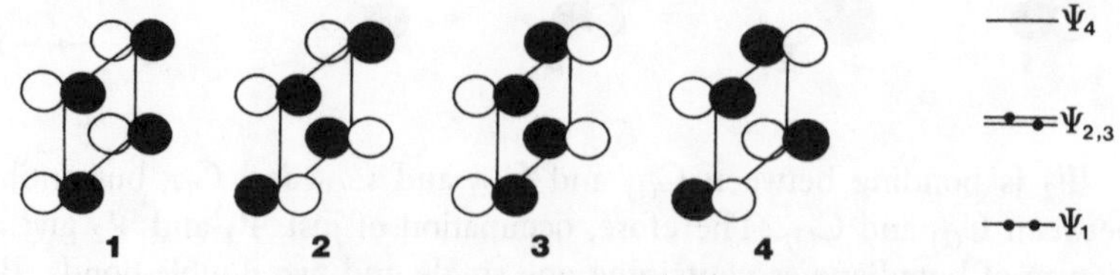

Two bonding and two antibonding interactions lead to **2** and **3** being overall non-bonding between the carbons.

This is an anti-aromatic, $4n\pi$ ($n = 1$) electron, system (Hückel's rule), and as such will be unstable and will suffer a Jahn–Teller type distortion to give a rectangular structure.

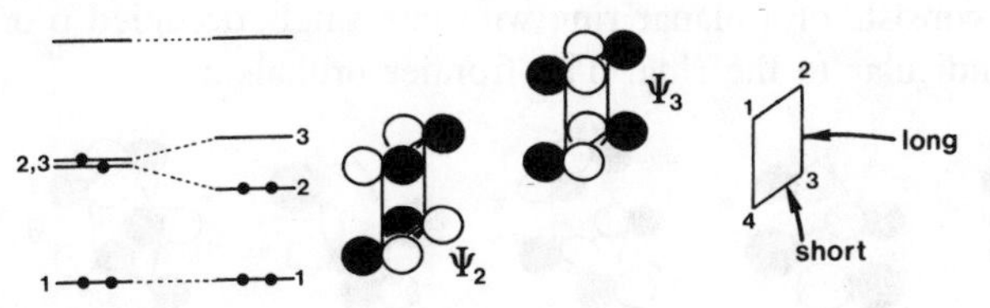

In the square structure, Ψ_2 and Ψ_3 were non-bonding (equal bonding and antibonding character) but when this distorts to a rectangle, Ψ_2 becomes bonding (bonding between $C_{(1)}$—$C_{(2)}$ and $C_{(3)}$—$C_{(4)}$ is now stronger than antibonding between $C_{(2)}$—$C_{(3)}$ and $C_{(1)}$—$C_{(4)}$) and Ψ_3 antibonding. In going from a square to a rectangle two electrons are thus taken from non-bonding orbitals and put into a bonding orbital, therefore stabilizing the system – this is the driving force for the distortion.

The presence of the low lying LUMO (Ψ_3) in the rectangular cyclobutadiene makes it very reactive – little energy is required for a nucleophile to donate electrons into Ψ_3 (a very low energy vacant orbital). Thus, although rectangular cyclobutadiene is more stable than the hypothetical square complex it is still very unstable. (Note that this is kinetic instability.)

When cyclobutadiene coordinates to a metal atom it adopts a side-on bonding mode and donates four electrons. There is σ donation from Ψ_1 (L → M) and π donation from Ψ_2 (L → M) into vacant metal orbitals of suitable symmetry and then π back donation into Ψ_3 (M → L). Thus, similarly to butadiene, these three contributions combine to make all C—C bond lengths in coordinated cyclobutadiene approximately equal.

The back donation into Ψ_3 means that we have three MOs on the ligand occupied and, essentially a 6π ($4n + 2$, $n = 1$) stable aromatic system. The stability is kinetic in origin and arises since any reaction of coordinated cyclobutadiene must begin with donation of electron density into a relatively high energy orbital (Ψ_4). Coordinated cyclobutadiene is stable enough so that we can do reactions with it, see page 253.

Cyclopentadienyl ligands and sandwich compounds

Cyclopentadiene (Cp), C_5H_5, is most often regarded as a 5e donor, adopting a face-on, η^5, bonding mode, as in, e.g. $[Cp(CO)_3Mo]_2$ (M—M bond results in each Mo having 18 electrons in its valence shell). Cp can also be regarded as a 6e donor if we consider Cp^-, a 6 π electron aromatic system, coordinated to a metal, i.e. M—Cp (5e donor) or M^+ Cp^- (6e donor). These two pictures are actually entirely equivalent and it does not matter which is used to describe the system as long as descriptions are never mixed.

C_5H_5 consists of a planar ring with one singly occupied p orbital on each C, perpendicular to the ring. The frontier orbitals are

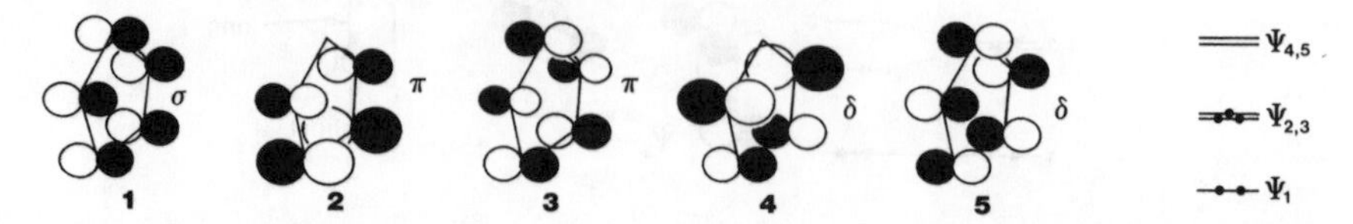

In free Cp, Ψ_2 and Ψ_3 are unsymmetrically occupied and we get a Jahn–Teller type distortion, lowering Ψ_2 and giving a system better represented as

When this coordinates to a metal we get σ donation from Ψ_1 (L → M), π donation from Ψ_2 (L → M) and π back donation into Ψ_3.

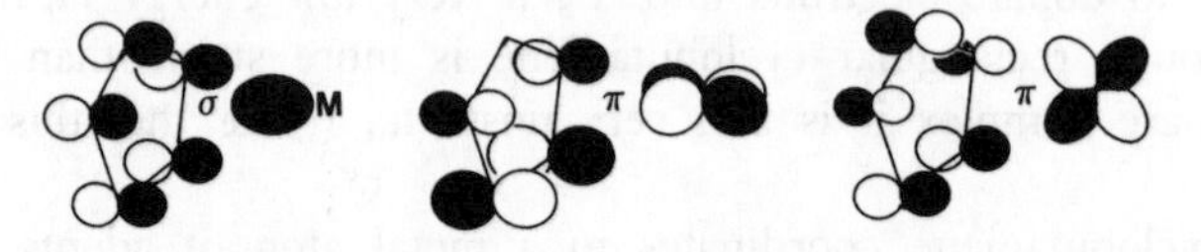

Thus, we are involving all of Ψ_1 to Ψ_3 in bonding and filling them with electron density so that we get, essentially, a six electron, $4n + 2$, $n = 1$, π aromatic system with all C—C bond lengths equal, i.e.

A particularly well studied group of compounds involving Cp rings is the sandwich compounds, such as ferrocene ($FeCp_2$). Here we have two parallel Cp rings sandwiching a metal atom. The bonding can be adequately described by an MO approach. The two Cp rings are treated as a single unit and group orbitals which can interact with metal d orbitals are generated, see opposite.

Note: it is also possible to get sandwich compounds, such as $CoCp_2$, with electron counts greater than 18 (19 electrons in this case). The MO diagrams of these sandwich compounds are quite complex; there are a few orbitals which are essentially non-bonding in character and therefore there should be no electronic preference for occupation or not of these orbitals. Thus, stability depends on some of the other factors discussed above (page 223/4). Such compounds are, however, generally fairly readily oxidized to an 18e system, e.g.

$$CoCp_2 \xrightarrow{CCl_4} [CoCp_2]^+Cl^-$$

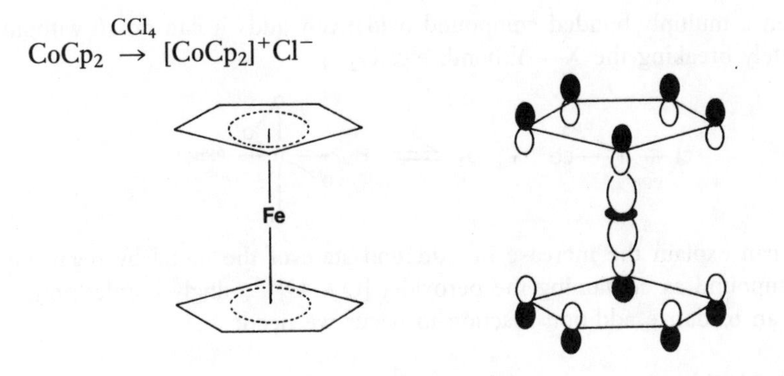

Benzene complexes

The bonding can be treated in the same way as for Cp complexes. The benzene usually adopts a side-on bonding mode, acting as a 6e donor, as in, e.g. $(CO)_3Cr(C_6H_6)$ (18 electrons).

Types of Organometallic Reactions

Oxidative addition

	$M^xL_n + XY \rightleftharpoons$	$M^{x+2}L_nXY$
Coordination number	n	$n+2$
Oxidation state	x	$x+2$

The reaction involves an increase in both the coordination number and the formal oxidation state of the metal by two units, e.g.

$$Ir^{I}Cl(CO)(PPh_3)_2 + HCl \rightarrow Ir^{III}HCl_2(CO)(PPh_3)_2$$

To work out the oxidation state of Ir, calculate the charge left on the Ir when all the ligands have been removed with closed shells of electrons, i.e. CO and PPh_3 are removed as neutral groups and Cl and H are removed as Cl^- and H^-. Thus, the oxidation state of Ir on the left-hand side is +1 and that on the right-hand side is +3. Therefore, although the actual mechanism of addition to the Ir^I complex may involve reaction with H^+ (see below) this does not affect the *formal* oxidation state of the final product.

Even though this reaction involves formal oxidation of the metal by two units the overall result is that the electron count for the complex increases by two electrons, i.e. from 16e to 18e in this case.

When a multiply bonded compound oxidatively adds it can do so without completely breaking the X—Y bond, e.g. O_2

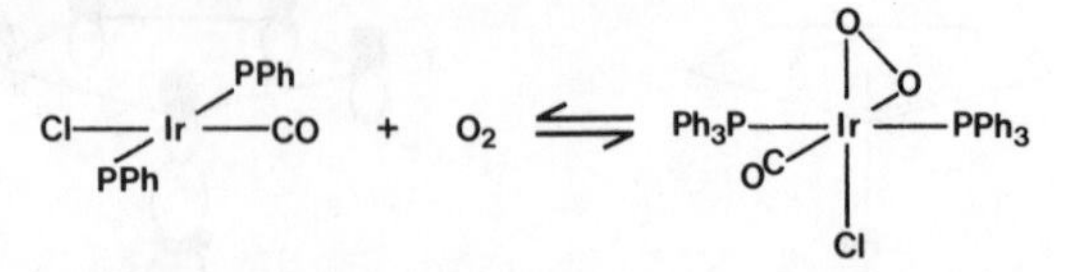

We can explain the increase in oxidation state of the metal by regarding this compound as containing the peroxide, $[O{-}O]^{2-}$, singly bonded ion.

For an oxidative addition reaction to occur we need:

1. A metal with two vacant coordination sites – if none exists they have to be generated by dissociation or change in bonding mode of ligands.
2. An unsaturated metal complex. Organometallic complexes generally obey the 18e rule, therefore we usually require a 16e (14e for square planar) complex before reaction can occur.
3. A metal with fairly stable oxidation states separated by two units.

The position of the equilibrium,

$$L_nM + XY \rightleftharpoons L_nMXY$$

depends on:

- the nature of the metal,
- the nature of the ligands (L) attached to the metal,
- the nature of XY, i.e. the strength of bonds formed/broken during the reaction,
- solvent effects.

Oxidative addition is favoured for:

1. A metal which has an easily accessible $(n+2)+$ oxidation state. Ease of oxidative addition increases down a transition metal triad (third row > second row > first row) as the ease of oxidation increases (ionization energy decreases down a triad, page 6).
2. Elements at the left-hand side of the transition metal series. Ionization energy increases from left to right (effective nuclear charge increases) and metals at the left show a much greater variety of positive oxidation states.
3. Good σ donor ligands. The donation of electron density to the metal increases the ease with which it can be oxidized, i.e. good σ donors stabilize higher oxidation states, e.g. PPh_3 is better able to stabilize higher oxidation states than is PF_3 or CO.

4. Hard ligands, e.g. F^-. They stabilize higher positive oxidation state metals (stronger electrostatic interaction with higher oxidation state metal).
5. Small ligands. The oxidative addition reaction involves an increase in coordination number, therefore the smaller the ligands already attached to the metal, the easier it is to bring another two in.
6. Large metals. The larger the metal atom the easier it is to coordinate two more ligands.
7. Stronger M—X and M—Y bonds and weaker X—Y bonds.

STEREOCHEMISTRY AND MECHANISM When a molecule such as O_2 undergoes oxidative addition it must occupy *cis* positions in the complex as the O—O bond is not broken (as in, e.g. the compound above). When the X—Y bond is broken, although initial addition could still be *cis*, rearrangement can occur and the final product will be the most thermodynamically favoured one, which could be *cis* or *trans*.

There are three main types of mechanism for oxidative addition reactions: concerted, S_N2 and S_N1. The mechanism adopted depends on the metal, the substrate, XY (including steric and electronic effects) and the solvent.

CYCLOMETALLATION

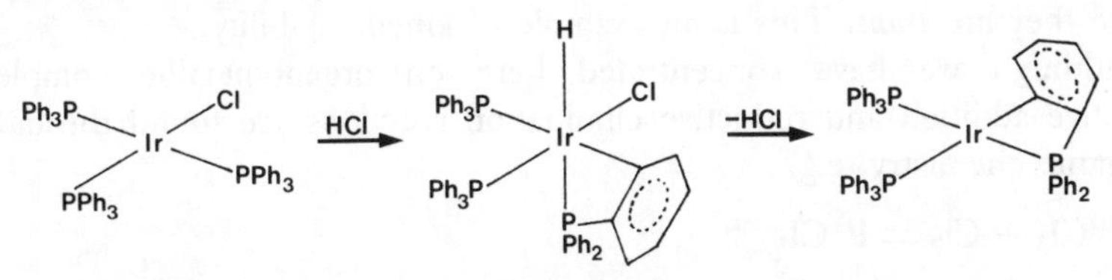

This is an *intra*molecular oxidative addition reaction and is a favourable reaction because:

- the group is already coordinated to the metal and there is very little decrease in entropy, when the other end coordinates, compared to a conventional oxidative addition reaction;
- the H atom is in close proximity to the metal atom, making initial attack very easy.

Reductive elimination

This is the reverse of oxidative addition, i.e.

$$L_nMXY \rightleftharpoons L_nM + XY$$

	L_nMXY	L_nM
Coordination number	$n+2$	n
Oxidation state	z	$z-2$

The above equilibrium will be favoured by factors which are opposite to those listed for oxidative addition.

Although this reaction type is not as widely studied as oxidative addition, it is a very important reaction, being the way in which many organic molecules disassociate themselves from transition metals in homogeneous catalysis reactions (see below).

The most common mechanism for reductive elimination appears to be a concerted one, e.g.

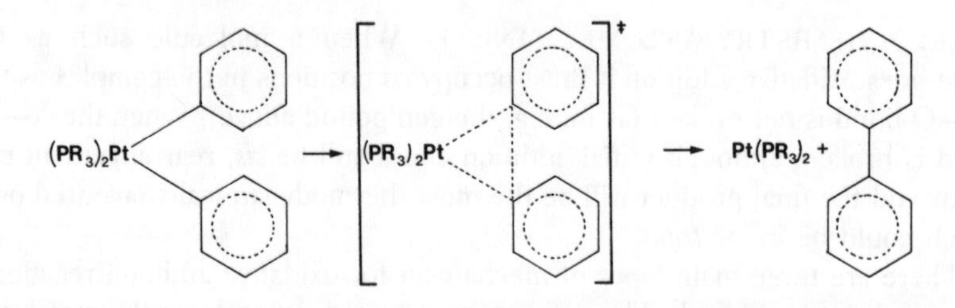

For this to occur the groups must be *cis* so that interaction can occur between them. This can be shown by the fact that complexes which have R (alkyl) and H groups in *cis* positions tend to be much more unstable than those where they are *trans*. This is an example of kinetic stability.

Although we have concentrated here on organometallic complexes, oxidative addition and reductive elimination reactions are found throughout inorganic chemistry, e.g.

$$P^{III}Cl_3 + Cl_2 \rightleftharpoons P^{V}Cl_5$$

Migration/insertion reactions

The general reaction is,

$$L_nM—X + AB \rightarrow L_nM—(AB)—X$$

Although the actual mechanism involves **migration** of a group, these reactions are more usually referred to as **insertions** since it appears that a group has inserted itself into an M—X bond.

We will concentrate on CO insertion. Consider the reaction

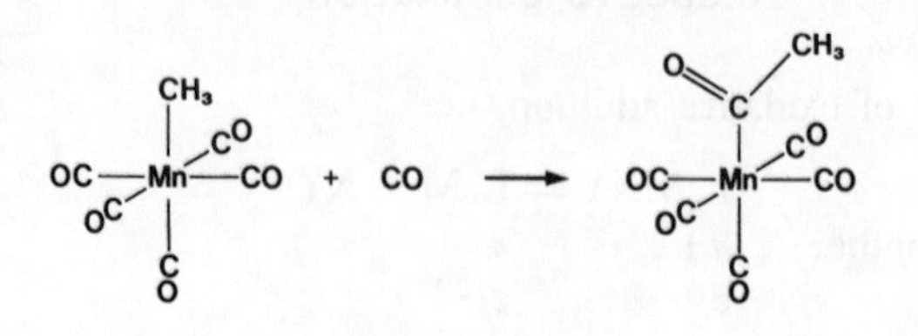

There are three points which have to be considered:

1. Is the group which migrates already attached to the metal?
2. Is it really a migration or an insertion?
3. Is the mechanism concerted?

Let us look at each point in turn.

Is the group which migrates already attached to the metal?
The answer to this appears to be 'yes'. We can show this in several ways.

(a) If we use ^{14}CO as the substrate we don't end up with any ^{14}C in the $COCH_3$.
(b) Varying the CO pressure has no effect on the reaction rate. This suggests that the initial, rate-determining step of the reaction is a rearrangement of the groups already on the metal, to give $COCH_3$ and a vacant cordination site. With a vacant site, coordination of an external CO is very easy and will be a fast step. Thus, the rate will not be affected by the CO pressure (concentration).
(c) Reaction with PPh_3 also results in the formation of $COCH_3$, i.e.

$$Mn(CO)_5(CH_3) + PPh_3 \rightarrow Mn(CO)_4(PPh_3)(COCH_3)$$

Points (a), (b) and (c) show that the reaction does not involve insertion of an *external* CO.

Is it really a methyl migration or an insertion of an internal CO?
It is thought to be a migration.

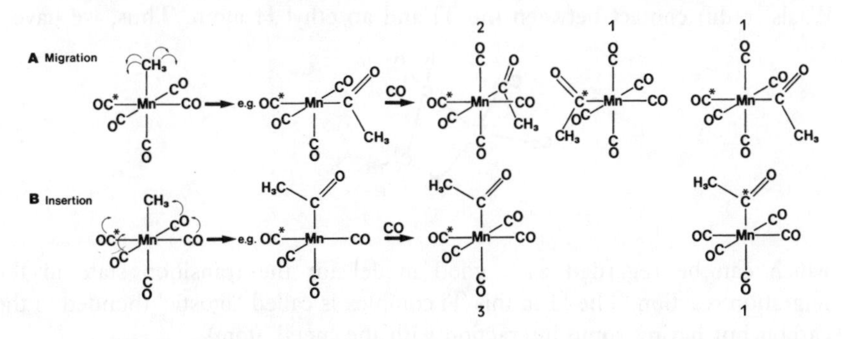

On the diagram we have assumed that we do not get migration from *trans* sites, which is quite reasonable if the reaction involves a concerted mechanism (see below).

We actually get products in the ratio 2:1:1 and, therefore mechanism **A** is correct, that is, methyl migration occurs.

Is the mechanism concerted?

This question cannot be answered by looking at the reaction above, instead a reaction involving a chiral alkyl group needs to be considered.

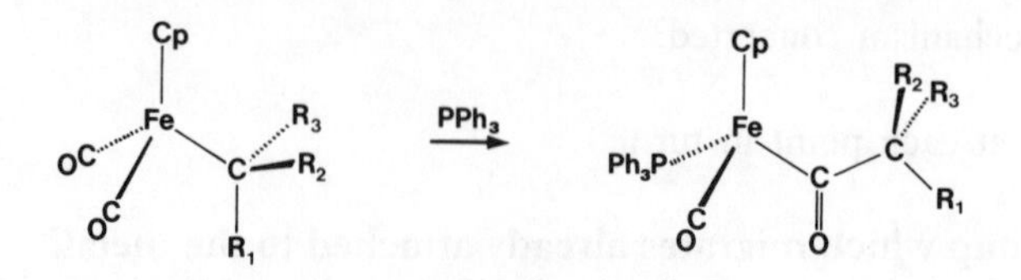

There is retention of configuration at the chiral centre, indicating a concerted mechanism.

The overall mechanism is

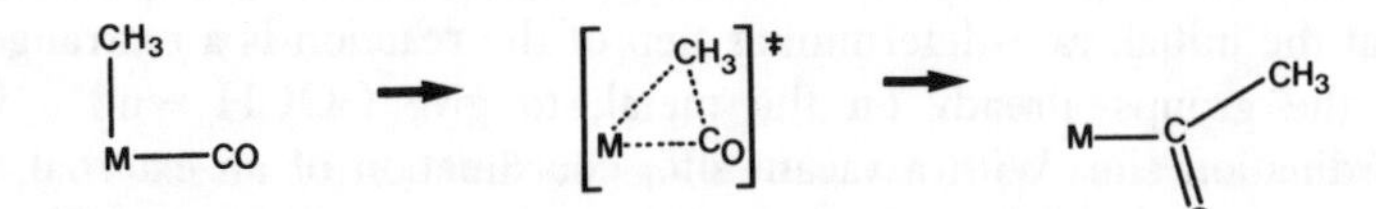

MIGRATION OF H TO ALKENES

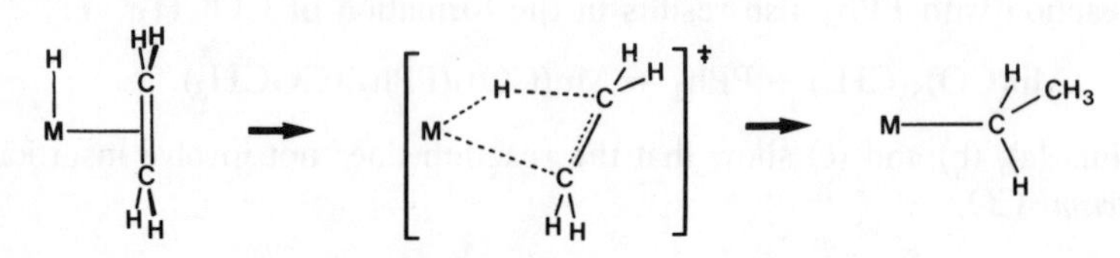

Evidence for a concerted mechanism comes from the fact that a complex, $[TiCp_2(Et)]$ exists, in which there is a short (less than the sum of the van der Waals' radii) contact between the Ti and an ethyl H atom. Thus, we have

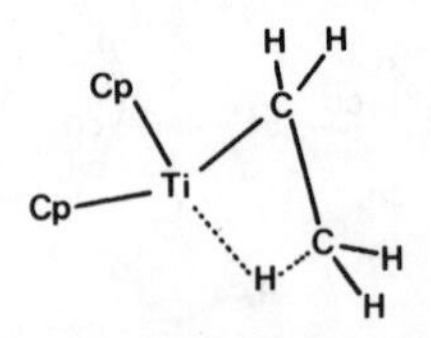

which can be regarded as a good model for the transition state in the migration reaction. The H in this Ti complex is called 'agostic' (bonded to the carbon but having some interaction with the metal atom).

β-Elimination (see, also, transition metal alkyls, below)

This involves transfer of a β-hydrogen (counting outward from the metal we have α, β, γ, etc., carbon atoms) from an alkyl group to the metal, which generates an M—H bond and an alkene. It is a particularly favourable, low activation energy, process because of the close proximity of the β—H to the metal atom.

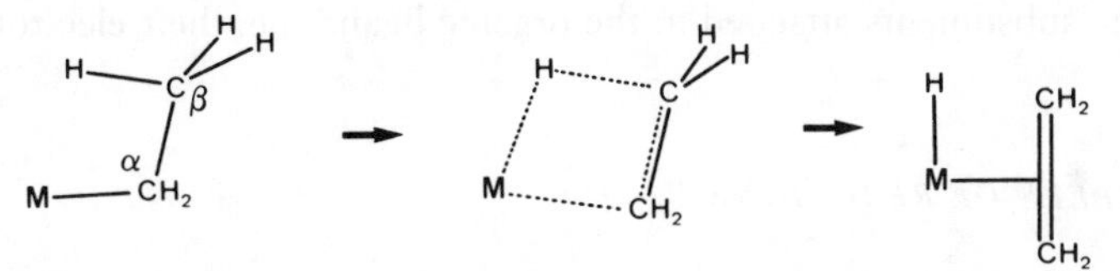

Less favourable are α- and γ-eliminations; for α-elimination, a great deal of strain would have to be introduced for the H to be close to the metal and, for γ-elimination, the hydrogen is further away, the chain is waving around more, therefore there is less chance of the H 'bumping into' the metal and a reaction occurring.

β-elimination does not occur for coordinatively saturated metal complexes, i.e. those with 18e and their full complement of ligands. This is because a vacant coordination site is required for initial attack of the H.

Specific reactions of organic ligands

When an organic ligand is coordinated to a transition metal atom its reactivity is radically affected; previously unreactive molecules can be made to undergo a variety of reactions. For example

$C_6H_5Cl + MeO^- \rightarrow$ No reaction at room temperature

But, under the same conditions,

$$[Cr(CO)_3(\eta^{6-}C_6H_5Cl)] + MeO^- \rightarrow [Cr(CO)_3(\eta^{6-}C_6H_5OMe)]$$

We have seen that organic ligands participate in synergic bonding with the metal. This changes the charge distribution in the ligand and, depending on the degree of M → L and L → M donation, it can react as a positively or negatively charged species, i.e. as a nucleophile or electrophile. What type of reaction we get depends on:

1. The oxidation state of the metal. For metals in high positive oxidation states there will be increased L → M σ donation and reduced M → L π back donation, which will result in the ligand becoming more positively charged, i.e. subject to nucleophilic attack. For metals in low oxidation

states the converse of this occurs, and the ligand becomes more negatively charged and more susceptible to electrophilic attack.
2. The number of d electrons on the metal, i.e. how much non-bonding electron density we have available for back donation to the ligands.
3. The other ligands attached to the metal, i.e. whether the other ligands increase or decrease the charge density on the metal, and hence on the target ligand.
4. The substituents attached to the organic ligand, i.e. their electronegativity.

EXAMPLES OF REACTIONS

1. Nucleophilic attack on an alkene, e.g.

$$\left[Cl_3Pt—\overset{CH_2}{\underset{CH_2}{\|}}\right]^- \xrightarrow{OMe^-} \left[Cl_3Pt{-}CH_2{-}CH_2{-}OMe\right]^{2-}$$

Nucleophilic attack at the C, which is made δ+ by increased σ donation to Pt^{2+} and reduced π back donation from Pt^{2+}.
2. Nucleophilic attack on benzene, e.g.

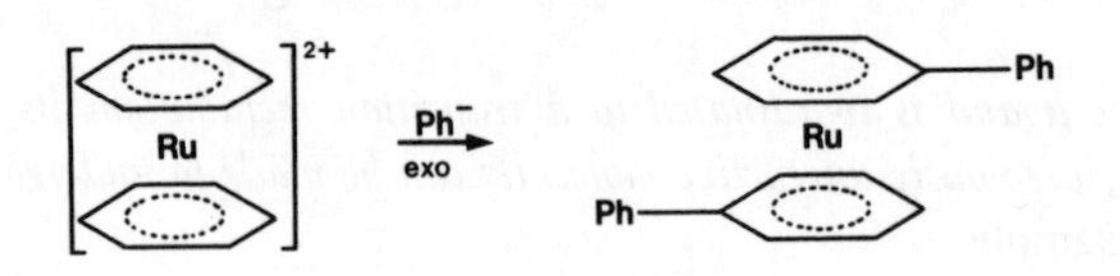

The positively charged metal withdraws electron density from the ring making it susceptible to nucleophilic attack.
Note: in organic chemical reactions, both alkenes and benzene usually react via electrophilic mechanisms.
3. Electrophilic attack on alkenes, e.g.

$$L_3Pt—\overset{CF_2}{\underset{CF_2}{\|}} \xrightarrow{HCl} L_3\underset{Cl}{Pt}{-}CF_2{-}CF_2H$$

The presence of very electronegative F atoms on the alkene causes it to be a stronger π acceptor, so that more electron density is donated from the metal into the C—C π* orbital. This renders the alkene more susceptible to electrophilic attack since the mechanism involves the proton (H^+) attacking the highest occupied molecular orbital, i.e. the

C—C π* orbital, which is partially filled due to back donation from the metal.

4. Electrophilic substitution in cyclobutadiene – analogous to the Friedel–Crafts reactions undergone by benzene, e.g.

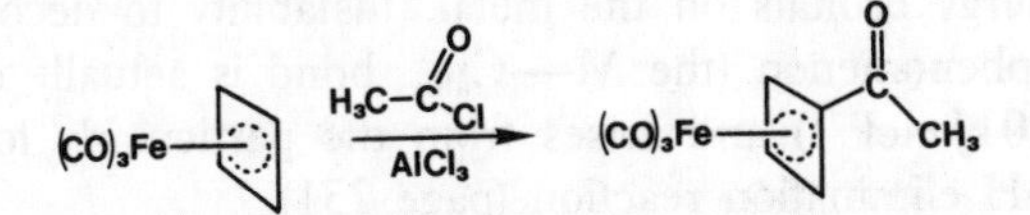

This reaction is facilitated by having the Fe in zero oxidation state and therefore better able to donate electron density to the cyclobutadiene and activate it towards electrophilic attack.

Exo/Endo Attack

Organic ligands can be attacked by a nucleophile/electrophile on the side away from the metal (exo attack) or on the same side as the metal (endo attack), i.e.

For endo attack to occur we must have prior coordination of the incoming ligand to the metal centre. This can be seen particularly well in

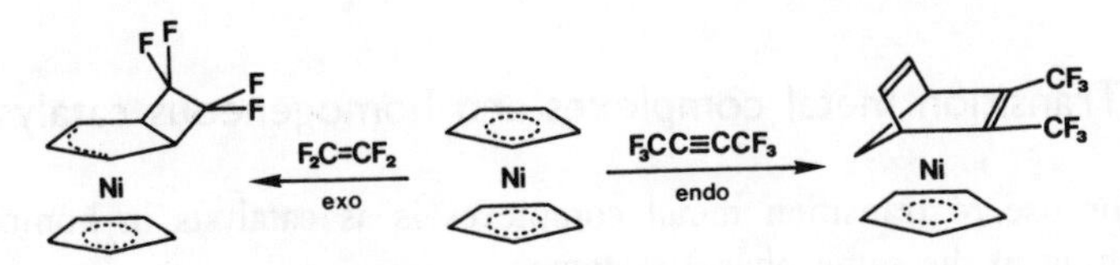

5. *Metathesis*, which basically is

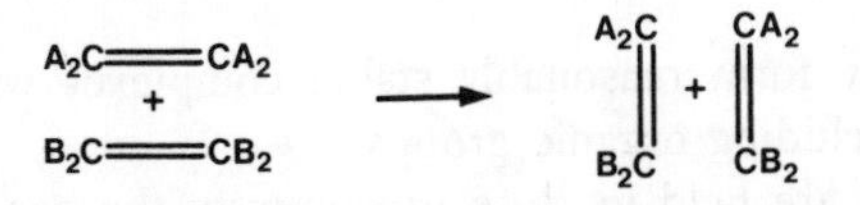

The reaction is catalysed by systems such as $WCl_6/RAlCl_2$ in the presence of small amounts of ethanol. The first stage is formation of a carbene complex, which reacts with the alkene to form a four-membered transition state,

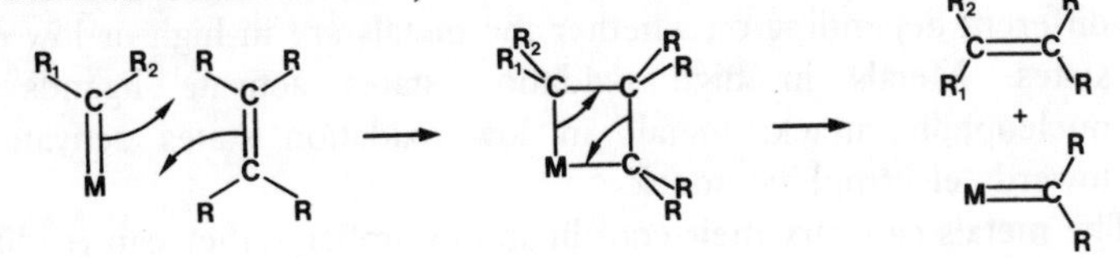

TRANSITION METAL ALKYLS This type of compound has a reputation for being unstable. Instability with respect to water and oxygen is thermodynamic and kinetic, and comes from the favourability of forming metal oxides and hydroxides, the polarity of the M—C bond ($M^{\delta+}$—$C^{\delta-}$) and the availabilty of vacant low energy orbitals on the metal. Instability to decomposition is mostly a kinetic phenomenon (the M—C_{alkyl} bond is actually quite strong, approximately 150 kJ mol^{-1}) and arises from the particularly low activation energy for the β-H elimination reaction (page 251).

We can increase the stability towards decomposition by blocking this reaction, which can be done in a number of ways:

1. By having a coordinatively saturated metal complex so that there is no site for the H to coordinate, e.g. $[Rh(NH_3)_5(C_2H_5)]^{2+}$, containing substitutionally inert, d^6 low spin Rh^{III}, is stable (see page 197).
2. By having no β-H, e.g. $Cr(CH_2CH_3)_4$ (β-H) decomposes at −80 °C whereas $Cr(CH_2SiMe_3)_4$ (no β-H) is stable at room temperature. (The stability of this second complex could also be due to the bulkiness of the ligands preventing any possible intermolecular decomposition routes.)

When β-H elimination is not possible, metal alkyls can decompose by homolytic fission (radical mechanism), reductive elimination or α-H elimination.

Transition metal complexes and homogeneous catalysis

A major use of transition metal complexes is as catalysts in homogeneous (everything in the same phase) systems.

What is it about transition metals that makes them so useful in catalysis?

1. They readily form reasonably stable complexes with a wide variety of ligands, including organic groups.
2. The ligands are held in close proximity in the coordination shell of the metal so that reaction can occur between them.
3. Ligands can be activated by coordination to the metal – σ and π donor/acceptor effects.
4. Transition metals have variable oxidation states – activation can be different depending on whether the metals are in high or low oxidation states. Metals in high oxidation states activate ligands towards nucleophilic attack; metals in low oxidation states activate ligands towards electrophilic attack.
5. The metals can vary their coordination number – they can readily switch

between, e.g. five and six coordinate complexes in catalytic cycles.
6. The steric and electronic properties of the 'non-participating' ligands can affect the activation of ligands and the products that are obtained.
7. They readily undergo oxidative addition and reductive elimination reactions so that substrates can be coordinated and products can dissociate.
8. The metal can play a part in reactions – reactions such as β-H elimination occur directly at the metal centre.
9. Ligands (e.g. allyl or NO) are able to vary their mode of coordination to a metal centre; this can readily generate vacant coordination sites.
10. What the transition metals do they do quickly.

Virtually all catalytic cycles proceed by elementary steps involving only 16 or 18 valence electron complexes. This is sometimes known as the '16 and 18 electron rule'.

We can illustrate most of the above points by considering the cobalt catalysed hydroformylation reaction.

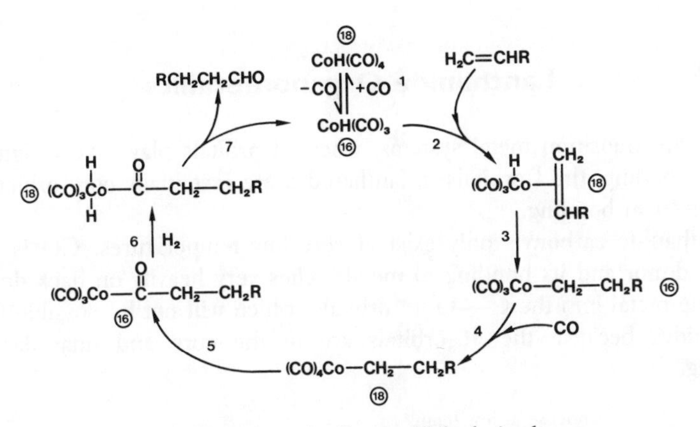

Only formation of a linear aldehyde is shown.

The overall reaction is

$$CH_2{=}CHR + CO + H_2 \rightarrow RCH_2CH_2CHO$$

The catalytic precursor, which is generated *in situ*, is $CoH(CO)_4$. This is an 18e complex and so will not be able to coordinate the substrate. Dissociation of a CO ligand must occur to give a 16e complex, $CoH(CO)_3$, which is the active catalyst (step 1). This can now coordinate a substrate molecule, a 2e donor, which brings the electron count of the complex up to 18e again (step 2). The alkene then 'inserts' into the Co—H bond to regenerate a 16e complex (step 3). This reaction is probably better regarded as a hydrogen migration – the reverse of β-H elimination.

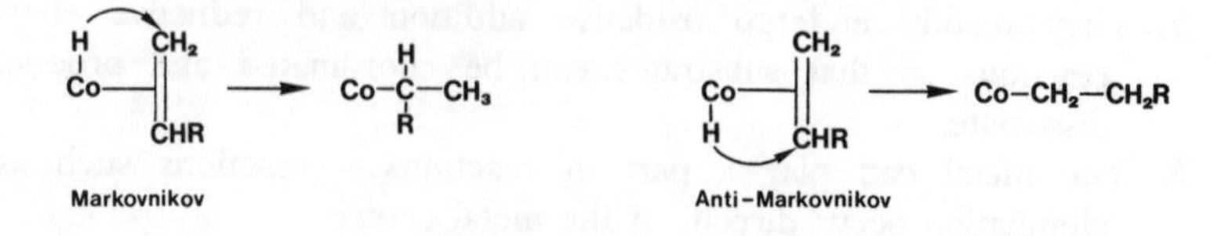

H migration can occur in two ways, Markovnikov or anti-Markovnikov, to produce branched or linear alkyls, respectively. The 16e complex coordinates another CO to form an 18e complex (step 4) which undergoes a CO insertion (alkyl migration) reaction to generate a 16e acyl complex (step 5). This 16e complex can coordinate H_2, via an oxidative addition reaction (step 6), to give an 18e species from which the aldehyde is reductively eliminated to regenerate the 16e active catalyst (step 7); this is then ready to coordinate another substrate.

Lanthanide Organometallics

Unlike the transition metal systems, where d orbitals play a very significant role in bonding, the f orbitals in lanthanides are essentially core orbitals and uninvolved in bonding.

Lanthanide carbonyls only exist at very low temperatures. CO is a very poor σ donor and its bonding to metals relies very heavily on back donation from the metal into the C—O π* orbitals, which will not be possible for the lanthanides because the 4f orbitals are in the core and unavailable for bonding.

$$Ln_{(g)} + CO \xrightarrow{\text{condense at low temp.}} Ln(CO)_n \quad (n = 1 \text{ to } 6)$$

The products decompose as the temperature is raised. There is some change in the C—O stretching frequency from that of free CO but this is only a few cm^{-1}, indicating a very small amount of π back donation (M → L).

Lanthanide alkyls/aryls can exist and are reasonably stable. Whereas in transition metal complexes there is a large covalent component in the M—L bonding, in the lanthanide equivalents, the bonding is almost totally ionic – a result of the unavailability of the f electrons. Thus, the compounds exhibit all the macroscopic properties associated with ionic compounds, high melting point, etc. (cf. transition metal alkyls, which are often low melting point, covalent compounds).

Despite the differences in bonding the lanthanide and transition metal organometallics actually undergo some similar reactions, e.g.

1. Insertion

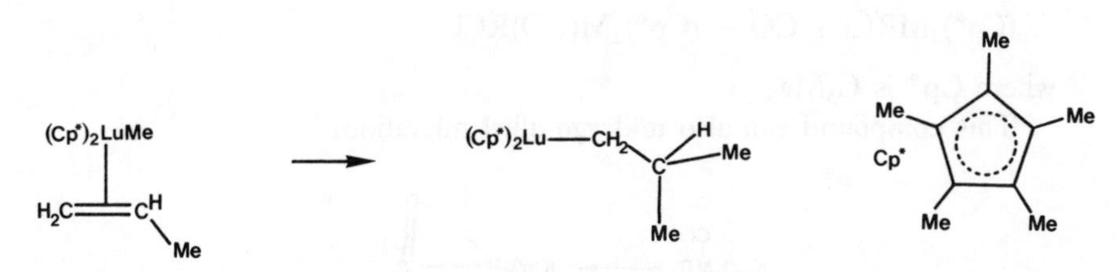

Methyls both on the same C since this gives less crowding at the Lu. Further reaction can occur to give a polymer.

2. C—H Activation, e.g.

$$(Cp^*)_2LuMe \xrightarrow{^{13}CH_4,\ \Delta} (Cp^*)_2Lu(^{13}CH_3) + CH_4$$

where $Cp^* = C_5Me_5$

The initial, weak, coordination of the $^{13}CH_4$ group involves a four membered transition state, i.e.

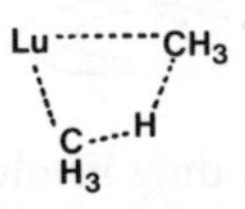

Actinide Organometallics

Actinide carbonyls, like their lanthanide analogues, are not very stable, again as a result of f electrons not being very available for back donation. However, the 5f orbitals are more available than the 4f in lanthanides and C—O stretching frequencies show a larger decrease from the free CO value, e.g. the CO stretching frequency in Cp_3UCO is 169 cm^{-1} less than that in free CO (cf. much larger decrease for transition metals complexes, page 232). This can be compared with a shift of just a few cm^{-1} in the stretching frequency when CO becomes coordinated to a lanthanide.

The bonding in actinide alkyls/aryls has more covalent character than that in lanthanide compounds but less than that in transition metal organometallics. The bonding is best regarded as mostly ionic but with some covalent contribution.

Like their lanthanide analogues, they are generally made by reaction between their halides and 'negatively charged' organic anions, e.g.

$$AnCl_3 + 3Cp^- \rightarrow AnCp_3$$
$$AnCl_4 + 3Cp^- \rightarrow AnCp_3Cl + RLi \rightarrow AnCp_3R$$

REACTION OF ACTINIDE ALKYLS WITH CO

$$(Cp^*)_2MRCl + CO \rightarrow (Cp^*)_2M(CO)RCl$$

where Cp* is C_5Me_5

This compound can also undergo alkyl migration, i.e.

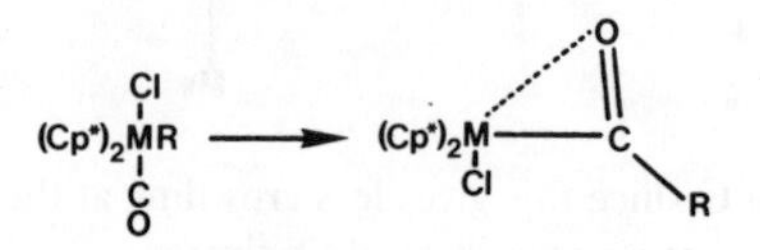

So there is interaction between the M and O, illustrating the desire of the actinides to form M—O bonds.

URANOCENE This is the most often encountered actinide sandwich compound,

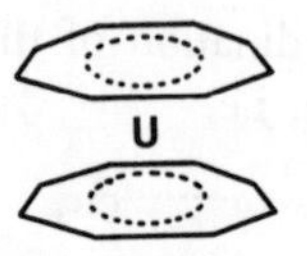

The rings are parallel and the bonding involves overlap of f orbitals on U with orbitals of the correct symmetry on the C_8H_8 ring (cf. ferrocene, $FeCp_2$, pages 243–5), e.g.

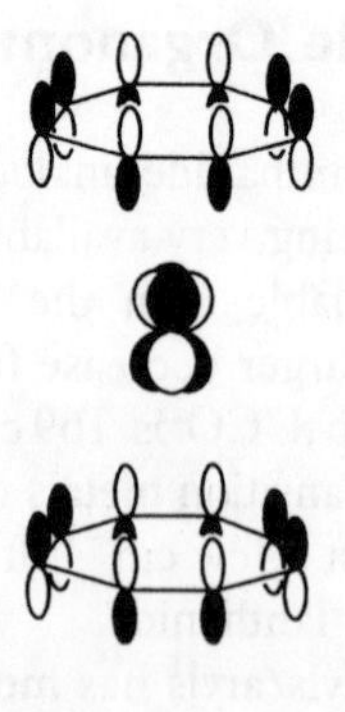

Summary

1. Transition metals in low oxidation states with coordinated organic ligands often have 18 electrons in their valence shell.

2. The electron count of a transition metal complex is only one of many factors that influence its stability.
3. The 18 electron rule can be used to predict the number of metal–metal bonds in di-, or tri-, etc., metal systems.
4. Transition metal fragments can be isolobal with organic fragments, i.e. the number, symmetry, extent in space, and approximate energy of frontier orbitals are similar.
5. The infrared stretching frequency of CO decreases when it becomes coordinated to a transition metal.
6. Bridging carbonyls are more commonly encountered for the first transition metal series than for the second or third series, and for anionic complexes rather than cationic ones.
7. When alkenes or alkynes bond to a transition metal the C—C bond length increases.
8. Coordination of an allyl, butadiene, cyclobutane or cyclopentadiene to a transition metal often results in the C—C bond lengths within each structure becoming more equal than in the free, uncoordinated, organic molecule.
9. Oxidative addition is most favoured for low oxidation state coordinatively unsaturated third row transition metals which have small hard ligands attached.
10. ‘Insertions’ are actually ‘migrations’.
11. When an organic ligand is coordinated to a transition metal its reactivity is often drastically affected. Low oxidation state transition metals activate organic ligands to electrophilic attack; high positive oxidation state transition metals activate them to nucleophilic attack.
12. Transition metal complexes can act as good homogeneous catalysts. Most catalytic cycles proceed by elementary steps involving only 16 or 18 valence electron complexes.
13. Lanthanide organometallics are predominantly ionic in character.
14. Actinide organometallics are more covalent than their lanthanide analogues but are still more ionic than equivalent transition metal compounds.

Questions

1 What are the oxidation states of the transition metals in each of the following compounds?

(a) $Ti(NEt_2)_4$

(b) $NiBr_3(PEt_3)_2$

Answer

- Treat the bonding between the transition metal and ligands as ionic.
- Remove all ligands with full outer shells for donor atoms.
- What is the charge left on the transition metal?

(a) $Ti(NEt_2)_4$: we remove 4 NEt_2^- ligands (cf. NH_2^-), which leaves a 4+ charge on the Ti since the overall complex is neutral. Therefore, the oxidation state of the Ti is +4.

(b) $NiBr_3(PEt_3)_2$: the P in PEt_3 already has a full outer shell, therefore we remove PEt_3 as a neutral molecule. Br is removed as Br^- and, as we have three of these, the charge left on the Ni is +3, i.e. the oxidation state of the Ni is +3.

2 Explain the variation in the following rates for the oxidative addition reaction between MeI and $IrX(CO)L_2$ as X and L are changed:

$$X = F > Cl > Br > I$$

$$\xrightarrow{\hspace{3cm}}$$

Rate decrease

$$\xrightarrow{\hspace{3cm}}$$

$$L = PMe_2Ph > PEt_3 > PEt_2Ph > PEtPh_2 > PPh_3.$$

Answer

The oxidative addition reaction is,

$$Ir^{I}X(CO)L_2 + MeI \rightarrow Ir^{III}X(CO)L_2(Me)(I)$$

Oxidative addition involves the Ir increasing its oxidation state by two units and adding two more ligands to its coordination shell. The reaction is favoured by ligands which stabilize higher oxidation states and by small ligands – less steric repulsion (repulsion between closed shells of electrons on adjacent ligands).

F can stabilize high oxidation states. Since F^- is a hard ligand the bonding in its compounds has a high degree of ionic character and is going to be stronger when the metal is in a higher oxidation state – electrostatic interaction is proportional to the product of the charges. F^- is also hard to oxidize and will not be oxidized to F_2 by higher oxidation state metals.

I is not good at stabilizing higher oxidation states. Since I^- is soft the bonding has a large proportion of covalent character and will not suffer the same strengthening upon increase in oxidation state. Also, I^- is fairly readily oxidized to I_2 and will tend to be oxidized by metals in high oxidation states (this is accompanied by reduction of the metal).

Therefore, the order is F > Cl > Br > I.

The variation in order with different phosphine ligands is a steric effect – the smaller the phospine, the easier it is going to be to add another two ligands to the coordination shell of the metal. PMe_2Ph is the smallest ligand with regard to the amount of space taken up, PPh_3 is the largest.

There is also going to be a small steric effect with variation in halogen – I is larger than F. This is not as important as the electronic effect.

3 Explain the following variation in IR C—O stretching frequencies in *fac*-$(R_3P)_3Mo(CO)_3$ complexes:

$\nu_{C—O}$ (cm^{-1})	2074,2026	2041,1989	1949,1835
R	F	Cl	Ph

Answer

The general structure is

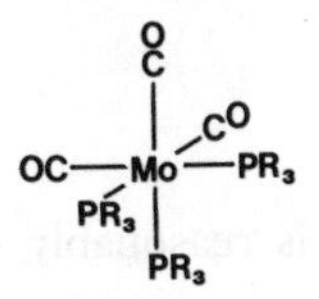

The weakening of the C—O bond is due to varying amounts of electron density being donated into the C—O π* orbital – more electron density donated to this orbital causes a greater weakening of the bond and a lower infrared stretching frequency.

There are two factors to be considered:

(a) π bonding
(b) σ bonding

Both rely on F being more electronegative than Cl, which, in turn, is more electronegative than Ph.

(a) The phosphine uses its d orbitals to accept π electron density from the metal. The phospine and the *trans* CO compete for π electron density in the same d orbital,

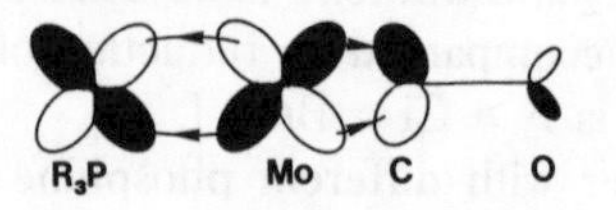

The electronegativity of F causes the d electrons on P to be lowered in energy so that they are better able to accept electron density from the metal. Cl and Ph are progressively less electronegative than the F, therefore the d orbitals are higher in energy and less available for π bonding. PF_3 is, thus, the strongest π acceptor, so that it competes most strongly with the *trans* CO for the metal's π electron density and reduces the electron density available for donation to the CO. Less electron density available for donation into C—O π* orbitals means that the C—O bond is stronger and the stretching frequency higher when CO is *trans* to PF_3.

(b) This can also be regarded as just a σ effect. The fluorines make the P to which they are attached more δ+, reducing σ donation from the P to the metal. Less electron density donated to the metal means that there is less tendency for the metal to donate electrons to the C—O π* orbitals, therefore, the C—O bond is stronger and the stretching frequency is higher.

The above effect is probably due to a combination of both σ and π factors.

4 Explain the following:

(a) $V(CO)_6$ exists and is reasonably stable but it is quite readily reduced to $[V(CO)_6]^-$.

(b) $CoCp_2$ reacts with alkyl halides (RX) to give $[CoCp_2]^+$ and $[Co(\eta^4 - C_5H_5R)Cp]$.

Answer

Both of these facts can be rationalized by use of the 18 electron rule.

(a) $V(CO)_6$ is a 17e species (5e from V and 2e each from 6 carbon monoxides) and would therefore be expected to dimerize to give $V_2(CO)_{12}$, with a metal–metal bond and 18e in the valence shell of each V. Dimeric $[V(CO)_6]_2$ does not, however, exist due to steric factors – the dimeric compound would involve seven coordinate V and a great deal of repulsion between CO ligands. However, carbonyl compounds do usually prefer to have 18e in the valence shell of the metal and, in this case, it can be obtained by reduction to the anion.

(b) $CoCp_2$ is a 19e complex (9e from Co and 5e from each Cp) but both the reactions involve generation of 18e species, i.e. $[CoCp_2]X$ contains $[CoCp_2]^+$ and $\eta^4 - C_5H_5R$ is a 4e donor giving a total electron count in $[Co(\eta^4 - C_5H_5R)Cp]$ of 9 + 4 + 5 = 18e.

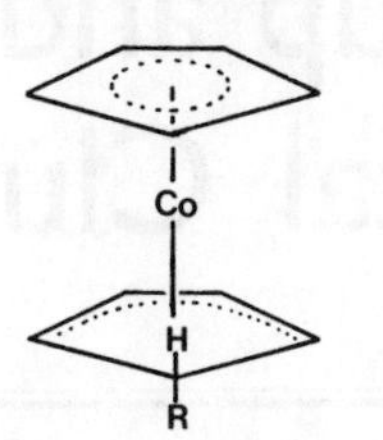

Thus, the reaction is more favourable due to the generation of the 'stable' 18e complexes.

CHAPTER 8

Main Group and Transition Metal Clusters

Boron Hydrides (Boranes)

B_2H_6 (diborane) versus C_2H_6 (ethane)

C_2H_6 has 14 valence electrons ($2 \times 4 + 6 \times 1$), i.e. sufficient electrons to form seven electron pair bonds, so that all the atoms can be held together by two centre–two electron (2c–2e) bonds, i.e.

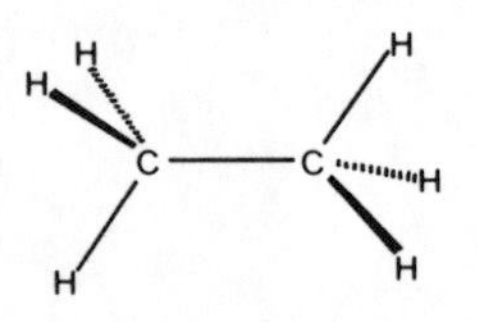

B_2H_6 has 12 valence electrons (2×3 from B + 6×1 from H) and as it is not possible to hold eight atoms together with six 2c–2e bonds diborane cannot adopt a structure analogous to that of ethane. *Diborane and all other boranes are electron deficient in that the atoms cannot be held together solely by 2c–2e bonds* and some sort of multi-centre bonding scheme must be invoked to rationalize their structures. *Boranes are not electron deficient in the sense that they have all bonding MOs filled and do not readily accept electrons*, e.g. B_2H_6 has no tendency to form $[B_2H_6]^{2-}$, isoelectronic with C_2H_6. When looking at borane structures it is most important to realize that the lines between atoms are simply that (they are there to give perspective to the shape), and not, as we have been used to, necessarily 2c–2e bonds.

The structure of diborane is

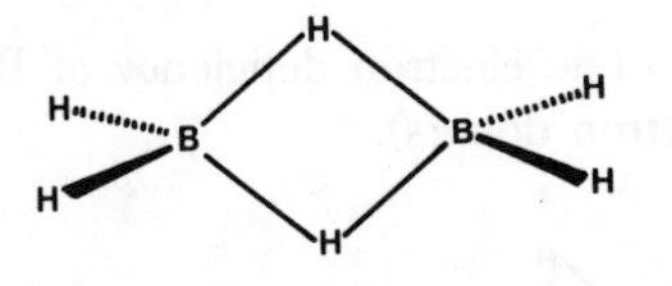

The boron atoms are approximately sp^3 hybridized, which explains the shape – **IT IS NOT PLANAR.**

Each B can be considered to use two sp^3 orbitals and two electrons in forming 'normal' 2c–2e bonds to the terminal hydrogens, leaving one electron and two orbitals which can be used to form 3c–2e bonds to the hydrogens, i.e.

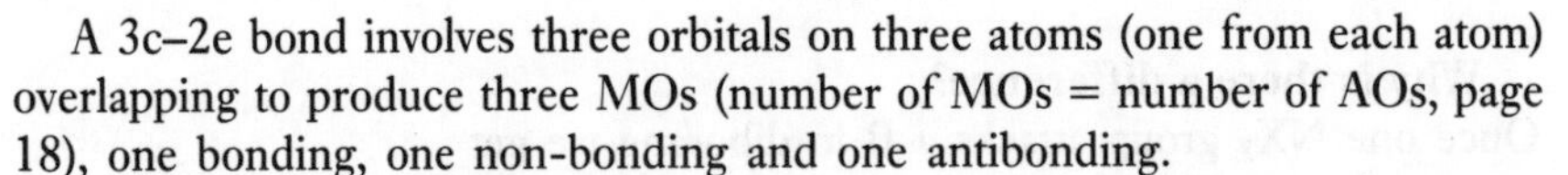

A 3c–2e bond involves three orbitals on three atoms (one from each atom) overlapping to produce three MOs (number of MOs = number of AOs, page 18), one bonding, one non-bonding and one antibonding.

In B_2H_6 we have 2e from the borons and 2e from the hydrogens which can fill up the two bonding MOs produced. A partial MO diagram for B_2H_6 is,

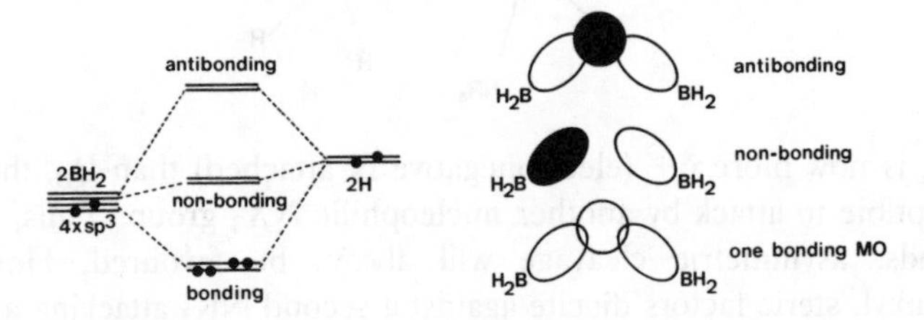

So, we can see here that, if we add two more electrons to B_2H_6, to generate $[B_2H_6]^{2-}$, the electrons are added to a non-bonding orbital; there is no energy advantage in doing this. Addition of two more electrons, however, increases the electron–electron repulsion in the molecule, i.e. there is an overall destabilization of the system and $[B_2H_6]^{2-}$ does not exist.

One line of evidence which supports this type of bonding in diborane is that the terminal B—H bonds are significantly shorter than the bridging B—H bonds. This is to be expected since, in the B—H_t bond there are 2e shared between two atoms but in the B—H—B unit there are 2e shared between three atoms, i.e. 1e per B—H interaction. Thus the B—H_t interactions are stronger and the bonds shorter.

Reactions of diborane

WITH LEWIS BASES The electron deficiency of B_2H_6 leads to it reacting with Lewis bases (electron donors).

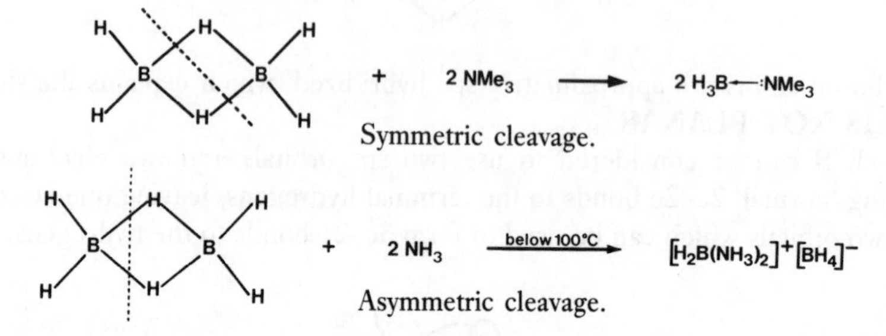

In the products of these reactions the B has eight valence electrons and it is no longer electron deficient with respect to 2c–2e bonding.

Why is there a difference?

Once one NX_3 group attacks a B in diborane we get

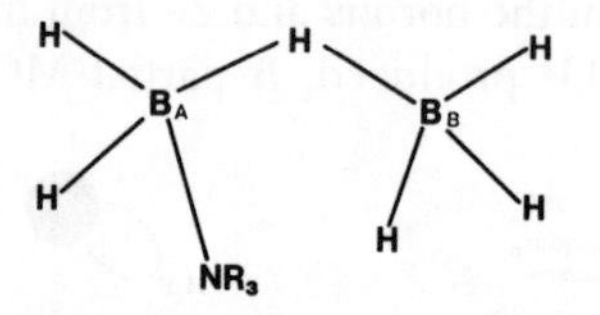

B_A is now more $\delta+$ (electronegative N attached) than B_B, therefore more susceptible to attack by another nucleophilic NX_3 group. Thus, on electronic grounds, asymmetric cleavage will always be favoured. However, when X = alkyl, steric factors dictate against a second NR_3 attacking a small B with a relatively large NR_3 already attached and therefore we get symmetric cleavage.

PYROLYSIS Pyrolysis (heating) in a sealed tube at temperatures above 100 °C leads to formation of a variety of higher boranes, such as B_4H_{10}.

Other boranes and Wade's Rules

All boranes can be considered as derived from closed B_n deltahedral units. (A deltahedron is a polyhedron with all triangular faces, e.g. an octahedron.) These boranes are called 'boron cluster compounds'. For example, B_5H_9 is derived from an octahedron by removal of a vertex,

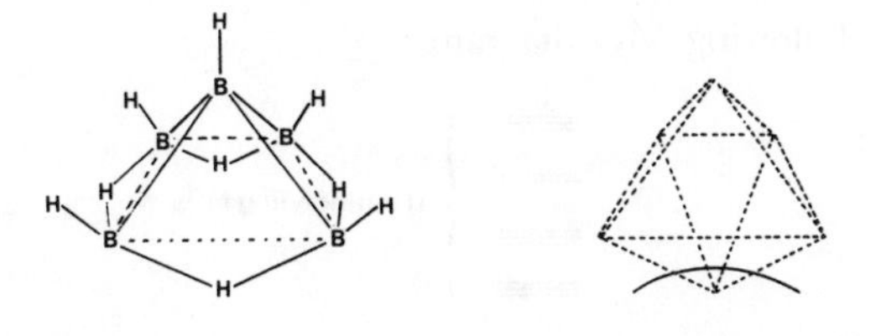

Wade's rules are a set of rules which were developed to explain the structures adopted by polyhedral boranes on the grounds of the number of valence electrons which are available for bonding.

Everything depends on the rule: *any* n *vertex* closo-*deltahedral structure requires* (n+1) *electron pairs for skeletal bonding*, i.e. (n+1) electron pairs are required to fill up all the skeletal bonding MOs and, therefore, to hold the cluster together. (A *closo*-deltahedron is one in which all vertices are intact, e.g. a complete octahedron.) The theoretical basis of this rule can be seen by a consideration of $[B_6H_6]^{2-}$.

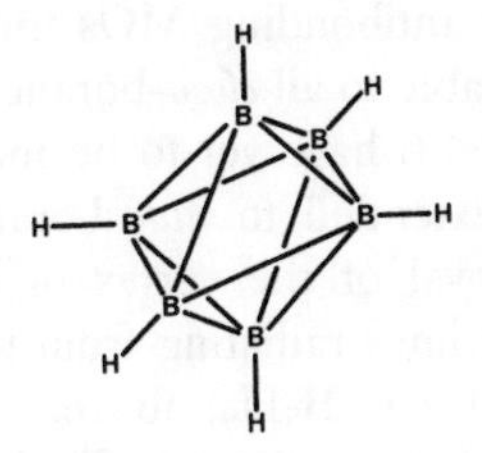

All boranes are considered as made up from BH units. The B can be considered as sp hybridized and uses one sp orbital and 1e to form a normal 2c–2e bond to the H. This leaves one sp hybrid, two p orbitals and 2e for skeletal bonding.

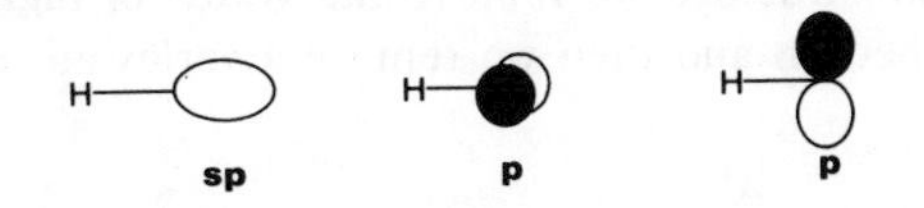

Thus, if we take six B—H units, as in $[B_6H_6]^{2-}$ we have a total of 18 AOs which combine to generate 7 bonding MOs and 11 antibonding MOs (remember that the number of MOs = number of AOs we started with). The bonding MOs are

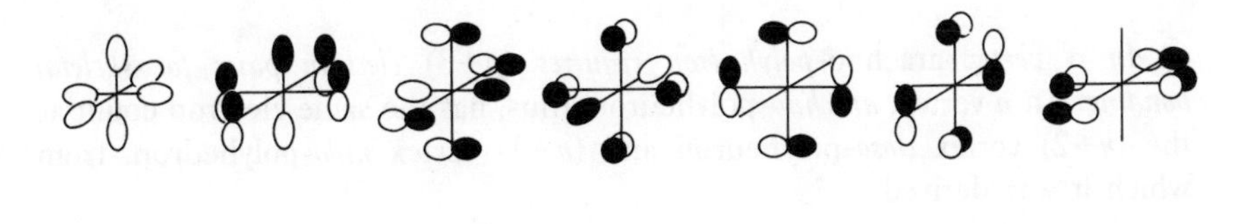

and we have the following MO diagram

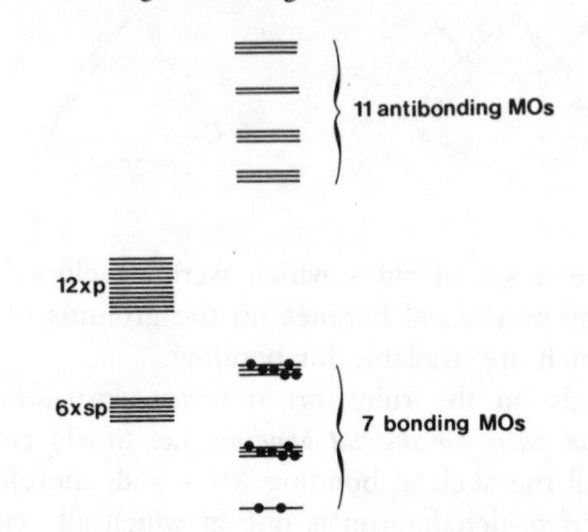

There is a total of 14 electrons available for cluster bonding (two from each BH and two from the negative charge); these are used to fill up all the bonding MOs leaving the antibonding MOs unoccupied.

This is generally applicable to all *closo*-boranes of formula $[B_nH_n]^{2-}$ ($n = 6$ to 12 – analogues with $n < 6$ have yet to be made).

Wade's rules can be extended to *nido*-boranes, which are derived from *closo*-boranes by the removal of the vertex of highest connectivity (i.e. the vertex which has the most lines radiating from it – the vertex which is joined to the most other vertices), e.g. B_5H_9, above.

An n *vertex* nido-*polyhedron requires* (n+2) *electron pairs for skeletal bonding.* The electron count for an *n* vertex *nido*-polyhedron is, thus, the same as that of the (n+1) vertex *closo*-polyhedron from which it was derived. *Arachno*-polyhedra are derived from *closo*-polyhedra by the removal of two vertices, or from the appropriate *nido*-polyhedron by the removal of one vertex. To generate the *arachno*-structure we remove the vertex of highest connectivity from the *closo*-polyhedron and then we remove a vertex adjacent to this, e.g. B_5H_{11}.

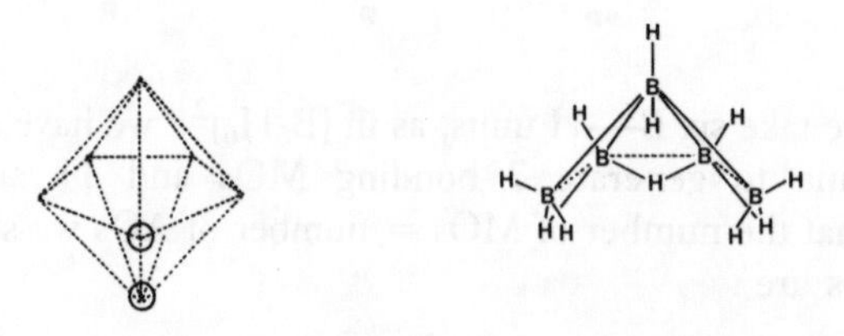

An n *vertex* arachno-*polyhedron requires* (n+3) *electron pairs for skeletal bonding.* An *n* vertex, *arachno*-polyhedron, thus, has the same electron count as the (n+2) vertex *closo*-polyhedron and (n+1) vertex *nido*-polyhedron, from which it was derived.

We could never remember which is *nido* and which *arachno* until we realized that n1do can be written thus, indicating 1 vertex removed.

HOW TO PREDICT STRUCTURES USING WADE'S RULES

1. Count the total number of electrons – three from each B, one from each H, add or subtract electrons for the charge (add for negative charge, subtract for positive).
2. Assume that each B forms *one* 2c–2e B—$H_{terminal}$ bond. This uses up 2e per BH, which cannot therefore participate in cluster bonding. So, $2n$ electrons are subtracted from the total (where n is the number of borons), which gives the total number of electrons available for skeletal bonding.

   Note: you must always assume that there is just *one* B—H_t per boron, no matter how many terminal hydrogens there really are. This is because any other H atoms can be regarded as part of the polyhedral cluster framework, i.e. bonding to the p orbital on the B, so that their electrons are used in cluster bonding.
3. For n B atoms, if the total number of electron *pairs* available for cluster bonding is $(n+1)$ then the structure is based on an n-vertex *closo*-polyhedron. $(n+2)$ Electron pairs gives a *nido*-structure and $(n+3)$ electron pairs, an *arachno*-structure.

n-VERTEX *CLOSO*-POLYHEDRA

n	POLYHEDRON
5	trigonal bipyramid
6	octahedron
7	pentagonal bipyramid
8	dodecahedron
12	icosahedron

Note: no tetrahedron is included here as the tetrahedron is anomalous – a *closo*-tetrahedron would be expected to have five electron pairs for skeletal bonding; a tetrahedron treated as a *nido*-trigonal bipyramid should have six electron pairs; theoretical calculations predict either six or four electron pairs. The problem is avoided in that there are no boranes with a tetrahedral structure (see below). There is one boron compound with a tetrahedral structure, B_4Cl_4, which has four electron pairs for skeletal bonding

$(4 \times 3 + 4 \times 1 - 4 \times 2 = 8e$, i.e. 4 pairs).
B   H   B—Cl

EXAMPLES OF THE PREDICTION OF SHAPES

B_5H_9 (n = 5)
Total number of electrons is $5 \times 3 + 9 \times 1 = 24$, i.e. 12 pairs.

Subtract 10 electrons, i.e. 5 pairs, for 5 B—H_t bonds, which leaves 14e, i.e. 7 electron pairs (n+2) for skeletal bonding. (n+2) Electron pairs means that we have a five vertex *nido*-structure based upon a six vertex *closo*-polyhedron (an octahedron).

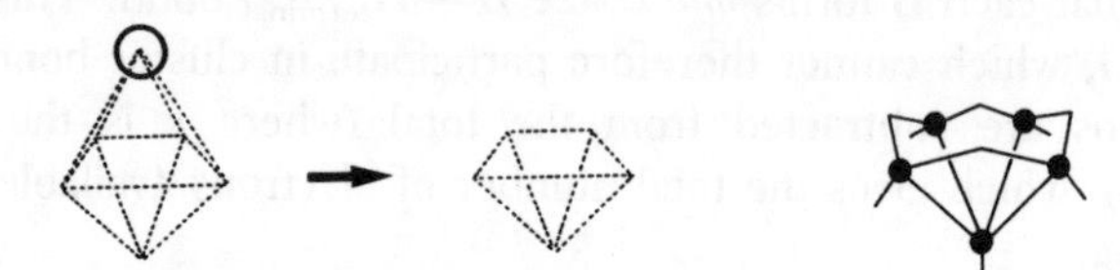

All vertices of an octahedron are of equal connectivity.

Note: once again, there are not enough electrons for all the lines in the diagram to be 2c–2e bonds – the compound is electron deficient.

Wade's rules cannot be used to predict the positions adopted by the H atoms. However, we can make a few generalizations based upon a study of existing structures:

1. Hydrogens tend to bridge the edges around open faces in *nido*- and *arachno*- polyhedra.
2. In *arachno*-polyhedra you often get two terminal hydrogens on a B, with one H pointing roughly towards the missing vertices.

The hydrogens interact with bonding MOs (full of electrons) which would have been used for bonding to the missing vertices. The hydrogens thus occupy regions of high electron density around the framework.

B_4H_{10} (n = 4)
Total number of valence electrons = $4 \times 3 + 10 \times 1 = 22$e.

Of these, 8e are used for B—H_t bonding, leaving 14e, i.e. 7 pairs (n+3) for skeletal bonding. (n+3) Electron pairs for skeletal bonding means that we have a four vertex *arachno*-structure based upon the six vertex *closo*-octahedron. The *arachno*-octahedron is generated by removal of two adjacent vertices, i.e.

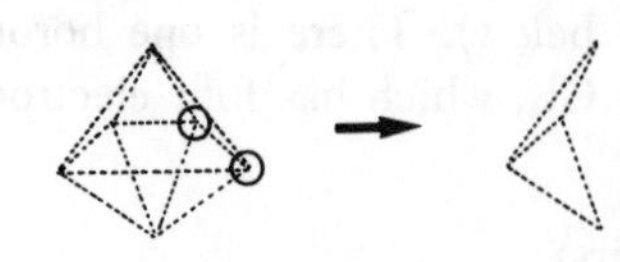

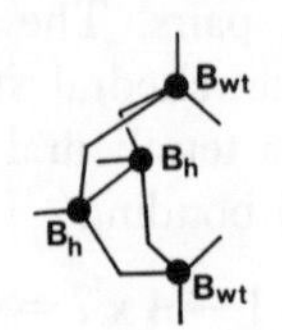

A 'butterfly' structure
wt = wing-tip
h = hinge

Four hydrogens bridge the edges around the open face, and one hydrogen on each of the wing-tip borons points roughly towards the missing vertices (see above).

B_5H_{11} ($n = 5$)
Total number of valence electrons $= 5 \times 3 + 11 \times 1 = 26$e.

Of the total, 10e are used for $5 \times$ B—H$_t$, leaving $26 - 10 = 16$e, i.e. 8 ($n+3$) electron pairs for skeletal bonding. The structure is thus a five vertex *arachno*-polyhedron based on a *closo*-polyhedron with seven vertices, i.e. a pentagonal bipyramid. We first remove the vertex of highest connectivity, i.e. an apical one, and then a vertex adjacent to this.

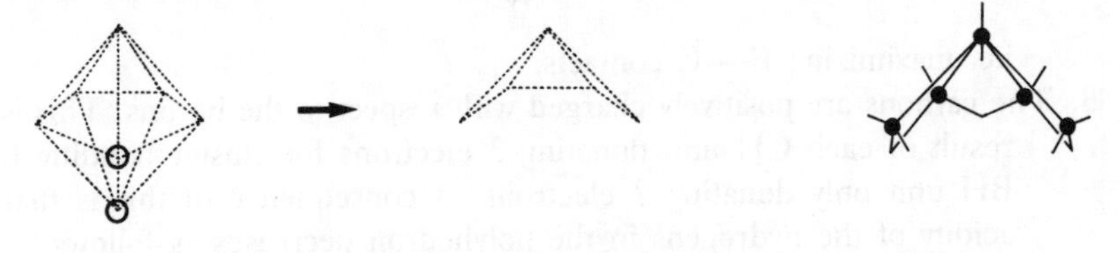

Carboranes (carbaboranes)

It is possible to incorporate CH units into borane frameworks, e.g.

$$B_{10}H_{14} + C_2H_2 \rightarrow C_2B_{10}H_{12} + H_2$$

CH and BH^- are isolobal (see page 227). These both have one sp orbital, two p orbitals and three electrons available for cluster bonding. Therefore, in principle, BH^- units, in a polyhedral borane, can be replaced by CH units without changing the structure. This generates the series of compounds $[CB_{n-1}H_n]^-$ and $[C_2B_{n-2}H_n]$, equivalent to $[B_nH_n]^{2-}$. The structure of carboranes can be worked out using Wade's rules, in exactly the same way as for boranes, e.g.

$C_2B_3H_5$ ($n = 2 + 3 = 5$)
C B

Total number of valence electrons $= 2 \times 4 + 3 \times 3 + 5 \times 1 = 22$e.

Of these, 10e are used in $3 \times$ B—H$_t$ and $2 \times$ C—H$_t$ bonds, leaving $22 - 10 = 12$e, i.e. 6 ($n+1$) pairs of electrons for skeletal bonding. ($n+1$) Electron pairs means that the structure is a *closo*-polyhedron with five vertices,

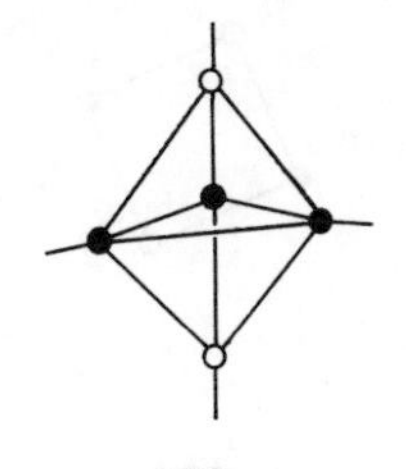

SOME GENERALIZATIONS ABOUT THE STRUCTURES OF CARBORANES

1. The carbons go for the positions of lowest connectivity.
2. When there is more than one C in a compound, the thermodynamically most stable product involves them being as far apart as possible,

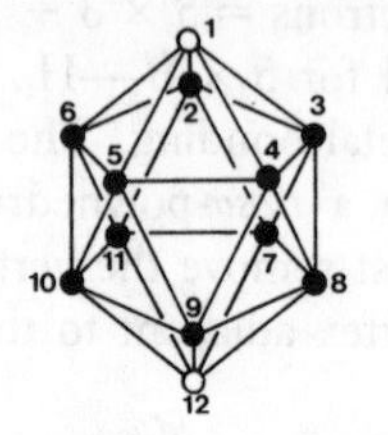

i.e. maximizing B—C contacts.

3. The carbons are positively charged with respect to the borons. This is the result of each CH unit donating 3 electrons for cluster bonding but a BH unit only donating 2 electrons. A consequence of this is that the acidity of the hydrogens in the polyhedron decreases as follows,

$$\mu_2\text{-H} > \text{C-H} > \text{B-H}$$

The bridging hydrogens are less strongly held and more electron density is pulled away from them by two main group atoms, making them more $\delta+$. C being more positive and more electronegative than B, pulls more electron density away from its H, making it more protonic.

REARRANGEMENT IN CARBORANES Consider the reaction sequence,

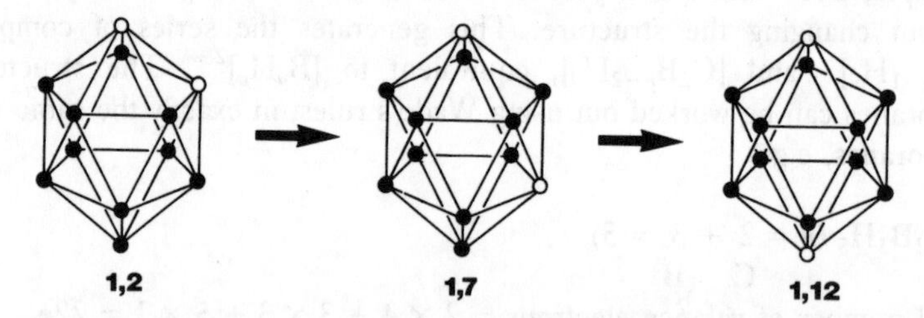

The Diamond–Square–Diamond (DSD) mechanism involves an icosahedron going through a cuboctahedral transition state to give another icosahedron.

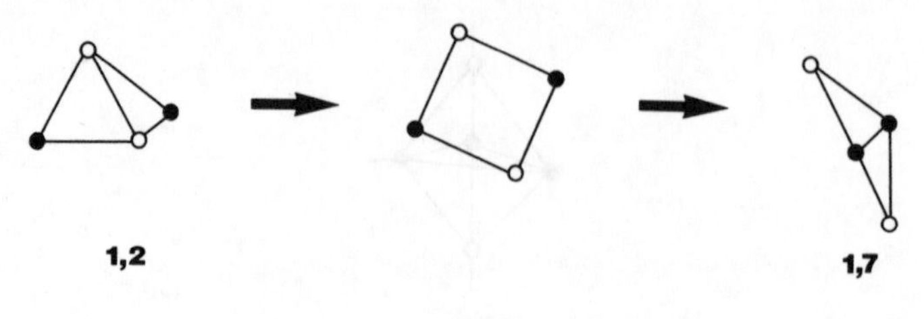

This mechanism can rationalize the rearrangement from 1,2 to 1,7 but it cannot explain the transformation from 1,7 to 1,12. Other, more complex, mechanisms, which can rationalize both rearrangements, have been suggested.

Metalloboranes

$Fe(CO)_3$ is isolobal with BH

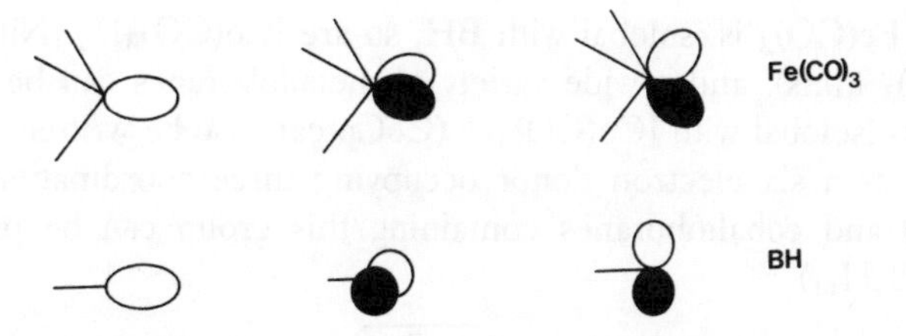

Thus, we can replace BH units in polyhedral boranes by $Fe(CO)_3$ without changing the structure.

	$C_2B_3H_5 + Fe(CO)_5 \rightarrow$	$C_2B_3H_5Fe(CO)_3 \rightarrow$	$C_2B_3H_5(Fe(CO)_3)_2$
isostructural-borane	$[B_5H_5]^{2-}$	$[B_6H_6]^{2-}$	$[B_7H_7]^{2-}$

Wade's rules can be used to predict the structures of metalloboranes. The electrons are counted in the usual way for the BH and CH units; for the $Fe(CO)_3$ unit there are 8e from the Fe and 2e donated by each CO, i.e. a total of 14e; we assume that 3 pairs of electrons (6e) are used in three Fe—C_{CO} bonds and a further three pairs occupy the t_{2g} set on Fe and play no part in cluster bonding. There are thus 14 − 12 = 2e available for skeletal bonding, i.e. the same number as a BH unit.

EXAMPLE

$C_2B_3H_5Fe(CO)_3$ ($n = \underset{C}{2} + \underset{Fe}{3} + \underset{B}{1} = 6$)

Total number of valence electrons $= \underset{2C}{2 \times 4} + \underset{3B}{3 \times 3} + \underset{5H}{5 \times 1} + \underset{Fe}{8} + \underset{3CO}{3 \times 2}$

$= 36e$

$2 \times C—H_t$	−4e
$3 \times B—H_t$	−6e
$3 \times Fe—C_{CO}$	−6e
$Fe_{t_{2g}}$	−6e
Total no. of electrons for cluster bonding	14e

There are, thus, 7 (n+1) electron pairs available for cluster bonding and the structure is a six vertex *closo*-polyhedron, i.e. an octahedron,

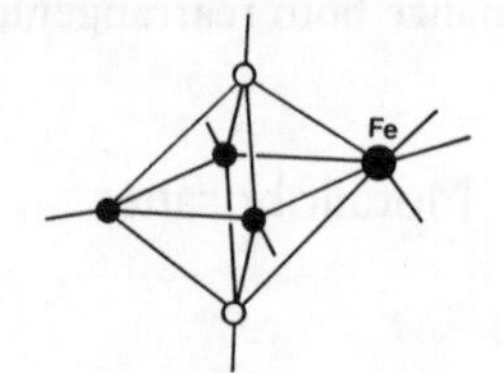

Just as $Fe(CO)_3$ is isolobal with BH, so are $[Co(CO)_3]^+$, $[Ni(CO)_3]^{2+}$ (all d^8 $M(CO)_3$ units), and a wide variety of metalloboranes can be made.

CoCp is isolobal with $[Co(CO)_3]^+$ (CoCp can also be written as Co^+ Cp^-, with Cp^- as a six electron donor occupying three coordination sites – see page 243) and cobaltoboranes containing this group can be prepared, e.g. $CpCo(C_2B_9H_{11})$.

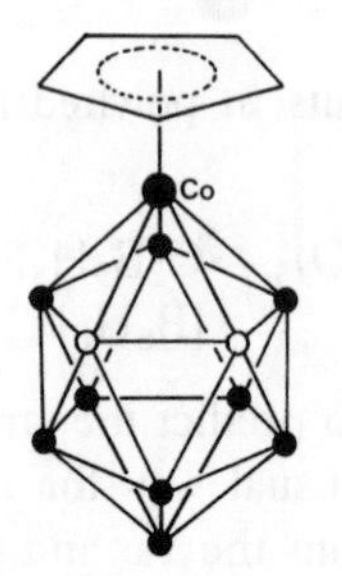

This is obviously a metalloborane analogue of $[B_{12}H_{12}]^{2-}$ but, perhaps less obviously, it is also an analogue of cobaltocene ($CoCp_2$), with the $[B_9C_2H_{11}]^{2-}$ unit isolobal with a Cp^- group (see page 243).

Why are Cp^- and $[B_9C_2H_{11}]^{2-}$ isolobal?
$[B_9C_2H_{11}]^{2-}$ presents a planar pentagonal face to the Co, with frontier orbitals of the approximate form,

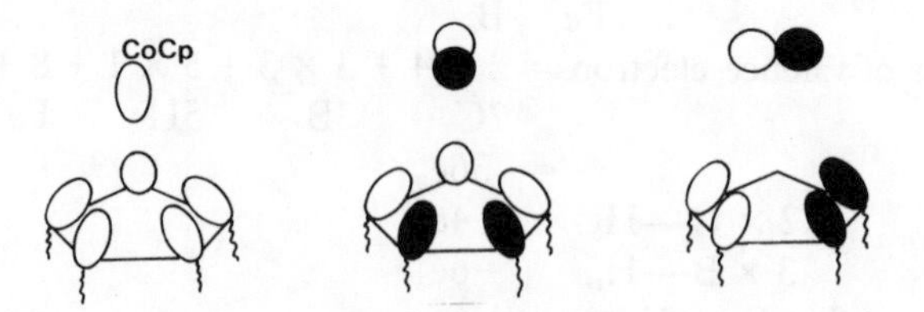

That is, these frontier orbitals are approximately the same as the three frontier orbitals of Cp^- and match the three frontier orbitals of the CoCp unit (Chapter 7). The borane fragment is a 6e donor because the three orbitals shown are involved in skeletal bonding and are filled with electrons.

Coordination of transition metals further towards the right-hand side of the Periodic Table to a borane fragment results in 'slip distortion' complexes, where the metals are coordinated to fewer atoms around the open face, e.g. $[Cu(C_2B_9H_{11})_2]^{2-}$.

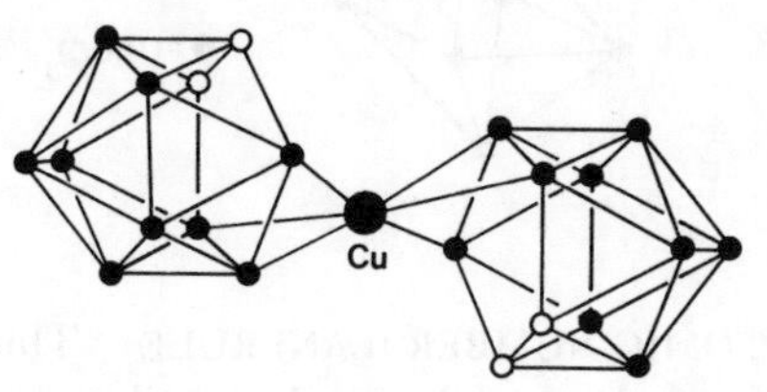

This can be understood in two ways:

1. Metals towards the right of the transition metal series require fewer electrons to make up their electron count to 18. The open face of the borane can donate electrons from one σ and two π orbitals, therefore just like Cp^- (see page 219) it has the ability to adopt η^1, η^3 and η^5 bonding modes, where it acts as a 2, 4 or 6e donor respectively.
2. Transition metals towards the right have more electrons available for cluster bonding, therefore there are too many electrons for a *closo*-structure to be adopted and the polyhedra open up to give *nido*- and *arachno*-structures based upon higher *closo*-polyhedra.

Transition metal clusters

BH is isolobal with CH^+, which is isolobal with $M(CO)_3$ (M = Fe, Ru, Os).

BH units in a borane can be totally replaced with $M(CO)_3$ units to give transiton metal polyhedral clusters. This could not be done with CH^+, to produce carbon–hydrogen clusters, because of the build-up of positive charge (*closo*-$[C_6H_6]^{4+}$ does not exist). Thus, just as $[B_6H_6]^{2-}$ can be made, so can $[Os_6(CO)_{18}]^{2-}$, which has the same structure.

Wade's rules may be used to work out the shapes of transition metal clusters:

$Os_6(CO)_{18}]^{2-}$ ($n = 6$)

Total number of valence electrons = $6 \times 8 + 18 \times 2 + 2 = 86$e
(Os, CO, charge)

Subtract 12e for each metal (i.e. assume 6e in t_{2g} and 6e used in 3M—L bonds. Note that this is always done, no matter how many ligands are attached to the metal.)

Therefore there are $86 - (6 \times 12) = 14$e, i.e. 7 ($n+1$) electron pairs left for skeletal bonding and the structure is a *closo*-polyhedron with six vertices, i.e. an octahedron,

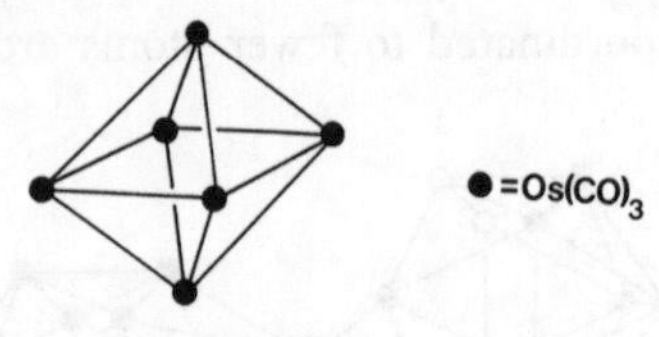

THE EFFECTIVE ATOMIC NUMBER (EAN) RULE This can be used to work out the number of metal–metal bonds in small ($n < 6$) clusters. Whereas Wade's rules employed a delocalized MO approach to bonding, the EAN rule is essentially a localized 2c–2e valence bond approach to cluster structures, relying on the 18e rule.

To work out the number of M—M bonds, m,

$$m = \frac{18 \times n - c}{2}$$

where n is the number of metal atoms and c is the total number of electrons in the valence shell of the metal.

EXAMPLES

$[Os_5(CO)_{15}]^{2-}$

$$m = \frac{18 \times 5 - 72}{2} = 9$$

Thus, the structure is a trigonal bipyramid with 9 M—M bonds,

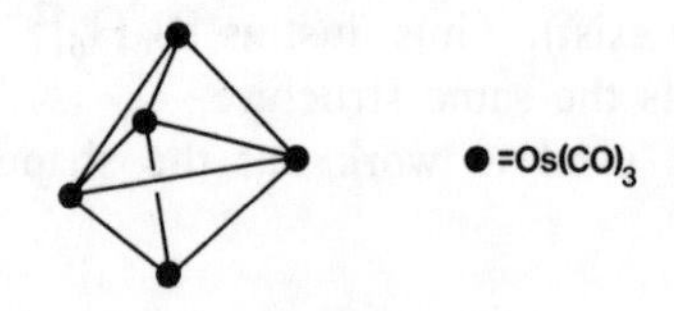

$Os_6(CO)_{18}$

$$m = \frac{6 \times 18 - 84}{2} = 12$$

Thus, a structure with 12 M—M bonds is predicted, and such a structure is actually obtained, but not the most obvious, octahedral, one. $Os_6(CO)_{18}$ is a bicapped tetrahedron,

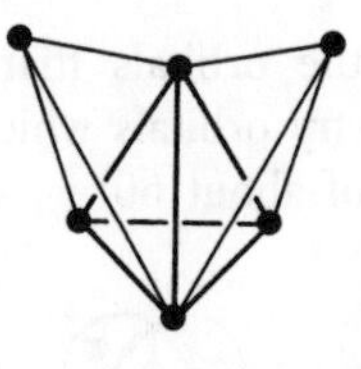

$[Os_6(CO)_{18}]^{2-}$

$$m = \frac{6 \times 18 - 86}{2} = 11$$

A structure with 11 metal–metal bonds is predicted but the actual structure is a regular octahedron with 12 M—M edges. For transition metal clusters containing six or more metal atoms the EAN rule loses its usefulness and the bonding is much better described using a delocalized approach where polyhedral edges do not necessarily correspond to bonds. Even for $n < 6$, perhaps it is better to regard the bonding as delocalized rather than consisting of 2c–2e bonds along the edges.

Wade's rules are more widely applicable than the EAN rule. Using that approach $Os_3(CO)_{12}$ is an *arachno*-trigonal bipyramid; $[Os_5(CO)_{15}]^{2-}$ is *closo*-trigonal bipyramid; $[Os_6(CO)_{18}]^{2-}$ has been discussed above; but what about $Os_6(CO)_{18}$?

$Os_6(CO)_{18}$ ($n = 6$)

No. of valence electrons $= 6 \times 8 + 18 \times 2 = 84$e

Subtract $6 \times 12 \Rightarrow 12$e, i.e. 6 electron pairs

Thus a structure based on a *closo*-polyhedron with five vertices would be predicted. A five vertex structure is predicted but there are six Os atoms and so the structure must be a *capped* trigonal bipyramid, which is the same as the bicapped tetrahedron, above.

THE CAPPING PRINCIPLE A capped polyhedron is a *closo*-structure with extra vertices sitting over one or more face of the cluster. *The electron count for a capped cluster is the same as that for the uncapped parent cluster.*

A way of thinking about this is to consider, e.g. $Os_6(CO)_{18}$, as made up of *closo*-$[Os_5(CO)_{15}]^{2-}$ and an $[Os(CO)_3]^{2+}$ unit. The number of electrons that an $[Os(CO)_3]^{2+}$ unit can use for cluster bonding is $8 + 3 \times 2 - 12 - 2$(charge) $= 0$ and, therefore a capping group contributes no electrons to cluster bonding and the structure can be regarded as
$[Os_5(CO)_{15}]^{2-} \rightarrow [Os(CO)_3]^{2+}$ – cf. a dative covalent bond.

TETRAHEDRAL OR SPHERICAL? Tetrahedral and capped transition metal clusters are quite common but there are no tetrahedral or capped boranes.

This is a consequence of the orbitals that are used in bonding. For an $M(CO)_3$ group the π symmetry orbitals which participate in bonding are dp hybrids, giving a σ-π angle of about 60°.

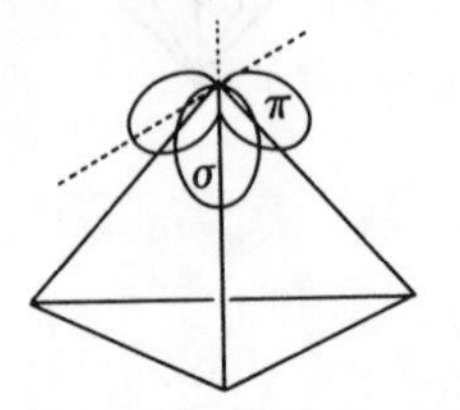

This angular arrangement favours a tetrahedral geometry, the angles in a tetrahedron (60°) being such as to give good interaction between the $M(CO)_3$ groups.

For BH, the π symmetry orbitals are purely p, and the angle between these and the sp, σ symmetry, orbital is 90°. These tangential π orbitals favour spherical polyhedra.

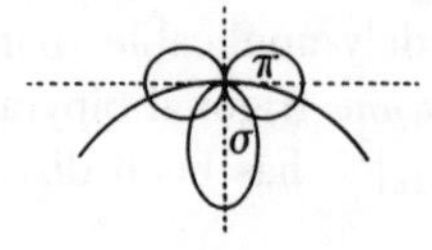

Zintl ions and main group clusters

These are clusters of bare (no ligands) main group atoms and can be cationic or anionic.

ANIONIC MAIN GROUP CLUSTERS

$$\text{Preparation: } x\text{Na} + n\text{M} \xrightarrow{(\text{liq. NH}_3)} [\text{Na}^+]_n[\text{M}_x]^{n-}$$

e.g. $[Sn_5]^{2-}$, $[Ge_9]^{2-}$, $[Ge_9]^{4-}$.

To explain the bonding, the Group 14 atom is considered as sp hybridized, with two electrons available for cluster bonding in an sp orbital pointing towards the centre of the cluster, two electrons in an sp orbital pointing away from the cluster (lone pair, not used in cluster bonding) and two vacant p orbitals perpendicular to these.

A bare Group 14 atom is thus isolobal with BH and $M(CO)_3$ (M = Fe, Ru, Os). Therefore $[Sn_5]^{2-}$ is isoelectronic (in terms of valence electrons available for cluster bonding) and isostructural with $[Os_5(CO)_{15}]^{2-}$ (see above).

CATIONIC MAIN GROUP CLUSTERS If E is a Group 15 atom then E^+ is isolobal with BH and $M(CO)_3$, having 2e available for cluster bonding (E^+ has 2e in a lone pair pointing out of the cluster, 2e in an sp hybrid orbital pointing into the cluster and two vacant p orbitals perpendicular to these – 2e and 3 orbitals available for cluster bonding).

Closo-clusters of Group 15 atoms have the general formula,

$[E_n]^{(n-2)+}$ e.g. $[Bi_5]^{3+}$, trigonal bipyramidal

and, *nido*-clusters,

$[E_n]^{(n-4)+}$ e.g. $[Bi_9]^{5+}$, tricapped trigonal prismatic

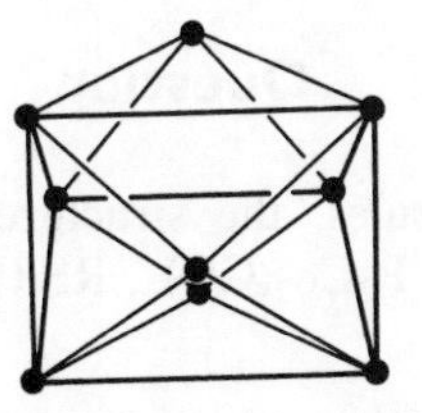

The elements on the right-hand side of the Periodic Table are quite electronegative and as the positive charge on a cluster is increased the orbitals become very contracted and less well able to participate in good E—E bonding at the extended E—E distances involved in these clusters. Thus, the highly charged *closo*-cluster, $[Bi_9]^{7+}$, is unknown.

Summary

1. B_2H_6 is electron deficient with respect to forming 2c–2e bonds and does not have a planar structure.
2. Reaction of B_2H_6 with NH_3 results in asymmetric cleavage of the borane whereas reaction with NR_3 results in symmetric cleavage.
3. *Closo*-deltahedral borane clusters require $(n+1)$ electron pairs for bonding. *Nido*-deltahedral borane clusters require $(n+2)$ electron pairs for bonding. *Arachno*-deltahedral borane clusters require $(n+3)$ electron pairs for bonding.
4. A *nido*-borane is derived from a *closo*-borane by removal of the highest connectivity vertex. An *arachno*-polyhedron can then be generated from a *nido* by removal of a vertex adjacent to the one first removed.
5. CH^+, BH and $Fe(CO)_3$ are all isolobal.
6. The number of metal–metal bonds in a small ($n < 6$) cluster is given by the equation:

$$\text{number of M—M bonds} = \frac{18 \times \text{number of metal atoms} - \text{total no. of valence electrons}}{2}$$

7. The total number of cluster electrons is the same for a capped and an uncapped cluster.
8. Boranes prefer spherical geometries whereas their transition metal analogues often prefer geometries based upon the tetrahedron.
9. B_nH_n, $(Fe(CO)_3)_n$, $(Ge_n)^{2-}$ and $(Bi_n)^{(n-2)+}$ are isostructural.

Question

Predict, using Wade's rules, the structures of the following cluster compounds: $Co_4(CO)_{12}$, $P_2Co_2(CO)_6$, $RSiCo_3(CO)_9$, P_4.

Answer

$(CO)_3Co$, P and RSi are isolobal.

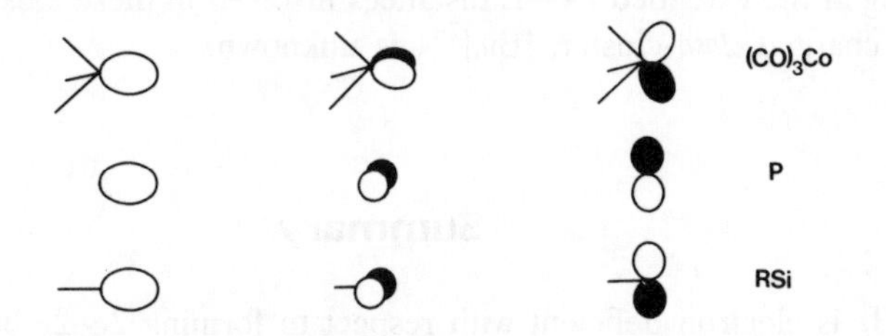

For the $(CO)_3Co$ unit the total number of electrons is 15, of which six are used for Co—CO bonding and six occupy the t_{2g} set of orbitals; this leaves it with three electrons for cluster bonding, which occupy the frontier orbitals as shown.

P has a total of five electrons in its valence shell; two electrons constitute a lone pair (sp hybrid) pointing away from the cluster, and so are unavailable for skeletal bonding; the remaining three electrons are available for cluster bonding.

An SiR unit has five electrons; two are used in an SiR σ bond, leaving three for skeletal bonding.

Therefore in each cluster compound there are $4 \times 3 = 12$ electrons (6 pairs) available for skeletal bonding. All the clusters have four vertices ($n = 4$) and there are $(n+2)$ pairs of electrons for cluster bonding – the structures are all *nido*, based on a *closo* five vertex deltahedron (a trigonal bipyramid) – all the clusters are tetrahedral.

● = $Co(CO)_3$, P, RSi

CHAPTER 9

Electrical Conduction in the Solid State

Band Theory

Electrical conduction in solids can be explained can be explained using **band theory**, which can be considered as an extended form of MO theory. A band is a very closely spaced group of MOs delocalized over the whole solid. The orbitals are so close in energy that there is effectively no energy gap between them.

Consider the build-up of a linear chain of atoms; let's start with two atoms, with one AO on each – we get a bonding and an antibonding MO; three atoms give a bonding, a non-bonding and an antibonding MO. As the chain is built up so that it is essentially infinite, a whole series of MOs is obtained, ranging in character from totally bonding to totally antibonding.

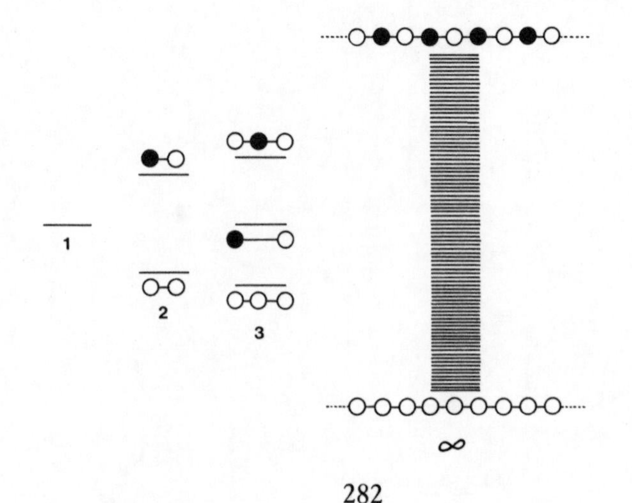

A band is made up of n MOs (n is the number of atoms) and is therefore filled by $2n$ electrons.

For the 3-D case we could have 1s orbitals overlapping to give a 1s band, 2s orbitals giving a 2s band and 2p orbitals giving three 2p bands, etc.; these are all separated by a **band gap**. The highest filled band is called the **valence band** and the lowest unoccupied, the **conduction band**.

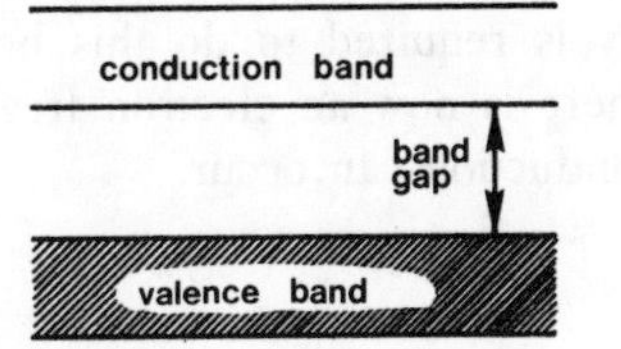

The width of the bands and the size of the band gap depend on the degree of interaction between the orbitals; the weaker the interaction, the narrower the band and the smaller the band gap.

Consider the 1-D case, the lowest, and highest, energy MOs in the band are:

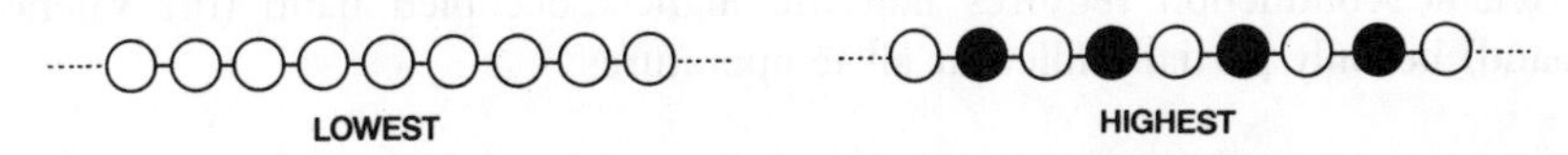

If the interaction between adjacent orbitals is small there is going to be only a small energy difference between the highest and the lowest MOs, i.e. a narrow band – cf. two AOs giving a bonding and an antibonding MO.

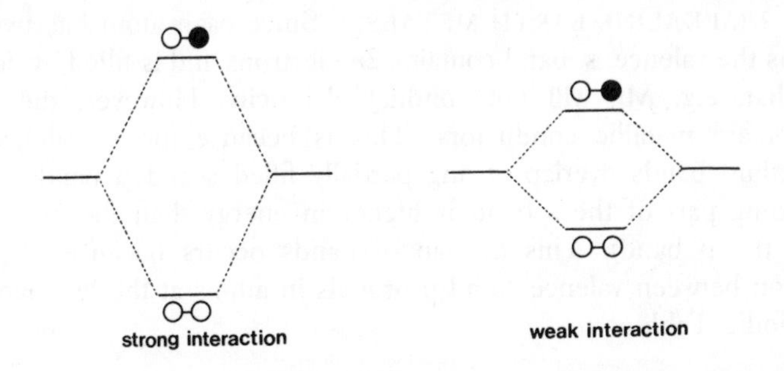

In the 3-D case, if we have three bands derived from three p orbitals then the lowest band is going to have, on average, the greatest amount of bonding character and the next highest band, being higher in energy, is going to contain more antibonding interactions. Thus, if the interaction between orbitals is strong the energy gap between successive bands is large and there is a large band gap (cf. two atomic orbitals overlapping to give two molecular orbitals – the stronger the interaction the larger the energy difference between the bonding and antibonding orbitals).

Electrical conductivity requires:

- an infinite, non-molecular, delocalized structure; and
- *a partially filled band.*

When a potential difference is applied to a solid an electron in a partially filled band can jump up to a vacant molecular orbital in the same band; very little (negligible) energy is required to do this because of the continuous nature of the band. There is now an electron free to move throughout the delocalized MO and conduction can occur.

Metals, Semiconductors and Insulators

Metals

Metallic conduction requires that the highest occupied band (the valence band) be only partially filled at all temperatures.

GROUP 1 (ALKALI METALS) Each atom has one s electron, therefore the s (valence) band contains *n* electrons, i.e. it is half-filled and electrical conduction can occur.

GROUP 2 (ALKALINE EARTH METALS) Since each atom has two valence electrons the valence, s, band contains $2n$ electrons and is filled, which should mean that, e.g. Mg will not conduct electricity. However, the Group 2 elements are metallic conductors. This is because the s (valence) and p (conduction) bands overlap, giving partially filled s and p bands (the more antibonding part of the s band is higher in energy than the more bonding part of the p band). This overlap of bands occurs because of the small separation between valence s and p orbitals in atoms at the left-hand side of the Periodic Table.

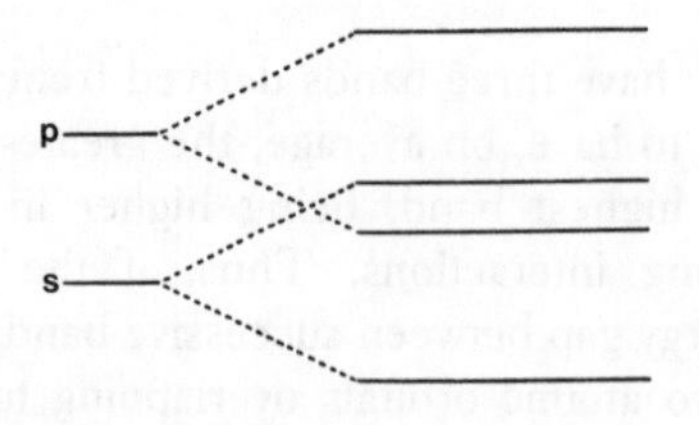

(It also occurs for Group 1 but there it is not essential for conduction to occur as the s band is already partially filled.)

GROUP 13 Aluminium and the elements below it in the group have partially filled p bands, i.e. they exhibit metallic conduction.

Metallic conduction decreases with increasing temperature. If we think of electrons travelling through the solid, their progress is inhibited by scattering due to vibrating nuclei; as the temperature is raised there is more vibration, therefore more scattering of the electrons and lower conductivity.

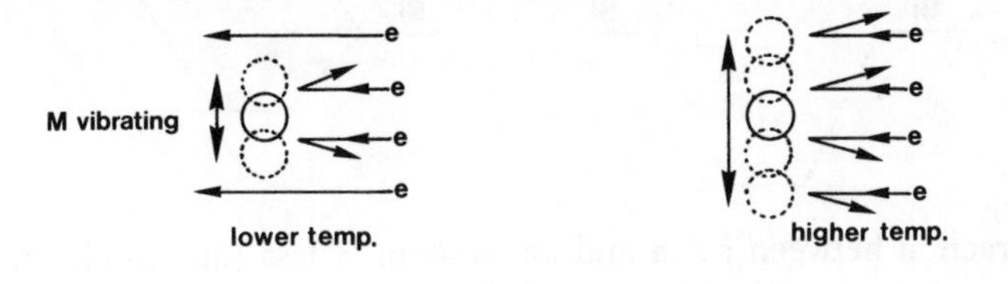

Semiconductors

In the ground state, semiconductors have a filled valence band and an empty conduction band; thus, they would be expected to be insulators. However, the gap between valence and conduction bands is relatively small and, at room temperature, thermal energies are large enough to promote electrons to the conduction band. This generates two partially filled bands and electronic conduction is then possible.

For example, $Si_{(s)}$ is a semiconductor. This is because, with each Si atom having four valence electrons, there is a total of $4n$ electrons available to just fill the 3s and lowest 3p band, but the energy gap to the next (empty) 3p band is relatively small and electrons can be promoted by light and thermal energies at room temperature to produce partially filled bands.

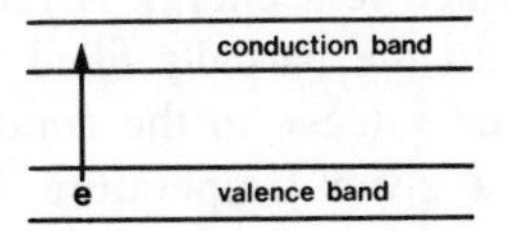

The electrical conductivity of semiconductors increases as the temperature is raised (opposite to metallic conduction) since more electrons are promoted to the conduction band, i.e. there are more electrons available for conduction.

DOPING OF SEMICONDUCTORS The conductivity of a Group 14 semiconductor, such as Si, can be increased by incorporating a small percentage of either Group 13 or Group 15 atoms into the structure; this is known as **doping**. The size of the doping atom should be as close as possible to that of the Group 14 atom to cause the minimum amount of disruption in the lattice structure.

The interaction between Ga and Si atoms is going to be different from that between Si atoms, which results in the production of orbitals in the band gap, i.e. if we consider a localized interaction between just atomic orbitals:

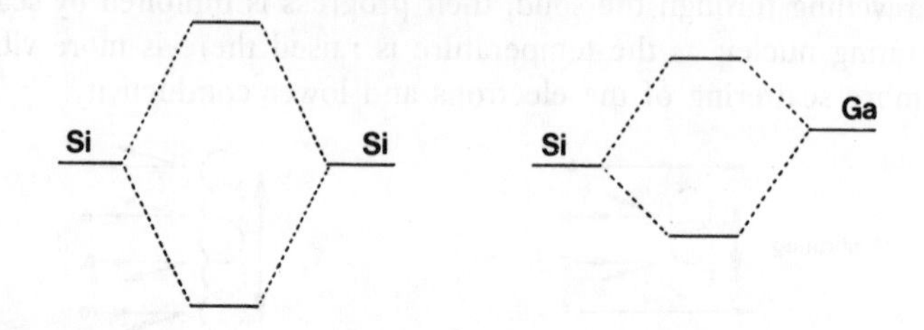

Interaction between a Ga and an Si atom is less (due to electronegativity difference between Si and Ga, and Ga being a fourth period element, i.e. $4p_{Ga}$–$3p_{Si}$ overlap is worse than $3p_{Si}$–$3p_{Si}$ overlap) than that between two Si atoms and, consequently, the bonding molecular orbital is not lowered to such an extent. In the 3-D case, doping the structure with Ga generates a series of partially filled discrete orbitals just above the valence band (they are discrete because there is only a very small amount of Ga in the structure and hence there is no interaction between Ga atoms, i.e. 'neighbouring' Ga atoms are probably, on average, separated by several 100 Si atoms).

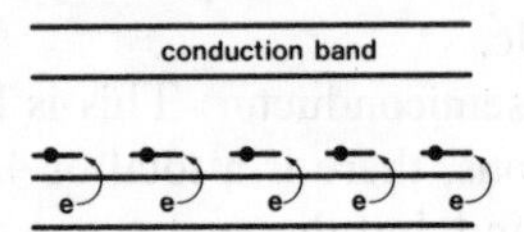

Excitation of electrons from the valence band into these partially filled orbitals results in the valence band being no longer completely filled, so that conduction is possible. Since less energy is required to promote electrons from the valence band into the partially filled orbitals, than is required to promote electrons from the valence to the conduction band in pure Si, the conduction is higher at a given temperature for SiGa than for pure Si. Conduction is higher at lower temperatures but still falls to zero at absolute zero because, at this temperature, there is no energy to excite electrons at all.

SiGa is a 'p-type' semiconductor since for each Ga a positive 'hole' is generated in the valence band due to the excitation of an electron to the discrete orbitals.

In the diagram, the movement of an electron to the left, to fill the positive hole, is equivalent to the hole moving to the right, i.e. we can regard conduction as the movement of positive holes through the solid, in the opposite direction to electrons. When this is the major mechanism for conduction we have a *p*- (positive) type conductor. 'Normal' conduction, simply regarded as the movement of electrons, is termed *n*- (negative) type.

Doping with Group 15 atoms

A Group 15 atom, such as As, has five valence electrons, i.e. one more than Si. There is an energy mismatch between the valence orbitals of As and Si, therefore interaction between them is smaller than that between two Si atoms (as above).

In the 3-D case partially filled discrete (only relatively few As atoms in the lattice) orbitals are produced just below the conduction band; relatively little energy is required to promote electrons from these orbitals into the conduction band and thereby generate a partially filled conduction band. The conductivity is again higher than for pure Si since, at a given temperature, less energy is required for promotion of electrons into the conduction band from these discrete orbitals than from the valence band. This type of conduction is termed n-type because it involves the movement of electrons through the solid.

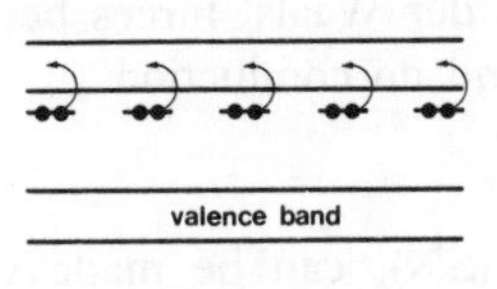

Insulators

Insulators have a filled valence band and a large energy gap to the empty conduction band. The gap is large enough so that at normal temperatures electrons will not be promoted to the conduction band, hence no conduction is observed.

Diamond (4*n* electrons) is an insulator since the 2s band and the lowest of the 2p bands are filled and then there is a large energy gap to the next 2p band.

Why is there a difference between diamond and silicon?

This is due to better interaction between the 2p orbitals in diamond (smaller inter-atomic distance and orbitals more concentrated) than between the 3p orbitals in Si, so that the band gap between the p bands is larger for diamond than for Si. Therefore a lot less energy is required for promotion of electrons into the conduction band for Si. The result of this is that at room temperature silicon is a semiconductor whereas diamond is an insulator.

I-D Conductors

$(SN)_x$ This is a near planar chain polymer with all S—N bond lengths approximately equal,

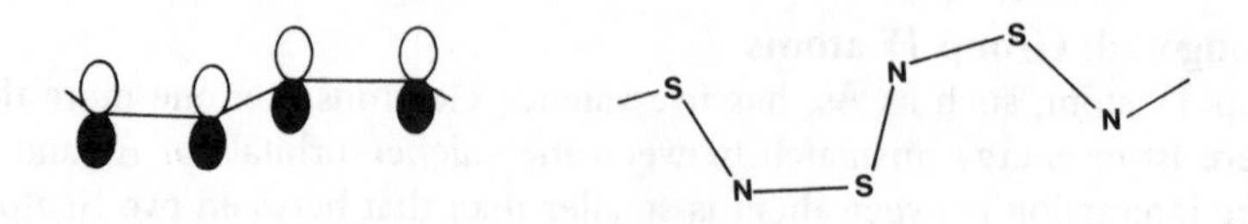

All atoms are assumed to be sp^2 hybridized, so that a delocalized band is formed by π overlap of the p orbitals, not used in σ bonding, along the chain; this band is capable of holding $2n$ electrons. Each S and each N uses 2e in two S—N bonds and 2e in a lone pair (so accounting for the three sp^2 orbitals), leaving S 2e and N 1e (an average of 1.5e per atom), for π bonding. There are, thus, $1.5n$ electrons over the whole chain and a partially filled band is obtained, making $(SN)_x$ a 1-D metallic conductor – conduction increases as the temperature is lowered and it even superconducts at very low temperatures. $(SN)_x$ is an insulator perpendicular to the chain direction because there are only van der Waals' forces between the chains, therefore there is no delocalization and no conduction.

Bromination of $(SN)_x$

Brominated derivatives of $(SN)_x$ can be made with stoichiometries in the range $(SNBr_{0.25})_x$ to $(SNBr_{1.5})_x$. These exhibit higher conductivities than the parent polymer. A possible explanation is that in $(SN)_x$, the delocalization of electron density along the chain is not complete because of the electronegativity difference between S and N. This makes the π cloud 'lumpy' (electron density tends to be more localized on the N), which reduces the conduction along the chain (ease of flow of electrons through the solid). In the bromine doped compound, the bromine atoms are believed to sit between the polymer chains. The electronegative bromines interact more strongly with the sulphurs (the greater electronegativity difference between S and Br than between N and Br leads to a stronger interaction – note that an N—Br bond is very weak) and electron density is partially withdrawn from the S, making it more electronegative. There is now greater similarity between the electronegativities of S and N so that the π cloud is less lumpy, i.e. the delocalization is more complete and the conductivity is higher.

PLATINUM CHAIN COMPOUNDS $K_2[Pt(CN)_4].3H_2O$ consists of a chain of $[Pt(CN)_4]^{2-}$ square planar units,

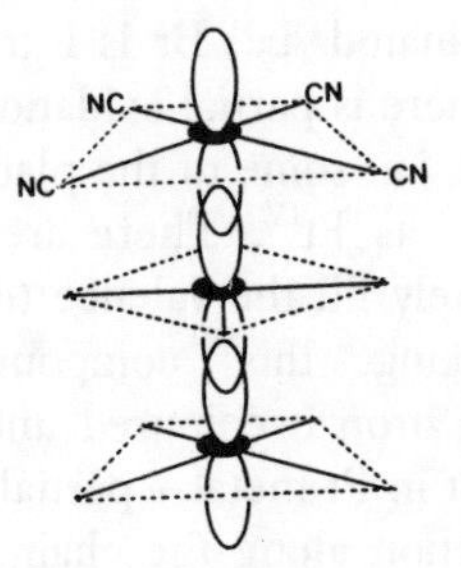

There is interaction between the d_{z^2} orbitals on adjacent platinums in the stack to give a band delocalized along the chain. The splitting diagram for Pt^{2+}, d^8 square planar, is

$d_{x^2-y^2}$

d_{xy}
d_{z^2}
d_{xz}, d_{yz}

so that the d_{z^2} is filled. Although the d_{xy} orbital is the highest occupied orbital there will be very little interaction between d_{xy} orbitals on adjacent Pt units along the chain (a face-on, δ symmetry, overlap is poor). The d_{z^2} orbitals on adjacent units point directly at each other producing a stronger interaction, i.e. a wider band, which overlaps the very narrow d_{xy} band and becomes the valence band.

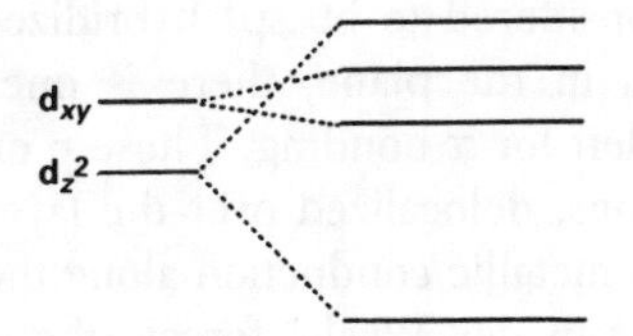

Another way of thinking about this is that stacking the Pt units in a chain produces a pseudo-octahedral geometry about the Pt and a splitting of orbitals such that the d_{z^2} is higher in energy than the d_{xy} orbital, leading to the d_{z^2} band being the valence band (see splitting diagrams, page 148).

The d_{z^2} band contains $2n$ electrons and is filled. Thus, $K_2[Pt(CN)_4]$ is a white non-conductor of electricity with an average Pt—Pt distance, within the chain, too long to suggest significant M—M interaction. {Consider the 1-D chain discussed at the beginning of this chapter: when a 1-D band is completely filled there is no net bonding interaction (as many antibonding as bonding orbitals are filled), so explaining the long M—M distance in the Pt chain.}

If the complex is brominated, i.e. Br is introduced into the lattice, as in $K_2[Pt(CN)_4]Br_{0.3}.3H_2O$, there is partial oxidation of Pt^{II} to Pt^{IV} (accompanied by reduction of Br to Br^-), i.e. some of the platinums along the chain can be regarded as Pt^{II} and some as Pt^{IV}. There are no longer enough electrons along the chain to completely fill the valence (d_{z^2}) band and a partially filled band is obtained, making this compound a metallic conductor. $K_2[Pt(CN)_4]Br_{0.3}.3H_2O$ is bronze coloured and has Pt—Pt distance along the chains approaching that in Pt metal – partial oxidation means that there is now a net bonding interaction along the chain, due to removal of electrons from the higher energy, antibonding orbitals in the band (see the previous diagram).

There are also other compounds, e.g. $K_{1.75}[Pt(CN)_4]$, which similarly exhibit metallic conduction, again due to partial oxidation of Pt^{II} to Pt^{IV}, in order to maintain electroneutrality.

2-D conductors

GRAPHITE (see also page 64)

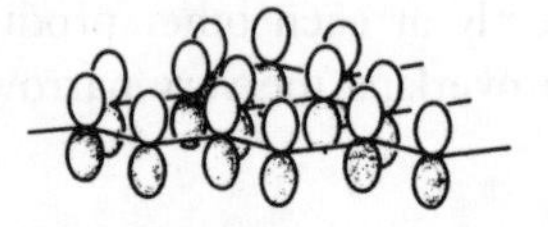

The C atoms can be considered to be sp^2 hybridized and after each has formed three 2c–2e bonds in the plane, there is one half-filled p orbital perpendicular to the plane left for π bonding. These p orbitals overlap to give a band, containing *n* electrons, delocalized over the layers. Since the band is half-filled, graphite exhibits metallic conduction along the layers. Between the layers there are only weak van der Waals' forces (the distance between the layers is much greater than the C—C distance within the planes), therefore there is no delocalization and graphite is an insulator perpendicular to the layers.

Intercalation Compounds of Graphite

The relatively large distance between the layers means that graphite can undergo reactions where atoms/ions are inserted between the layers. There are two extreme types of intercalation compound:

1. *The delocalized system is completely lost, e.g.* $(CF)_n$: this compound is completely saturated – all orbitals and electrons on C are used in forming 2c–2e σ bonds. Because there is no π bonding there is no

reason for the structure to be planar and the layers are buckled – steric factors pushing towards tetrahedral angles at the carbons.

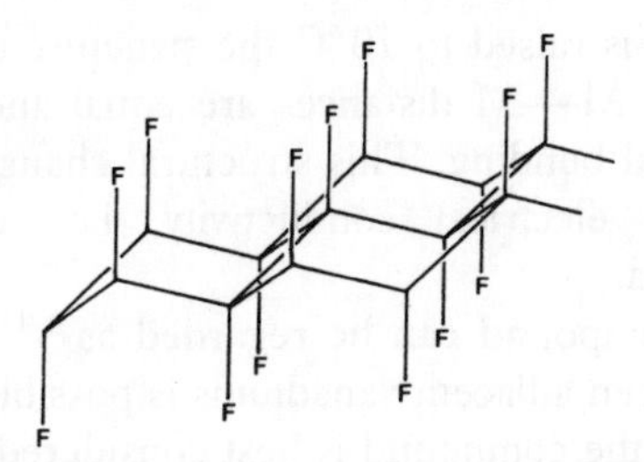

These compounds are colourless insulators – the extra electron contributed by F means that there is an even number of electrons in the valence shell of the C, therefore all bands are filled. This could also be explained by treating the bonding as just involving four localized σ bonds at each C – no delocalization means no conduction.

$(CF)_n$ is the extreme case of this type of compound and there are other less fluorinated derivatives of graphite, such as $(C_4F)_n$, where delocalization is only partially lost (only one in four carbons has four σ bonds, so that a delocalized description of the bonding is still valid).

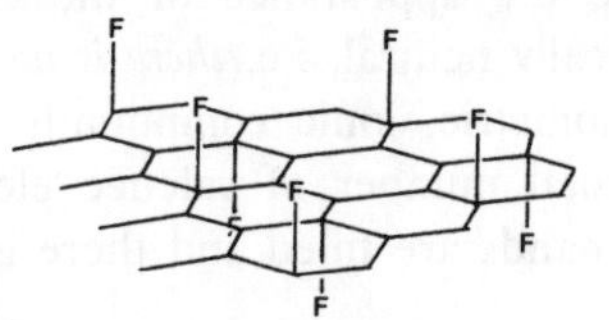

2. *The delocalized system is retained, e.g. C_8K*: the layers are planar but further apart than in graphite so that the heteroatom can be accommodated between them.

The compounds are coloured and conductors, with conductivity higher than in graphite. This is because, e.g. C_8K can be regarded as $[C_8]^-K^+$, that is with more electrons in the p band available for conduction, than in pure graphite.

3-D conductors

LnI_2, where Ln = La, Ce, Pr These are dark coloured metallic conductors. They do not contain the Ln^{2+} ion but, rather, are better described as $Ln^{3+} + 2I^- + e^-$, where e^- is an electron delocalized over the whole structure. There is one e^- per Ln, therefore a half-filled band is generated and metallic conduction is exhibited. Some LnH_2 compounds also exhibit metallic conduction for the same reason.

VO_2 At room temperature this has a distorted rutile structure, with alternating short and long M—M distances, indicating the presence of M—M bonds.

As the temperature is raised to 70 °C the structure changes to undistorted rutile in which all the M—M distances are equal and long, indicating that there is no metal–metal bonding. This structural change is accompanied by a dramatic increase in electrical conductivity, i.e. above 70 °C, metallic conduction is exhibited.

Below 70 °C the compound can be regarded as V^{IV} (d^1) + $2O^{2-}$, so that M—M bonding between adjacent vanadiums is possible by using the single d electron. Above 70 °C the compound is best considered as V^V (d^0) + $2O^{2-}$ + e^-, where e^- is a free electron delocalized over the whole of the solid. With no d electrons, V^V cannot participate in M—M bonding, therefore all M—M distances are equal and long.

Non-stoichiometry

Non-stoichiometric substances are ionic solids which have compositions differing slightly from what would be expected from the valencies of the ions, e.g. $Na_{0.98}Cl$. Despite the appearance of the chemical formula all these compounds are electrically neutral, i.e. *there is no overall charge on the solid.*

The parent, stoichiometric, ionic compounds are insulators in the solid state, e.g. KI – the total number of valence electrons is 16, i.e. an even number, such that all bands are filled and there is a large energy gap to the next band.

KI If KI is treated with K vapour (note that this is neutral and therefore the non-stoichiometric compound produced must also be electrically neutral), the following is produced,

K^+ I^- K^+ I^-
I^- K^+ e^- K^+
K^+ I^- K^+ I^-

i.e. the K sits in the lattice as $K^+ + e^-$, with the K^+ occupying a normal cationic site and e^- occupying a site which, in stoichiometric KI, would be taken by an I^- ion. Thus, we have generated $K_{1+x}I$ (x is small) containing anionic vacancies (I^- missing).

The electrons in the anionic sites are effectively surrounded by a sphere of positive charge (cations).

The surrounding shell of positive charge acts like a point positive charge at the centre of the sphere (electrostatic theory) and thus there is a situation analogous to that found in an H atom (positive charge at the centre plus $1e^-$).

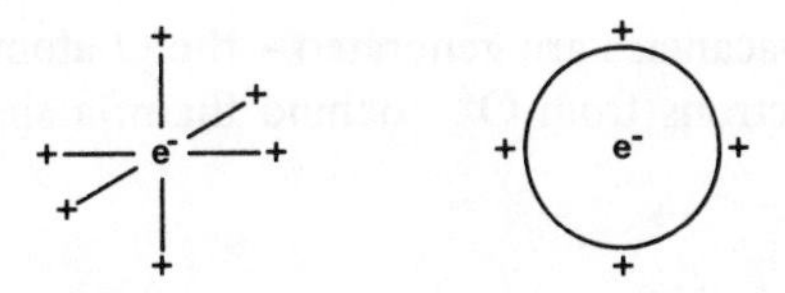

The electron here, like that in an H atom, can exist in only certain, allowed, energy levels; the energy of these levels comes in the band gap. Promotion of the electrons from these levels into the conduction band is easy, compared to promotion from the valence to the conduction band in KI. This generates a partially filled conduction band. Thus, compounds such as these are *n*-type semiconductors (cf SiAs, above).

The hydrogen-like electron energy levels in the band gap are the reason why many non-stoichiometric compounds are coloured, i.e. electronic transitions can occur at levels corresponding to absorption/emission of visible light.

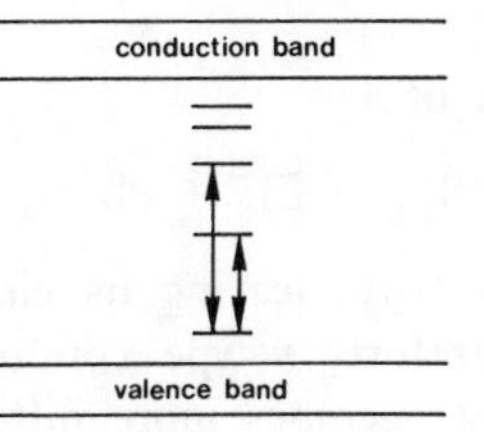

In the presence of I_2 vapour KI can pick up I to give $K_{1-x}I$, containing cationic vacancies. The I diffuses in, picks up an electron from the valence band, generating I^-, which takes up an anionic position in the KI lattice. Thus there are spaces left in the lattice where K^+ should be, i.e. cationic vacancies.

$K^+ \; I^- \; K^+ \; I^- \; K^+ \; I^- \; K^+ \; I^-$
$I^- \; K^+ \; I^- \; K^+ \; I^- \; \square^+ \; I^- \; K^+$

Since electrons are picked up from the valence band (I can be considered as supplying acceptor orbitals in the band gap), there are holes in the valence band, therefore we have a p-type conductor (cf. SiGa).

ZINC OXIDE If white ZnO, an insulator, is heated it becomes a yellow, *n*-type, semiconductor. This is because, as the temperature is raised, the equilibrium,

$$ZnO_{(s)} \rightleftharpoons O_{2(g)} + Zn_{1+\delta}O_{(s)}$$

moves to the right, i.e. we get,

$$Zn^{2+}{}_{latt.} + O^{2-}{}_{latt.} \rightarrow 1/2O_{2(g)} + Zn^{2+}{}_{latt.} + 2e^-{}_{latt.}$$

Thus anionic vacancies are generated – the O atoms depart (as O_2) leaving the other two electrons from O^{2-} behind them; a similar situation to that of $K_{1+x}I$, above.

LEAD II SULPHIDE (PbS)

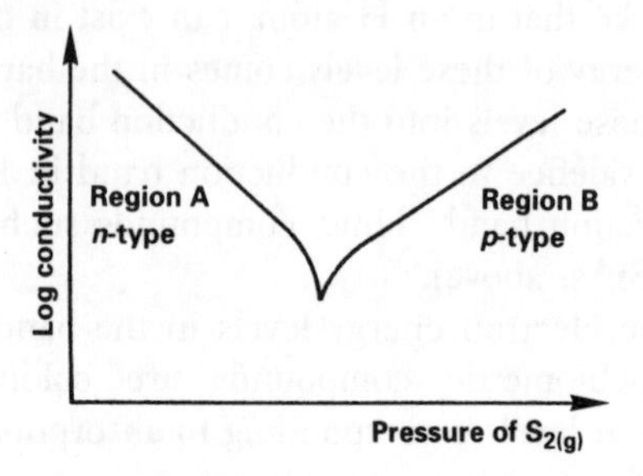

Region A: low pressure of S_2

$$Pb^{2+}{}_{latt.} + S^{2-}{}_{latt.} \rightarrow 1/2S_{2(g)} + \square^{-} + Pb^{2+}{}_{latt.} + 2e^{-}{}_{latt.}$$

S^{2-} departs the lattice, as $S_{2(g)}$, leaving its electrons behind, resulting in anionic vacancies and, therefore, *n*-type conduction. This decreases with increasing S_2 pressure as it becomes more difficult for the S to leave (Le Chatelier's principle).

Region B: high pressure of S_2

$$Pb^{2+}{}_{latt.} + 1/2S_{2(g)} \rightarrow Pb^{4+}{}_{latt.} + \square^{+} + S^{2-}{}_{latt.}$$

At higher S_2 pressure S atoms are absorbed. These take electrons from the valence band of PbS (equivalent to oxidizing Pb^{2+} to Pb^{4+}) to generate S^{2-} and cationic vacancies. With holes in the valence band, *p*-type conduction is exhibited. This increases with S_2 pressure as more S is absorbed into the lattice and more cationic vacancies are generated.

Summary

1. Electrical conduction in solids can be explained using band theory.
2. The band gap and band width depend on the degree of interaction between valence atomic orbitals on adjacent atoms in the solid; the larger the interaction, the larger the band width and band gap.
3. Electrical conduction requires a partially filled band.

4. Metals conduct electricity at all temperatures and conduction increases as the temperature is lowered. Semiconductors are insulators at absolute zero but their conductivity increases as the temperature is raised. Insulators do not conduct electricity under 'normal' conditions.
5. Doping $Si_{(s)}$ with either Ga or As increases its conductivity at a given temperature.
6. $(SN)_x$ is a 1-D metallic conductor; doping with Br_2 increases its conductivity.
7. $K_2Pt(CN)_4$ is an insulator but bromination produces a 1-D metallic conductor.
8. Graphite is a 2-D conductor, but $(CF)_n$ is an insulator. The conductivity of C_8K is higher than that of graphite.
9. $K_{1+x}I$ is an *n*-type semiconductor but $K_{1-x}I$ is a *p*-type semiconductor.

Questions

1 Silicon and gallium arsenide (GaAs) are insulators at absolute zero but as the temperature is raised GaAs becomes a conductor at a lower temperature than does Si. Explain these observations.

Answer

Si and GaAs are isoelectronic (considering valence electrons). In each case there are $4n$ electrons to occupy valence s and p bands. The s and lowest p band (valence band) are completely filled (each band accommodates a maximum of $2n$ electrons). Therefore, conduction must occur as the result of promotion of electrons from the filled valence p band to a vacant conduction band (next highest p band). At absolute zero all the electrons are in the valence p band and there is not enough energy to promote the electrons into the conduction p band, therefore both Si and GaAs are insulators. As the temperature is raised electrons can gain enough thermal energy to jump up to the conduction band, so that partially filled conduction and valence bands are generated, resulting in electrical conduction.

The temperature at which conduction occurs will depend on the magnitude of the band gap; smaller band gap, less energy is required to promote an electron, and conduction occurs at a lower temperature. The size of the band gap depends on the extent of interaction between adjacent atoms:

(a) The interaction between orbitals decreases as the electronegativity difference between atoms, and therefore the energy difference between interacting orbitals, increases. Interaction is strongest for orbitals of the same energy.

(b) The degree of interaction decreases down the Periodic Table as valence orbitals become more diffuse and interatomic distance increases; Ga and As are both in the fourth period and Si is in the third period.

Thus, the band gap is larger in Si than in GaAs and conduction begins at a higher temperature for Si.

2 Diamond is an insulator, Si is a semiconductor, Ge is a better semiconductor, the stable allotrope of Sn, at room temperature, is a metallic conductor, lead is a metallic conductor. Discuss.

Answer

If the bonding is considered in terms of band theory, then the valence orbitals in the atoms will give rise, in the 3-D solid, to one s and three p bands.

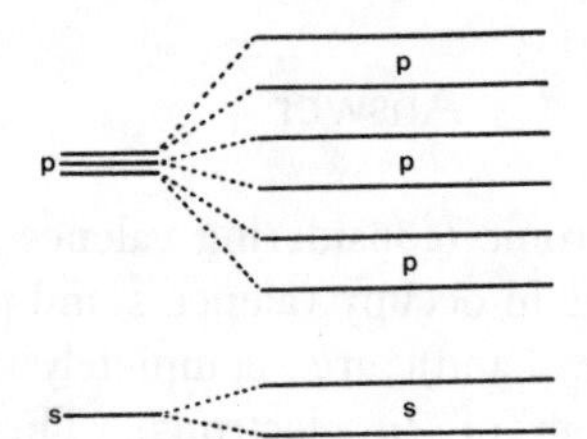

Each atom has four valence shell electrons, i.e. 4*n* valence electrons for the whole solid. Each band can accommodate 2*n* electrons, and so the s and lowest p band are completely filled. This lowest p band is the valence band and the next lowest p band is the conduction band.

The width of the bands and the size of the band gap between the valence and conduction bands is proportional to the degree of interaction between atomic orbitals on adjacent atoms; the stronger the interaction, the wider the band and the larger the band gap. The degree of interaction follows the order: 2p–2p > 3p–3p > 4p–4p > 5p–5p > 6p–6p (more diffuse orbitals and larger distance between interacting atoms down the group, cf. element–element bond energies decrease down a group, see pages 66/67), which results in the gap between conduction and valence bands being largest for diamond and smallest for lead. The band gap for Sn and Pb is small enough so that the valence and conduction bands overlap to produce two bands which are partially filled at all temperatures; this gives rise to metallic conduction.

The band gaps for Si and Ge are small enough so that electrons in the valence band can gain sufficient thermal energy, at room temperature, to be promoted to the conduction band producing two partially filled bands. Since promotion of electrons to produce partially filled bands requires energy, there is no promotion and, hence, no conduction at absolute zero, i.e. these substances are semiconductors. There is less interaction between the 4p orbitals on adjacent atoms in Ge than between the 3p orbitals in Si, therefore the band gap is smaller in Ge. This means that less energy is required to promote electrons to the conduction band and Ge will conduct at a lower temperature. At a given temperature there will be more electrons in the conduction band of Ge than in that of Si, therefore the conductivity of Ge is higher.

In diamond, the interaction between the 2p orbitals of adjacent carbon atoms is large, and the size of the band gap is large enough so that there is no promotion of electrons at normal temperatures, i.e. diamond is an insulator.

CHAPTER 10

Spectroscopy

Nuclear Magnetic Resonance (n.m.r.) Spectroscopy

Just as there is an electron spin angular momentum quantum number, s, there is likewise a nuclear spin angular momentum quantum number, I. I can take values 0, 1/2, 1, 3/2, . . ., depending on the nucleus. In general, for odd mass number, I is half integral and for even mass number, I is 0 or integral, e.g. ^{1}H, $I = 1/2$, ^{2}H, $I = 1$, ^{16}O, $I = 0$, ^{7}Li, $I = 3/2$.

A rotating charged particle generates a magnetic field so that a nucleus (positively charged) acts just like a tiny bar magnet. When an external magnetic field is applied a nucleus with nuclear spin I will be able to take up $(2I+1)$ orientations relative to this field, e.g. if $I = 1/2$ then the nucleus can be aligned either with, or against, the magnetic field, i.e. two orientations, which are given quantum numbers $m_I = \pm 1/2$.

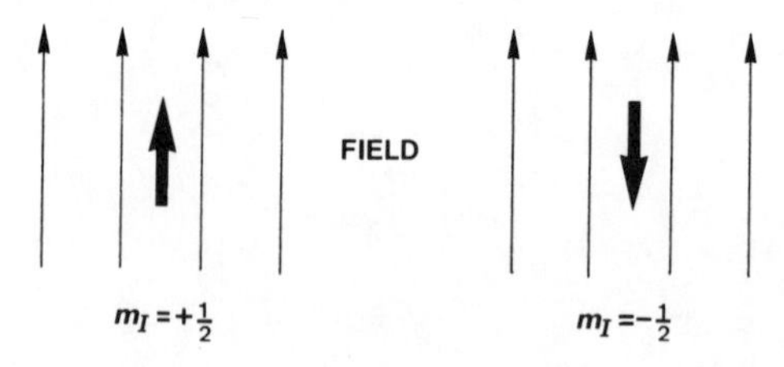

In general, for a given value of I the allowed values of m_I are I, $I-1$, $I-2$, . . ., $I+1$, $-I$, which always gives $2I+1$ values of m_I.

For the $I = 1/2$ case the nuclear spin aligned with the field is the ground state and when it is against the field this is a higher energy state. If a magnetic field is applied to a sample of nuclei with $I = 1/2$ n_w nuclei will align themselves with the field and n_a against it,

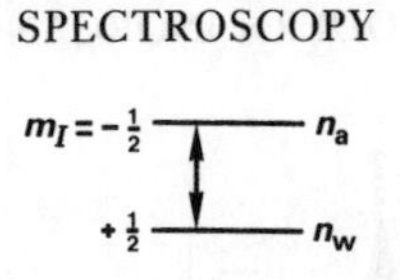

$n_w > n_a$ and there is a Boltzmann distribution between the levels,

$(n_w/n_a = \exp\{E/kT\})$

The energy difference between the levels, E, is very small compared to kT (i.e. $\exp\{E/kT\} \approx 1$), and so the states are approximately equally occupied, i.e. $n_w \approx n_a$. E depends on the magnetic field strength – as the field increases E increases.

By application of radio frequency radiation ($E = h\upsilon$) we can excite nuclei from $m_I = -1/2$ to $m_I = +1/2$. The frequency at which this occurs is the **resonance frequency**.

The selection rule for transitions between nuclear spin states is $\Delta m_I = \pm 1$.

There are basically two types of n.m.r. spectrometer; either the magnetic field is kept constant and the frequency of the radiation is varied to obtain resonance, or a fixed frequency is used and the field varied. For convenience, we will just talk in terms of a variable field but all the principles apply equally well to fixed field experiments.

The position of resonance is called the **chemical shift** and is given the symbol δ: it is expressed in units of p.p.m. (parts per million of the operating frequency). Thus δ is independent of the operating frequency of the machine used (since it is expressed as a *fraction* of this operating frequency). Lower δ corresponds to higher field, higher δ to lower field.

Nuclei cannot be continually excited to the upper state, there has to be some mechanism of *relaxation* back to the lower state. (We will not discuss relaxation here but it has many important applications in advanced n.m.r. spectroscopy.)

EXAMPLE

B_2H_6

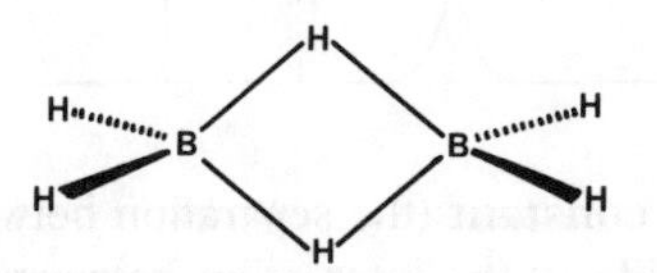

Let's break down the molecule into bits and build up the spectrum.

If the n.m.r. spectrum of isolated ^{1}H atoms ($I = 1/2$) could be measured a single line would be observed, corresponding to the excitation of nuclei from the lower to the upper state.

Similarly, ^{11}B ($I = 3/2$) n.m.r. would also give a single line. The ^{11}B nucleus can take up $(2I+1)$, i.e. $2 \times 3/2 + 1 = 4$, orientations with respect to the applied field,

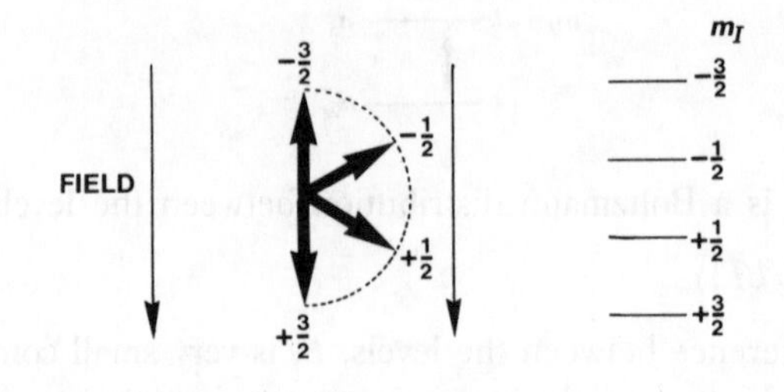

Since the selection rule for transitions between nuclear spin states is $\Delta m_I = \pm 1$ there are three possible transitions. These all occur at the same frequency, therefore there is only one line in the spectrum.

Now consider the B and H atoms joined together. Let us first look at the ^{11}B spectrum, i.e. considering transitions between nuclear spin states of ^{11}B. A B nucleus can be next to an H nucleus with $m_I = +1/2$ or $-1/2$,

$$\text{B—H}+\tfrac{1}{2} \qquad \text{B—H}-\tfrac{1}{2}$$

The magnetic field due to the H nucleus will affect the total magnetic field at the B nucleus. The magnetic field at B is affected in opposite but equal ways by the H in $m_I = \pm 1/2$ states. So the magnetic field due to the H nucleus changes the energy of the m_I states of the B nucleus, i.e. the resonance frequency for a B nucleus attached to an $m_I = +1/2$ H nucleus is different from that of a B attached to an $m_I = -1/2$ H nucleus – the B and H nuclei are said to **couple** together. As stated above, there are approximately equal numbers of molecules with H nuclei in $\pm 1/2$ states ($n_w \approx n_a$) and so the ^{11}B n.m.r. spectrum of the BH unit would consist of two lines of equal intensity, a **1:1 doublet**.

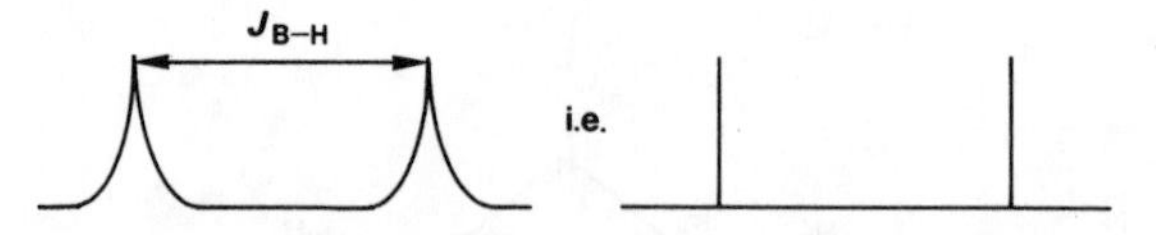

J is the **coupling constant** (the separation between, e.g. the two lines of a doublet) and is a guide to the interaction between two nuclei. J depends on three factors:

1. The distance between interacting nuclei – if two nuclei are directly joined then J is larger the shorter the bond length, i.e. for a particular pair of nuclei J is a guide to the strength of the covalent bond between them. For example, in diborane (page 265), J for B—H$_t$ (short bond) is larger than J for B—H$_b$ (long bond); this is true for boranes in general. J also

be transmitted.

2. The nuclei involved, e.g. $\mathcal{J}_{P—F}$ is greater than $\mathcal{J}_{B—F}$.
3. The degree of ionicity – as the degree of ionicity increases the coupling between nuclei decreases, e.g. $\mathcal{J}_{P—F}$ is greater than $\mathcal{J}_{Li—F}$ · Li—F bonds are much more ionic than P—F bonds.

In the ^{1}H spectrum (transitions between nuclear spin states of the H nucleus) of an isolated B—H unit, an H nucleus can be next to an ^{11}B nucleus in one of four states ($m_I = -3/2, \ -1/2, \ +1/2, \ +3/2$) with approximately equal probability (the four m_I states are approximately equally occupied). The B and H nuclei couple together, therefore four equal intensity, equally spaced, lines are observed in the ^{1}H spectrum, i.e.

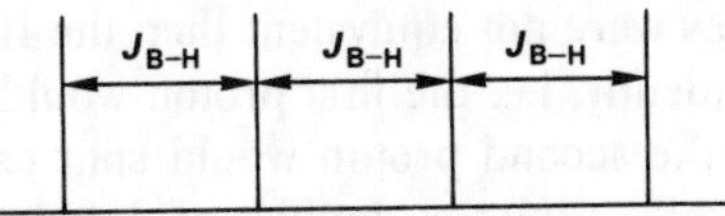

The equal spacing arises because there is a difference of $m_I = 1$ between each pair.

Now consider a B joined to two equivalent protons; the ^{11}B spectrum would consist of a 1:2:1 triplet, i.e.

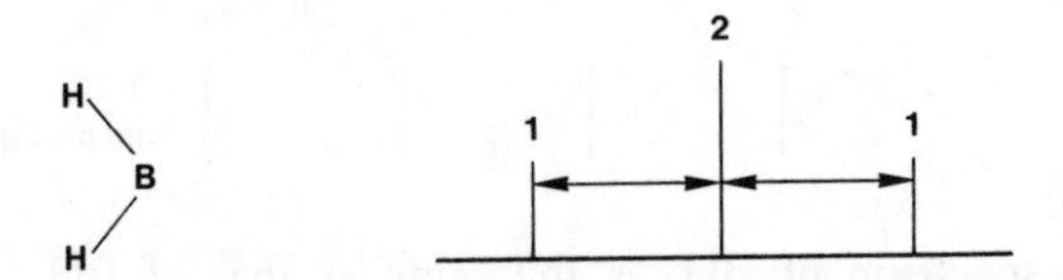

This comes about because in any BH_2 unit the H nuclear spins couple together to give a resultant M_I value.

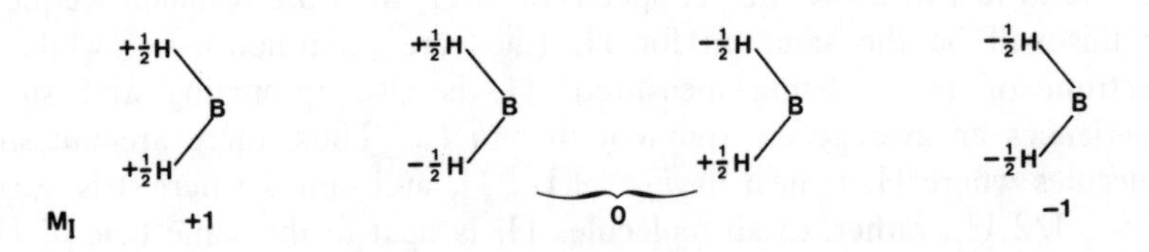

These three situations will all produce a different magnetic field at the B, i.e. a different resonance frequency, therefore there are three lines. They are equally spaced because $M_I = +1$ affects the B by the same amount as $M_I = -1$ but in the opposite direction. The intensities are 1:2:1 because there is twice the probability of having $+1/2, \ -1/2$ H nuclear spin states together.

In general, *a nucleus joined to* n *equivalent* I = *1/2 nuclei will give a spectrum split into* (n+1) *lines with intensities given by Pascal's triangle*, i.e.

n	
0	1
1	1 1
2	1 2 1
3	1 3 3 1
4	1 4 6 4 1
5	1 5 10 10 5 1

It has been stressed above that the H nuclei in the BH_2 unit had to be equivalent, i.e. in identical chemical and magnetic environments {the equivalent nuclei experience the same magnetic field due to other atoms in the molecule, their surroundings (i.e. other molecules, solvent) and the external field}. If they were not equivalent then the B signal would be split by each proton independently, i.e. the first proton would split the ^{11}B signal into a doublet and then the second proton would split each of these lines into a doublet, i.e. instead of a triplet, a doublet of doublets is obtained,

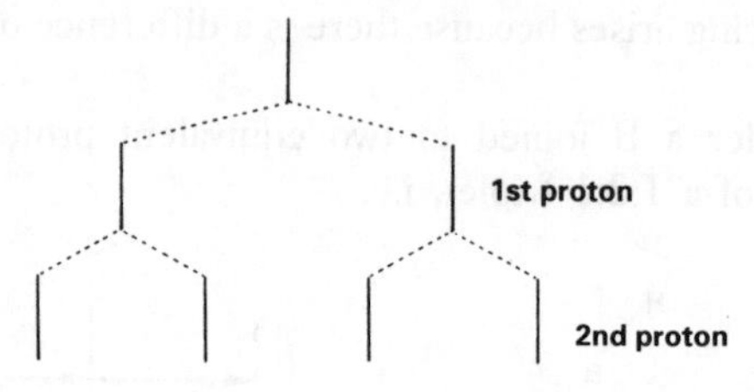

The ^1H spectrum of BH_2 is the same as that of BH (except twice the intensity, for equal number of units), i.e. the H nuclei are not splitting each other. In general *equivalent nuclei give rise to a single signal (not necessarily a single line), i.e. they do not split each other*. The reason for this is that if it were possible to just measure the ^1H spectrum of H_x then the resonant frequency for this will be the same as for H_y (they are equivalent), i.e. while the spectrum of H_x is being measured, H_y is also resonating and so H_x experiences an average environment due to H_y. Thus, there are not some molecules where H_x is next to $m_I = +1/2$ H_y and others where it is next to $m_I = -1/2$ H_y, rather, in all molecules H_x is next to the same type of H, a resonating H_y (constantly changing between $\pm 1/2$ states), therefore the H_x signal is not split.

For a

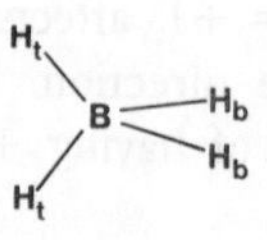

unit; if all the four hydrogens were equivalent a 1:4:6:4:1 quintet would be observed. However, B—H_t is shorter than B—H_b, therefore there are two sets of equivalent protons which must be considered separately.

Consider the ^{11}B spectrum: in the absence of hydrogens this will be just a single line, the hydrogens, however, cause it to be split. Coupling to the two equivalent H_t is considered first (since B—H_t is shorter than B—H_b, therefore $J_{B—H_t}$ is larger – see above) – this splits the single line into a 1:2:1 triplet. Now, coupling to the two equivalent H_b will split *each of these lines* into a 1:2:1 triplet (i.e. just as if the $B(H_t)_2$ unit were a single nucleus joined to two H_b units), i.e.

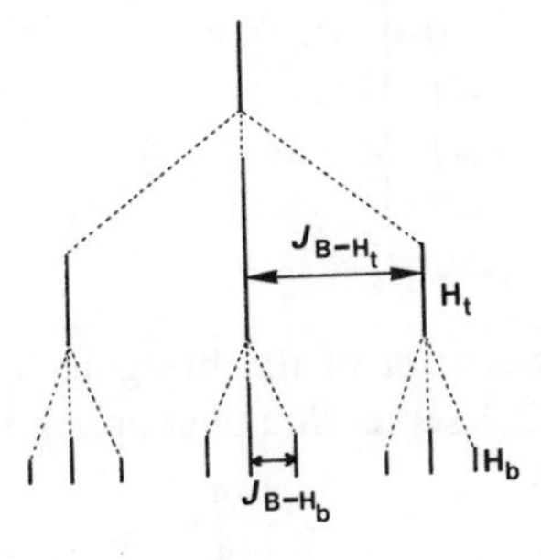

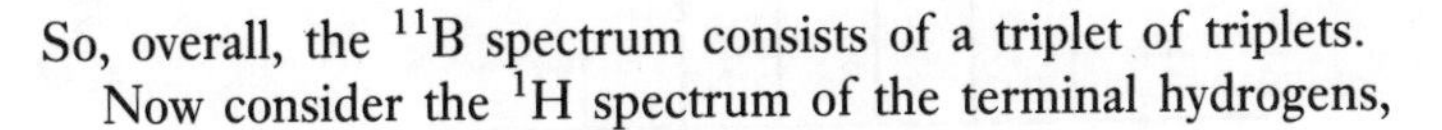

So, overall, the ^{11}B spectrum consists of a triplet of triplets.

Now consider the ^1H spectrum of the terminal hydrogens,

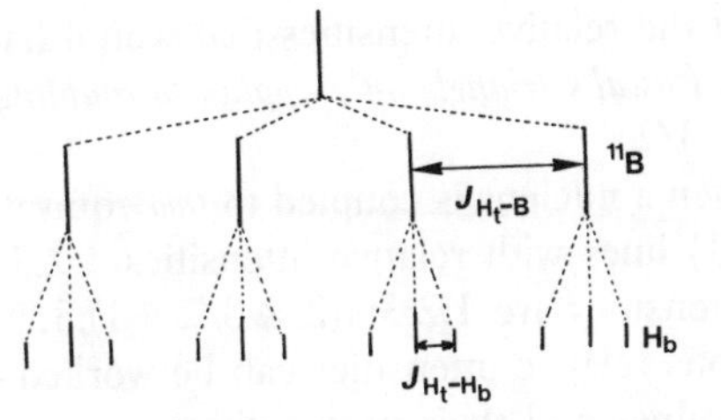

Therefore the ^1H spectrum consists of a quartet of triplets.

In B_2H_6 the H_b atoms are joined to two equivalent B atoms, i.e.

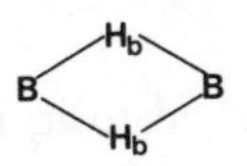

If we look at this unit by itself; B has $I = 3/2$ and the nuclear spins of the two borons couple together to give a total nuclear spin of $\Sigma I = (3/2 + 3/2) = 3$. This signal due to the bridging hydrogens is split by the borons into $(2\Sigma I + 1)$, i.e. $(2 \times 3 + 1) = 7$ lines. These correspond to the following M_I values:

$B_a(m_I)$	$B_b(m_I)$	M_I	Probability/intensity
+3/2	+3/2	3	1
+3/2	+1/2	2	2
+1/2	+3/2	2	
+3/2	−1/2	1	3
−1/2	+3/2	1	
+1/2	+1/2	1	
+1/2	−1/2	0	4
−1/2	+1/2	0	
+3/2	−3/2	0	
−3/2	+3/2	0	
−1/2	−1/2	−1	3
−3/2	+1/2	−1	
+1/2	−3/2	−1	
etc.			

Thus, the n.m.r. spectrum of the bridging hydrogens in the BHHB unit consists of a 1:2:3:4:3:2:1 septet due to coupling to the two equivalent borons.

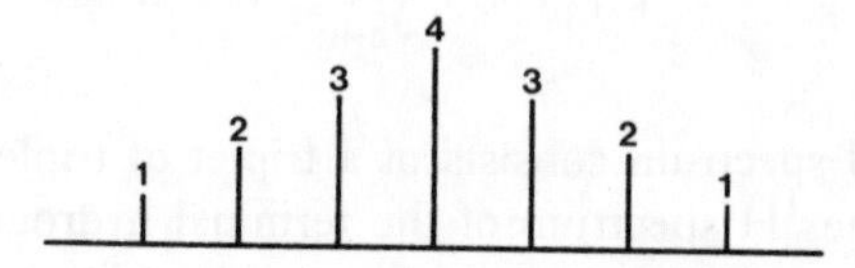

These are not the relative intensities that would have been predicted from Pascal's triangle. *Pascal's triangle only applies to coupling to n equivalent $I = 1/2$ nuclei.* B has $I = 3/2$.

In general, when a nucleus is coupled to *two* equivalent nuclei (nuclear spin I) we get $(2\Sigma I+1)$ lines with relative intensities 1:2:3:. . .$(2I+1)$. . .:3:2:1. e.g. for B the intensities are 1:2:3:4$(2 \times 3/2 + 1)$:3:2:1.

For all situations relative intensities can be worked out by writing down all the possible M_I values and their probabilities.
The

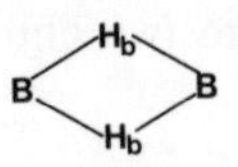

unit has four equivalent H_t attached to it, i.e.

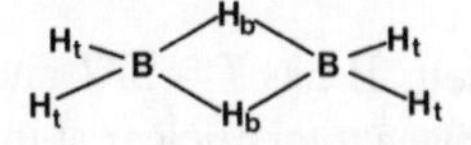

which means that the septet, obtained above, will be further split. Four equivalent hydrogens gives a total, $\Sigma I = 4 \times 1/2 = 2$, i.e. $2\Sigma I+1 = 5$ lines.

Intensities are given by Pascal's triangle, therefore a 1:4:6:4:1 quintet is observed. Thus, the spectrum of the bridging hydrogens in B_2H_6 consists of a septet of quintets,

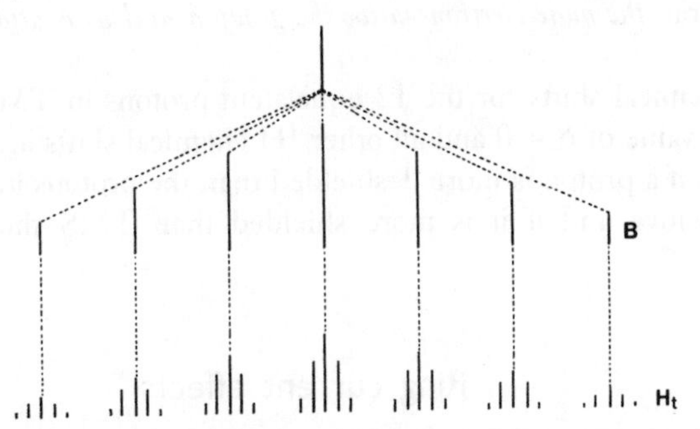

The resolution of a real instrument is unlikely to be sufficient to separate all the lines and so this will often be described as a 'complex multiplet'.

Now consider B_2H_6 as a whole:

1. ^{11}B spectrum – the borons are in equivalent environments and do not split each other; the ^{11}B spectrum just consists of a triplet of triplets.
2. ^{1}H spectrum – the terminal hydrogens give rise to a quartet of triplets; the bridging hydrogens give a septet of quintets. However, there are twice as many terminal hydrogens as bridging ones and so there is a more intense signal due to the terminal hydrogens – *the area under the peaks in the n.m.r. spectrum is proportional to the number of nuclei in a particular environment.* This would usually be written as, 'the integrated intensities for the terminal and bridging hydrogens are in the ratio 2:1'.

Where the H signals come relative to each other in the spectrum, i.e. their δ values, depends on the amount of electron density at the nucleus. *Circulating electron density in a magnetic field generates a magnetic field in opposition to the applied field* (Lenz's Law). Consider an isolated H atom, H attached to an electronegative group such as O in C_2H_5OH and H attached to an electropositive metal atom, as in, e.g. $[Pt(PMe_3)_3H]^+$.

The isolated H atom would resonate for a particular applied field, i.e. $\delta = \delta_H$. The electronegative O withdraws electron density from the H so that the field set up in opposition to the applied field is lower (less electron density circulating at the H nucleus), therefore the applied field has to be lower to cause the same resultant field at the H nucleus – lower field is equivalent to higher δ, i.e. a downfield shift, and the H nucleus is said to be **deshielded**.

A hydrogen attached to a metal is hydridic (has a small negative charge) – the electron density at the H nucleus is larger and the field set up in opposition to the applied field is greater. Therefore the applied field has to be increased to give the same resultant (i.e. resonant) field at the H nucleus. Higher field is equivalent to lower δ, i.e. an upfield shift and the H nucleus is **shielded**.

In general, *the more electronegative the group a nucleus is attached to, the higher the δ value.*

The chemical shifts for the 12 equivalent protons in TMS ($Si(CH_3)_4$) are assigned a value of $\delta = 0$ and all other ^1H chemical shifts are taken relative to this. Thus, if a proton is more deshielded than the protons in TMS its δ value will be positive and if it is more shielded than TMS the δ value will be negative.

Ring current effects

A hydrogen attached to an aromatic ring resonates at a lower field (higher δ) than one attached to a similar non-aromatic group; this is due to **ring current**. The circulation of charge around the delocalized ring reinforces the applied field at the H, i.e.

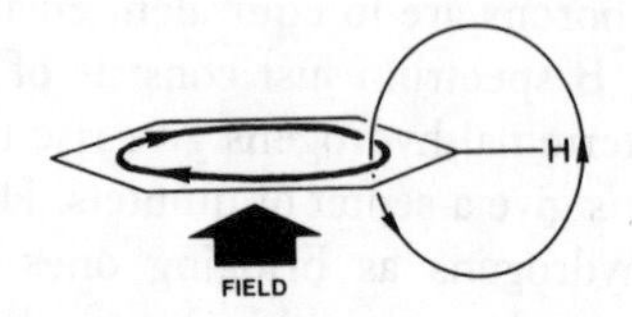

Therefore, a lower external field has to be applied to cause resonance.

Decoupling

Spectra can be simplified by decoupling experiments. If, for example, the ^{11}B spectrum of diborane is measured, it shows coupling to H, and is complex. However, if the sample is irradiated with radiation (this will only apply to a spectrometer where the field is fixed and we vary the frequency of the radiation) at the resonance frequency of the bridging and terminal hydrogens, just one signal is observed.

Consider the terminal hydrogens; these would normally split the ^{11}B spectrum into a triplet because there will be some molecules with each of the following permutations,

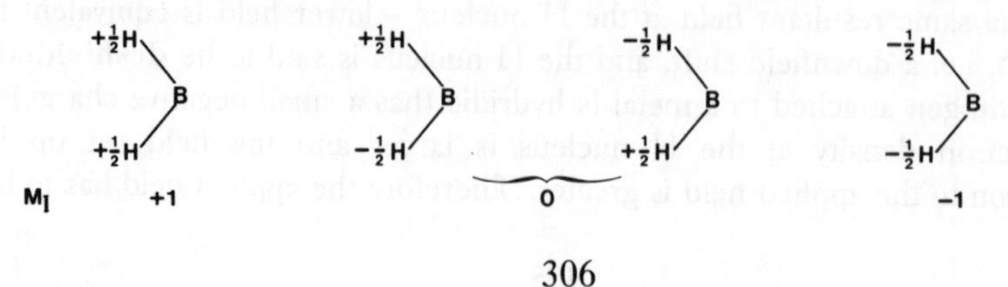

Each of these gives a different signal in the ^{11}B spectrum and there are therefore three lines. If the sample is being irradiated at the H resonance frequency then each H is constantly changing between $+1/2$ and $-1/2$ states so that all borons just experience an average field due to the H nuclei. Therefore there is only one B environment and only one signal in the spectrum.

EXAMPLES

PF_5

The structure by VSEPR theory is

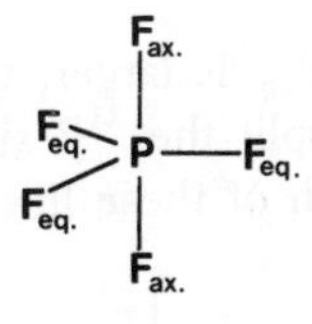

^{19}F, $I = 1/2$; ^{31}P, $I = 1/2$.

^{19}F spectrum

There are two F environments, $F_{ax.}$ and $F_{eq.}$

1. $F_{ax.}$: the two axial fluorines are equivalent so that they do not split each other. The $F_{ax.}$ atoms are first split by the P (they are joined directly to this, therefore J is larger) into a doublet and then each line of this is split into a 1:3:3:1 quartet ($2\Sigma I+1 = 4$ and the intensities are given by Pascal's triangle for $I = 1/2$ nuclei) by the three equivalent ($I = 1/2$) $F_{eq.}$ (coupling through two bonds, therefore J is smaller).

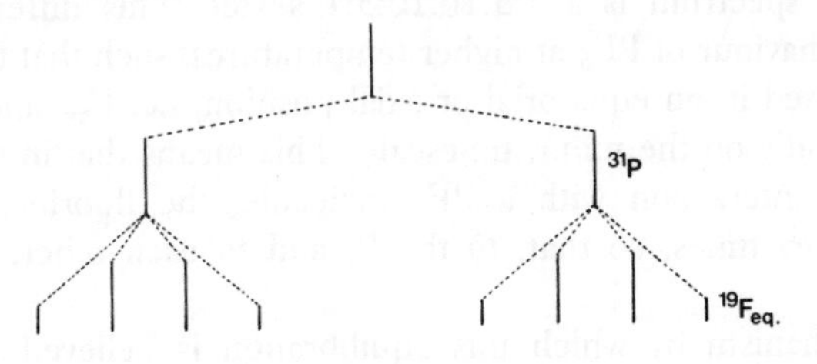

Therefore a doublet of quartets is observed, with integrated intensity 2.

2. $F_{eq.}$: there are three equivalent $F_{eq.}$ atoms. The signal is first split into a doublet by the P and then each line of the doublet is split into a 1:2:1 triplet by the two equivalent $F_{ax.}$ atoms.

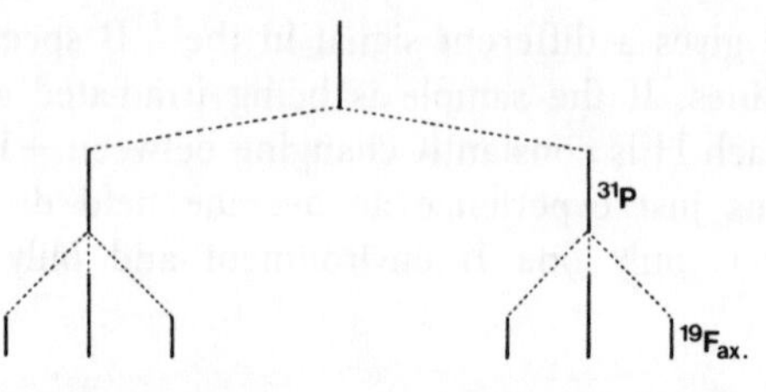

Therefore a doublet of triplets, of integrated intensity 3, is observed.

$\mathcal{J}_{P—F_{eq.}}$ is larger than $\mathcal{J}_{P—F_{ax.}}$ because the P—F$_{eq.}$ bond is shorter than the P—F$_{ax.}$ bond.

^{31}P spectrum

As stated above, coupling to F$_{eq.}$ is larger, therefore this is considered first. Three equivalent F$_{eq.}$ atoms split the ^{31}P signal into a 1:3:3:1 quartet; two equivalent F$_{ax.}$ atoms split each of these lines into a 1:2:1 triplet.

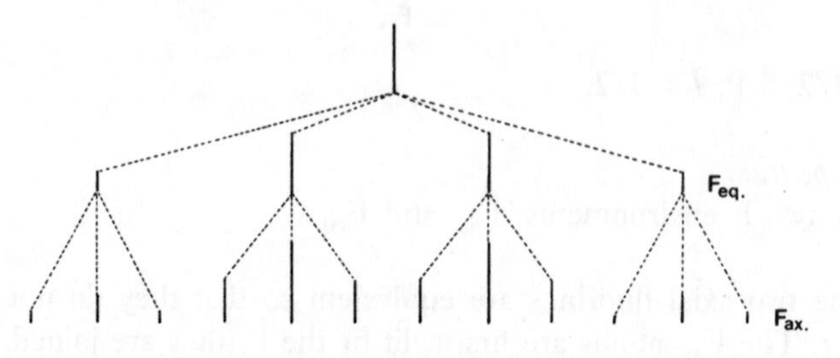

Therefore the ^{31}P spectrum consists of a quartet of triplets.

The spectra given above are for PF_5 at low temperature; at room temperature the ^{19}F spectrum is just a 1:1 doublet, of integrated intensity 5, and the ^{31}P spectrum is a 1:5:10:10:5:1 sextet. This difference is due to **fluxional** behaviour of PF_5 at higher temperatures, such that the fluorines are not rigidly fixed in an equatorial or axial position, i.e. F$_{ax.}$ and F$_{eq.}$ exchange positions rapidly on the n.m.r. timescale. This means that in the time it takes for a single interaction with a PF_5 molecule, the fluorines have changed positions many times, so that, to the P, and to each other, they all appear equivalent.

The mechanism by which this equilibration is believed to occur is the **Berry pseudorotation**.

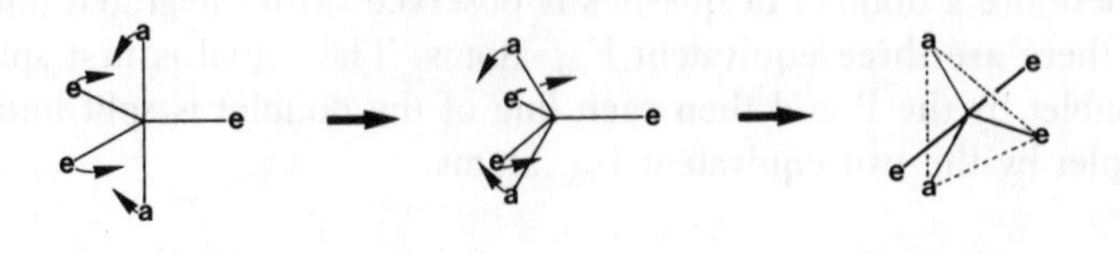

i.e. trigonal bipyramid → square based pyramid → trigonal bipyramid.

It is a pseudorotation since it appears that the whole molecule has been rotated through 90°. Other mechanisms have been proposed but this is by far the most widely accepted.

With five equivalent $I = 1/2$ fluorines the ^{31}P spectrum consists of a sextet {total $\Sigma I = 5 \times 1/2 = 5/2$ and the number of lines is given by $(2\Sigma I + 1) = 6$} with intensities given by Pascal's triangle (since $I = 1/2$). Since all the fluorines are equivalent they do not split each other, therefore the only splitting in the ^{19}F spectrum is due to coupling with the ^{31}P, giving a doublet.

Getting a doublet in the ^{19}F spectrum implies that the fluxional equilibration of the fluorines must be intramolecular (within a molecule) since coupling to the P is retained. In the presence of F^-, however, the spectrum collapses to a singlet, as there is equilibration by

$$PF_5 + {}^*F^- \rightleftharpoons PF_4{}^*F + F^-$$

which is rapid on the n.m.r. timescale. The chemical shift for F here is the weighted average of an F attached to a P and a free F^-. The coupling to P is lost since, if we take a particular F, this may be first joined to a P with $m_I = +1/2$, then it dissociates to give F^- (not joined to anything), which can then join on to another P, which could have $m_I = -1/2$. This happens very many times while the spectrum is being measured (rapidly on the n.m.r. timescale) and so the F appears to experience an average P environment.

It is possible to estimate activation energy for intramolecular fluxional processes. This is done by observing the temperature at which a fluxional process is 'frozen out', i.e. the temperaure below which the fluxional process ceases to operate.

$HPt(PMe_3)_3$

This is square planar,

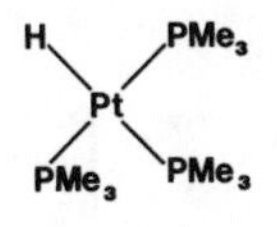

Nucleus	I	Natural abundance (%)
^{195}Pt	1/2	34
^{1}H	1/2	100
^{13}C	1/2	1
^{31}P	1/2	100

It is assumed that all other isotopes of Pt have $I = 0$.

We will consider the ^{1}H spectrum.

H_a, the metallic hydride

There are two possibilities for the H joined to the Pt; it can either be joined to the ^{195}Pt ($I = 1/2$), in which case the signal will be split into a doublet, or it can be joined to some other isotope of Pt ($I = 0$) so that the signal remains as a singlet. In 34 per cent of the molecules in the sample the H will be joined to ^{195}Pt, therefore the total intensity of the doublet is 0.34. In the remaining 66 per cent of the molecules there will be no ^{195}Pt, therefore the singlet has intensity 0.66. Thus, the spectrum appears to be an approximately 1:4:1 triplet, but is actually a doublet and a singlet.

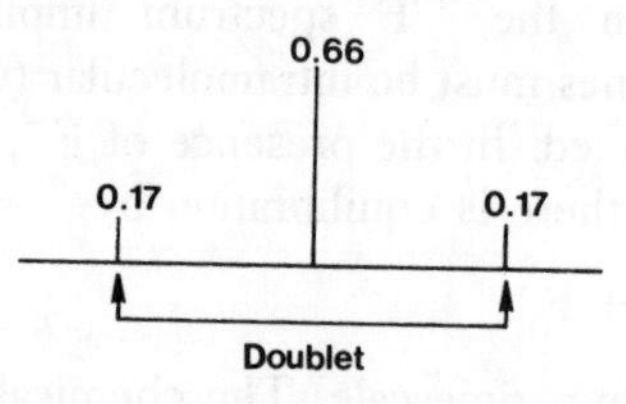

The doublet is centred on the singlet and is called a **satellite** signal.

The signals are then all further split by coupling to the P atoms. There are two different P environments, *cis* or *trans* to the H. *Trans coupling is always larger* – the H and the *trans* P interact with the same orbital on the Pt, therefore they are going to have a greater effect on each other.

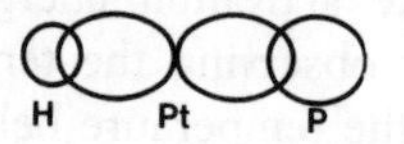

Trans coupling is considered first and splits each line of the spectrum into a 1:1 doublet. Coupling to the two equivalent *cis* P atoms then splits each of these lines into a 1:2:1 triplet.

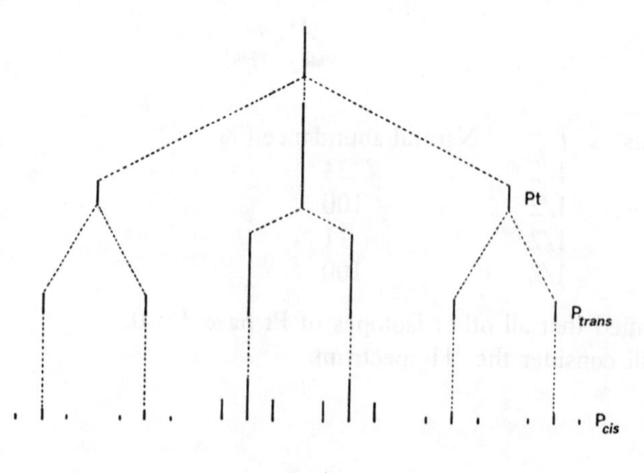

This signal comes at negative δ since the H is more electronegative than the Pt and is more shielded than TMS (H is δ−, quite hydridic). Virtually all H atoms directly attached to metal atoms have negative chemical shifts.

There are other hydrogens in this molecule:

H atoms in $P(Me)_3$ groups

There are nine equivalent hydrogens attached to the carbons on the *trans* P and 18 equivalent hydrogens in the Me groups on the *cis* P atoms. Coupling to ^{13}C ($I = 1/2$) can be ignored since the natural abundance is so low, i.e. only 1 per cent of H atoms are attached to a ^{13}C. There will, however, be weak coupling (through two bonds) to P, which will split the H signals into doublets. Coupling to Pt is possible in theory but this is through too many bonds (three) to be observed in practice. Thus, there are two signals; a doublet of intensity 9 and one of intensity 18. The chemical shifts of these are going to be very similar because the H atoms are in very similar environments.

In this case δ is positive since the P is more electronegative than H and the nuclei are deshielded. The entire spectrum thus consists of a multiplet, of intensity 1, a doublet of intensity 9, and a doublet of integrated intensity 18.

As the temperature is raised above room temperature the two doublets merge to give a doublet of intensity 27 as the *cis* and *trans* P atoms exchange positions rapidly on the n.m.r. timescale.

Alkene rotation

In transition metal–alkene complexes does rotation occur about the C—C axis or perpendicular to it?

This can be answered by a study of the ^{13}Cn.m.r. of the following module,

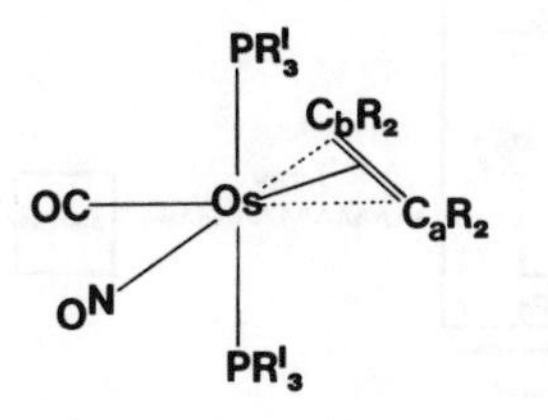

At low temperature there are two ^{13}C n.m.r. signals for the alkene; C_a and C_b are in different environments (*trans* to different ligands) and, therefore, give different signals. If mode 2 were the mechanism of rotation then the

carbons should remain inequivalent as the temperature is raised. However, as the temperature is increased the two signals merge into one, indicating that the propeller-like mode of rotation must be rapid on the n.m.r. timescale, i.e. C_a spends equal amounts of its time *trans* to CO and *trans* to NO, as does C_b; therefore one, average, signal is obtained.

Use of n.m.r. to study reaction rates and positions of equilibria

A good example of this is

$$PCl_3 + P(OEt)_3 \rightleftharpoons PCl_2(OEt) + P(OEt)_2Cl$$

Each compound has a different ^{31}P n.m.r. signal (different chemical environments) and although the products on the right-hand side of the equilibrium cannot be isolated, their rate of production and the position of equilibrium can be studied by looking at how the intensities of the various ^{31}P signals in the n.m.r. spectrum vary with time.

Mössbauer Spectroscopy

Mössbauer spectroscopy involves observation of transitions between different nuclear states (transitions between energy levels in the nucleus); these are well separated in energy and transitions involve absorption/emission of high energy, γ-radiation (cf. electronic spectroscopy, which involves observation of transitions between orbitals – absorption/emission of IR to X-ray frequencies of radiation).

Mössbauer spectroscopy can be carried out on a variety of heavy nuclei and a different source is required for each one. If Fe is taken as an example, the source is ^{57}Co, which is radioactive and slowly decays to give $^{57}Fe^*$, which has the Fe nucleus in an excited state. This nucleus decays quickly to the ground state, ^{57}Fe, with emission of a γ-ray, which is used in the experiment.

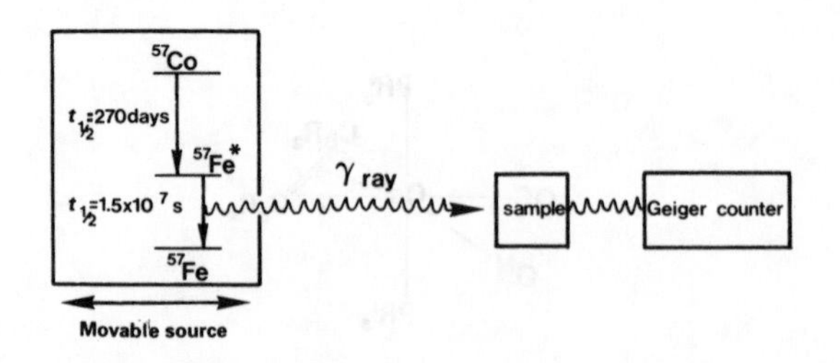

If the compound under investigation contains ^{57}Fe atoms with the same spacing between nuclear energy levels as in the ^{57}Fe source, then total absorption of the γ-rays occurs (Mössbauer spectroscopy can be considered as the 'nuclear version of electronic spectroscopy'). In order to get γ-rays of just one frequency (monochromatic radiation) it is necessary to have a source which is in the solid state and kept at a very low temperature. This is because a γ-ray has a large momentum associated with it [momentum = $h\upsilon/c$ (De Broglie relationship) where υ is large for the high frequency γ-rays], therefore when a γ-ray is emitted from the nucleus there is a large recoil (conservation of momentum) of the nucleus. If this nucleus (i.e. atom) is free to move, as in a gas or a liquid, then the recoil will be large, i.e. part of the energy given out when the excited nucleus decays to the ground state is converted to kinetic energy of the atom, therefore the γ-ray is of lower energy/frequency than expected. In the solid state the atoms are coupled together and so the body from which the γ-ray is emitted is effectively the whole solid, which has a very large mass, leading to very small recoil and very small energy loss from the γ-ray. The source is kept cold so as to reduce vibration of atoms – movement of atoms would cause γ-rays to be emitted with a range of frequencies – the Doppler effect.

The separation between nuclear energy levels in an atom in the compound under study depends on the environment of the atom. The frequency of radiation experienced by the sample can be varied by moving the source towards or away from it, at constant speed – the Doppler effect. When the source moves towards the sample, the sample experiences higher frequency radiation, and when the source moves away the frequency is lower. The result of a Mössbauer experiment is expressed as an **isomer shift** in units of millimetres per second ($mm\,s^{-1}$). The isomer shift is an indicator of the difference in chemical environment between a particular nucleus in the source (^{57}Fe, in this case) and a nucleus of the same element in the sample.

The isomer shift is dependent on the electron density at the nucleus. Only s electrons give rise to electron density at the nucleus as other orbitals have nodes there. However, p,d, etc., electrons and the overall charge on an atom do have some effect, in that screening can affect the s electron density at the nucleus.

Uses of Mössbauer

1. It can show differences in oxidation states, e.g. $FeCl_2$ versus $FeCl_3$; the isomer shift for the former is $+1.3\ mm\,s^{-1}$ and that for the latter is $+0.5\ mm\,s^{-1}$. The values cannot be used quantitatively.
2. Formal oxidation states can be shown to be quite different from actual charges on atoms, e.g. $[Fe(CN)_6]^{3-}$ and $[Fe(CN)_6]^{4-}$ have very similar isomer shifts, although the formal oxidation states of the Fe are +3 and +2, respectively.
3. The effect of ligands on the electron density at the nucleus can be

demonstrated, e.g. the isomer shifts for $FeCl_2$ and $[Fe(CN)_6]^{4-}$, both of which contain Fe^{II}, are $+1.3$ mm s^{-1} and -0.05 mm s^{-1}, respectively.
4. Different chemical environments in the same molecule can be shown, e.g. there are two different isomer shifts in $Fe_3(CO)_{12}$.

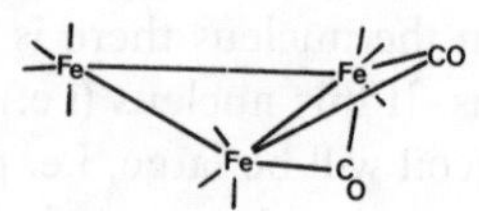

5. It can be used to give information about the symmetry of compounds. Often, ground and excited nuclear states have different values for the nuclear spin quantum number, I, e.g. ^{57}Fe and $^{57}Fe^*$ have $I = 1/2$ and 3/2, respectively. Therefore, on going from the excited to the ground state, there is a change in the charge distribution within the nucleus; if the environment of the Fe nucleus is unsymmetrical this will be seen as quadrupole splitting (see below). For example, there is no quadrupole splitting in the symmetrical $[Fe(CN)_6]^{4-}$, but it is observed in $[Fe(CN)_5(NO)]^{2-}$.

Although all the examples given here are for Fe Mössbauer spectroscopy, experiments can also be carried out for many other elements, e.g. Au, Sn, Eu.

Nuclear Quadrupole Resonance (n.q.r.) Spectroscopy

Nuclei with $I > 1/2$ are non-spherical (either egg- or tangerine-shaped). As the nuclei are centrosymmetric they do not give rise to dipoles; they do, however, act as quadrupoles and have quadrupole moments. A simple quadrupole would be,

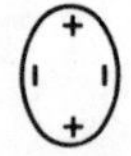

A (−) in a nucleus represents a deficiency of positive charge.

The nucleus can take up different orientations in the electric field due to the electrons in the atoms. If the field at the nucleus is spherically symmetrical then these orientations will all be equivalent, e.g. $[Fe(CN)_6]^{4-}$. However, if the field is asymmetric (i.e. an electric field gradient at the nucleus), e.g. $[Fe(CN)_5(NO)]^{2-}$, then the different orientations will correspond to different energies and the nucleus can be excited between the energy levels. The separation of these levels corresponds to the energy of radio frequency radiation and excitation between the levels is the basis of n.q.r. spectrocopy.

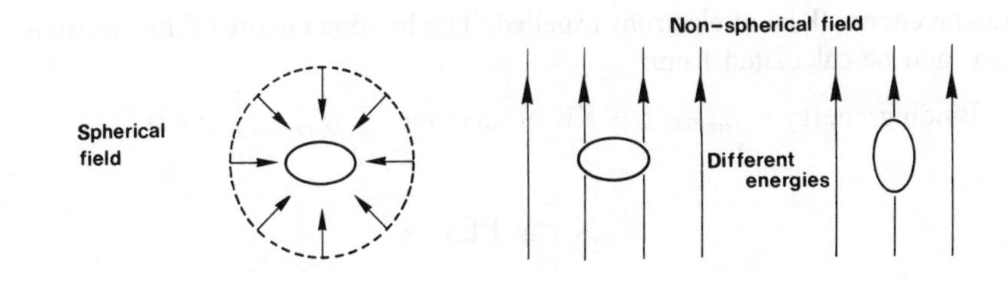

Consider the Cl^- ion ($I = 3/2$, therefore it has a quadrupole moment); this has a closed shell electronic configuration and is spherically symmetrical, so there is no field gradient at the nucleus and it exhibits no nuclear quadrupole coupling. This is going to be true of any compound in which the bonding is totally ionic (if there were such a thing). However, once some covalency is introduced, as in, e.g. Cl_2, the electrons involved in the covalent bond are going to lie more to one side of an atom and there is, thus, a large asymmetry in the electronic distribution, and a large *nuclear quadrupole coupling constant* (this is the product of the electronic charge, the nuclear quadrupole moment and the electric field gradient at the nucleus). Thus, *the nuclear quadrupole coupling constant is dependent on the degree of ionic character in a compound.*

Nuclear quadrupole coupling constants are also dependent on hybridization. Since an s orbital is spherically symmetrical a field gradient can never result from its occupation. Partial occupation of a p orbital (if the orbital is used in a covalent bond), however, results in a large electric field gradient and large nuclear quadrupole coupling.

The nuclear quadrupole coupling constant, Q_{X-Y}, can be related to the hybridization ($S = s$ character) and the degree of ionicity (i) by the equation,

$$Q_{X-Y} = (1-S)(1-i)Q_{X-X}$$

where, e.g. X = Cl and Y = I. Substitution of appropriate values for ICl gives a value for the ionicity of about 28 per cent. For $CrCl_3$ the value is closer to 90 per cent and so the numbers are as we expect from a qualitative consideration of electronegativity differences (I—Cl = 0.5, Cr—Cl = 1.5).

Hybridization and *s* character are not really easily assignable quantities and so this treatment can only be qualitative and comparative.

Photoelectron Spectroscopy (PES)

There are two types of photoelectron spectroscopy: ultraviolet (UV) and X-ray. Photoelectron spectroscopy involves irradiating a sample with monochromatic (one frequency) radiation of known frequency and measuring the kinetic energy (KE) of electrons expelled. The binding energy of the electrons can then be calculated from:

$$\text{Binding energy} = E_{\text{radiation}} - \text{KE of electron} \qquad (E_{\text{radiation}} = h\upsilon)$$

X-ray PES

X-rays are of sufficiently high energy (high frequency/short wavelength) to cause ionization of **core electrons** from atoms in molecules. The binding energy of an electron in a particular orbital of an atom of a particular element (e.g. an electron in the 1s orbital in C) is related to the charge on the atom – higher positive charge gives rise to a larger binding energy.

Consider 1,1,1-trichloroethane, CCl_3CH_3; attachment of three electronegative Cl atoms to one of the carbons results in a higher positive charge on this, than on the other, C atom. The binding energy of the C 1s electrons is larger, for the C with the Cl atoms attached and the X-ray pholoelectron spectrum is of the form,

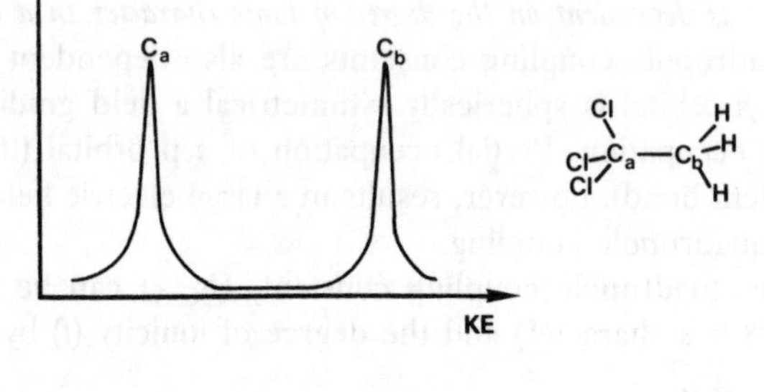

UVPES

Radiation in the UV region is of lower energy than X-rays and is only able to cause the ionization of electrons from frontier molecular orbitals (the molecular equivalent of valence atomic orbitals) in molecules. UVPES provides information about the relative ordering of molecular orbitals.

Consider the UV photoelectron spectrum of methane, CH_4: this shows two lines, one of intensity 1, and the other of intensity 3, i.e.

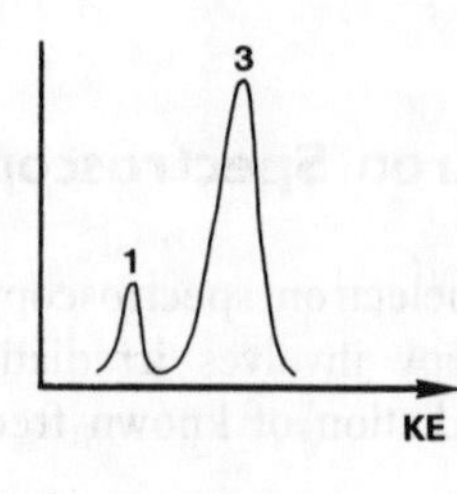

This provides support for the MO approach to the bonding in methane (page 43). A valence bond approach would consider the C atom as sp^3 hybridized, giving rise to four equivalent C—H bonds and, therefore, one line of intensity 4 in the photoelectron spectrum; this is not observed.

Electron Spin Resonance (e.s.r.) Spectroscopy

This shows the presence of unpaired electrons in a compound. The type of systems which have been studied are, e.g. O_2 (two unpaired electrons), NO (one unpaired electron), ions with partially filled d or f shells (transition metals, lanthanides and actinides). Electron spin resonance is analogous to n.m.r. in that a spinning unpaired electron acts like a tiny bar magnet, so that it can take up orientations aligned either with or against a magnetic field; radio frequency radiation can cause the electron to flip between these orientations.

An unpaired electron on an isolated metal atom with nuclear spin quantum number $I = 0$ will give a single signal in the spectrum, but signals in real complexes often exhibit **hyperfine splitting** (splitting of the signal in to many other lines). This is the result of coupling between nuclear and electron spin, i.e. coupling to a nucleus with nuclear spin I will give $(2I+1)$ lines in the spectrum. A nucleus with spin I can take up $(2I+1)$ orientations in a magnetic field (m_I values); nuclei in different orientations will give rise to different magnetic fields at the unpaired electron – see n.m.r., above. In a complex ML_n, if an unpaired electron on M spends some time on a ligand atom which has a nucleus with $I > 0$ then we get coupling. When this occurs it can provide evidence for covalency in bonding, since a 'purely ionic' compound would have no delocalization of electron density on to the ligands.

For example, the e.s.r. spectrum of UF_3 shows hyperfine splitting due to the interaction of unpaired 5f electrons on U^{3+} ($5f^3$) with the ^{19}F ($I = 1/2$) nucleus; such splitting is not observed in the lanthanide compound NdF_3. This indicates some covalency in the bonding in the actinide compound (not purely U^{3+} $(F^-)_3$) but very little in the lanthanide one (see pages 184 and 187).

Vibrational Spectroscopy

Infrared (IR) and Raman spectroscopy are the main methods of investigating the vibration of molecules in their electronic ground states. The experiments are easily performed and can be applied to all states of matter (solid, liquid, gas).

All molecules vibrate. The number of vibrational modes and whether they are infrared or Raman (or both) active is indicative of the overall symmetry of a molecule. The higher the symmetry of a molecule, the lower the number of allowed vibrations.

The number of bands in a spectrum is not necessarily dependent on the number of different environments for a particular group but rather depends on the overall symmetry. Thus, an $M(CO)_3$ unit,

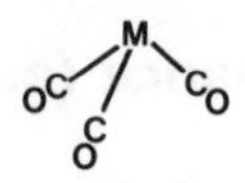

containing three identical carbonyl groups shows two C—O stretching bands at different frequencies, corresponding to asymmetric and symmetric stretches,

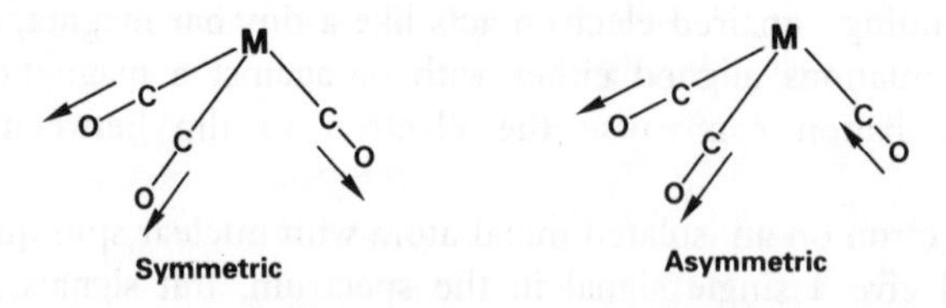

Infrared involves direct absorption of radiation and can only occur when the vibrational motion involves some *change in the dipole moment* of the molecule. If this is so then a vibration is said to be infrared active.

Raman is a second order effect resulting from the scattering of radiation by a molecule, rather than direct absorption. For a vibrational mode to be Raman active it must involve some *change in the polarizability* of the molecule.

The rule of mutual exclusion states that *if a molecule has a centre of symmetry then there cannot be any vibrational modes which are both IR and Raman active.* This is a good way of detecting a centre of symmetry in a molecule. For example, *cis–trans* isomers can be distinguished using the rule of mutual exclusion – only the *trans* isomer has a centre of symmetry, and hence no vibrational modes can be IR and Raman active, e.g. *cis*- and *trans*-$[Pt(NH_3)_2Cl_2]$ can be distinguished.

It is important to realize that no evident IR and Raman bands at the same frequency does not necessarily imply the presence of a centre of symmetry – the bands may be unobserved. However, coincident IR and Raman bands mean that the molecule definitely has no centre of symmetry.

Isotopic substitution

Replacing an atom by a different isotope changes the vibrational frequency. The force constant (strength) of a bond remains the same (isotopes are chemically identical) upon isotopic substitution, therefore since the mass which is vibrating changes, the stretching frequency changes – greater mass, lower stretching frequency. By observation of changes in stretching frequencies upon isotopic substitution we can assign bands to particular pairs of atoms. For example, M—H and C—O stretching frequencies in metal carbonyls often come in the same region making it difficult to assign bands. If, however, the H is replaced by a D ($^{2}_{1}H$) then the M—H bands will be shifted to lower frequency and so the bands can be assigned by comparison.

Solvent shifts

The vibrational frequency depends on the solvent in which the spectrum is measured. By correlation with known data we can assign bands to a particular pair of atoms on the grounds of variation of stretching frequencies between different solvents.

Applications of IR/Raman

STRUCTURAL STUDIES The number of bands in the IR/Raman of $N(SiH_3)_3$ is indicative of a planar rather than a pyramidal structure (page 110). In the spectra of NMe_3 and $P(SiH_3)_3$ there are more bands, due to deviations away from planar towards the lower symmetry pyramidal structure.

ELECTRONIC EFFECTS AND THE FORCE CONSTANT The force constant, and hence υ, is a measure of the strength of a chemical bond, i.e. the amount of electron density between the nuclei, e.g.

	$Ni(CO)_4$	$[Co(CO)_4]^-$	$[Fe(CO)_4]^{2-}$
$\upsilon_{C—O}$ (cm^{-1})	2060	1890	1790

More electron density being donated into the CO π^* antibonding orbitals, as the negative charge on the metal increases, weakens the C—O bond and lowers the stretching frequency (page 232).

BOND LENGTHS The energy required to excite molecules between rotational energy levels comes in the microwave region of the electromagnetic spectrum. Transitions between levels can be seen in specific microwave spectroscopic experiments or as 'rotational fine structure' in high resolution infrared spectra.

The energy difference between rotational levels depends on the moment of inertia of a molecule and once this has been worked out it is possible to deduce values for bond lengths in simple molecules, such as B_2H_6.

X-ray Diffraction

X-rays can be diffracted by planes of atoms in crystals (like light is diffracted by a diffraction grating). An X-ray diffraction experiment consists of placing a single crystal in a monochromatic X-ray beam and then measuring the position and intensity of the diffracted (scattered) rays; this is known as 'single crystal X-ray crystallography' and is the most important method for determining molecular structure. A single crystal is a regular periodic structure containing no faults or dislocations. The *X-rays are scattered by electrons in the crystal.*

Results of X-ray analysis

1. The chemical formula can be determined although it is more usual to carry out structural studies on a crystal of known composition.
2. The molecular structure can be obtained – this is the most important method for doing this and, in general, if a single crystal can be grown the molecular structure can be determined.
3. Bond lengths, bond angles and torsion angles can be obtained to a high degree of accuracy – this is really the only method (apart from neutron diffraction, which is not a readily accessible technique – see below) by which accurate bond lengths and angles can be determined.
4. Ionic and covalent radii can be derived from bond length data. Very accurate work can give ionic radii directly. This is done by using the experimental results to construct an electron density contour map and then taking the minima in this map to represent the boundary between ions.
5. The absolute configuration of chiral molecules can be determined.
6. Ground state structural/bonding effects can be studied, e.g. *trans* influence, steric effects, intermolecular contacts, hydrogen bonding.
7. Solid state structures obtained from X-ray analysis can be used to give an indication of possible solution reaction pathways, e.g. the agostic hydrogen in $TiCp_2ClEt$ (page 250).

Problems with X-ray analysis

1. A single crystal is required and sometimes it can prove very difficult to grow such a crystal.
2. X-rays are scattered by electrons and the scattering by a particular atom depends on the electron density around the atom. Thus, X-rays are scattered much more by, e.g. Hg than by H, making it very difficult to

detect light atoms in the presence of heavy ones, e.g. in many transition metal cluster compounds H atoms cannot be directly located.

3. X-rays can be absorbed by atoms (energies are such as to promote core electrons to higher orbitals) and in some cases, depending on the frequency of the radiation and the atoms involved, this can be acute and severely reduce the accuracy of results.

IR versus X-ray

In diborane (B_2H_6) the B—H bond lengths from IR spectroscopy are 1.20 Å and 1.32 Å but from X-ray analysis they are 1.08 Å and 1.25 Å.

The difference arises because we are measuring different things, i.e. X-ray measures the distance between centres of electron density in a molecule but IR spectroscopy gives information about internuclear distances. The difference is not usually very large as the maximum electron density is, in most cases, associated with the core electrons, which are centred on the nucleus. However, H has no core electron density and the maximum electron density is in the B—H bond, i.e. the maximum of electron density is displaced towards the B, therefore the bond length 'appears' shorter.

Neutron Diffraction

Neutron diffraction is similar to its X-ray counterpart in that molecular structures, bond lengths, etc., can be obtained from single crystal neutron diffraction experiments. However, neutrons are scattered by nuclei and not by electrons. The neutron scattering ability of nuclei varies irregularly by a factor of about four over the whole of the Periodic Table – it is not related to the size of the nucleus. Neutron diffraction is a good method for gaining information about light atoms in the presence of heavy ones, e.g. the H atoms in $H_2Os_3(CO)_{12}$ were accurately located by neutron diffraction but not found at all in X-ray experiments.

Problems with neutron diffraction

1. The crystal must be much bigger than for X-ray experiments – such single crystals often prove difficult to grow.
2. Because of large scattering by H atoms it can actually be quite difficult to solve structures of large organic molecules.
3. Experiments are very expensive; a nuclear reactor is required and there are few places where experiments can be performed.

Mass Spectroscopy

There are two basic types of mass spectroscopy: electron impact (EI) and fast atom bombardment (FAB). In both cases a vaporized sample is bombarded with high energy particles (electrons or small atoms), to produce positive ions. These are accelerated in an electric field and the resulting high energy stream of ions is passed between the poles of a large magnet. The charged particles are deflected to different degrees depending on their mass/charge (*m/z*) ratio. The highest mass peak observed is the molecular ion, $[\text{molecule}]^+$. The major use of mass spectroscopy is to find out the molecular mass of a compound. Other information can be obtained from mass spectroscopy, by observation of fragmentation and isotopic patterns.

1. *Fragmentation patterns*: from these we can obtain information about molecular structure. For example, consider an osmium carbonyl compound – the mass spectrum of this shows a molecular ion peak at 906 and a further 12 peaks, each separated by 28 mass units. The lowest observed peak in the spectrum is at 570 mass units.

   An osmium carbonyl contains just osmium, carbon and oxygen, the relative atomic masses of which are 190, 12 and 16, respectively. The peaks separated by 28 mass units correspond to loss of CO groups and thus this compound appears to have 12 carbon monoxides. The peak at 570 corresponds to an Os_3 unit, therefore the compound is $Os_3(CO)_{12}$ (as a check this has a molecular ion of 906 mass units). Now that the basic formula is known, from a knowledge of the 18 electron rule (page 220) a structure based on an Os_3 triangle with four CO units attached to each Os could be predicted.

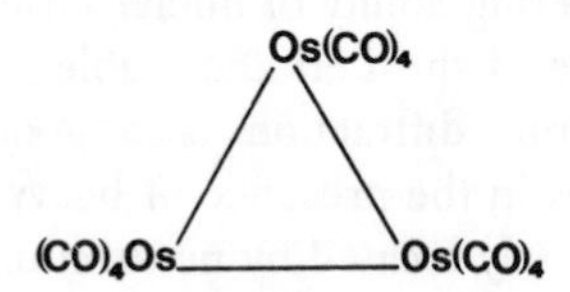

   This is not the only structure that can be derived from this formulation and to check our prediction we would have to use other techniques such as infrared spectroscopy. (What is the overall symmetry? Are there any bridging carbonyls?)

2. *Isotopic patterns*: from these we can obtain information about molecular composition. For example, we have a compound of empirical formula PNFBr. The highest observed peaks in the mass spectra are:

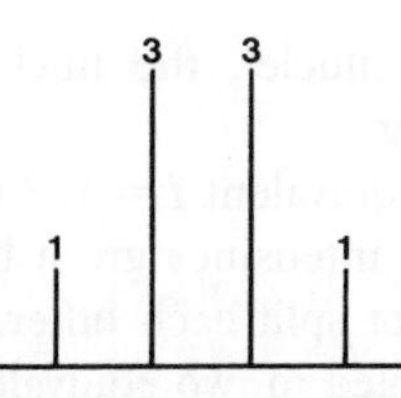

The intensities of the peaks are shown. Successive peaks are separated by 2 mass units.

P, N, and F all have only one common isotope, relative atomic masses 31, 14 and 19, respectively. However, Br has two isotopes with relative atomic masses 79 and 81 and natural abundances approximately 50 per cent each.

If we have two isotopes with relative abundances a and b, and n of these atoms in a compound, there will be $(n+1)$ molecular ion peaks with intensities given by the various terms in the expansion of $(a+b)^n$. In this case four peaks tell us we have three bromines. The relative intensities of the peaks are given by expanding $(0.5 + 0.5)^3$, i.e $(0.5)^3 + 3(0.5)^2(0.5) + 3(0.5)(0.5)^2 + (0.5)^3$, i.e. peaks in the ratio 1:3:3:1.

Electron Diffraction

This is the only method of structure determination in the gas phase. It involves diffraction of electron waves by interaction with electric potential fields in the molecule and provides information about inter-atomic distances. The bond lengths/angles obtained by this technique are much less accurate than those obtained by X-ray diffraction in the solid state.

Summary

Nuclear magnetic resonance spectroscopy

1. Nuclei with spin quantum number I can take up $2I+1$ orientations in a magnetic field (m_I values). The energy levels so generated are approximately equally occupied.
2. The selection rule for transitions between nuclear spin states is $\Delta m_I = \pm 1$.
3. Nuclear spins can couple; the size of the coupling depends on the

distance between the nuclei, the nuclei involved and the degree of ionicity in the bonding.

4. A nucleus joined to n equivalent $I = 1/2$ nuclei will give a spectrum split into $(n+1)$ lines, with intensities given by Pascal's triangle.
5. Equivalent nuclei do not split each other.
6. When a nucleus is coupled to two equivalent nuclei (nuclear spin I) there are $(2\Sigma I+1)$ lines with relative intensities 1:2:3:. . .$(2I+1)$. . .:3:2:1.
7. The more electronegative the group that a nucleus is attached to, the higher the δ value.

Mössbauer spectroscopy

8. Mössbauer spectroscopy involves observation of transitions between nuclear energy levels.
9. The sample in a Mössbauer experiment must be in the solid state at a low temperature.
10. The Doppler effect is employed to vary the frequency of radiation experienced by a nucleus.
11. The isomer shift depends on the electron density at the nucleus.

Nuclear quadrupole resonance spectroscopy

12. Nuclei with $I > 1/2$ are non-spherical and possess quadrupole moments.
13. If the electric field at the nucleus is asymmetric, nuclei with $I > 1/2$ can take up different orientations, which correspond to different energy states.
14. The nuclear quadrupole coupling constant depends on the degree of ionicity in a bond.

PES

15. X-ray PES involves excitation of core electrons in atoms and gives information about atomic charges within molecules.
16. UVPES involves excitation of valence electrons and provides information about the relative ordering of molecular orbitals.

Electron spin resonance spectroscopy

17. Electron spin resonance spectroscopy shows the presence of unpaired electrons.
18. Coupling of electron and nuclear spins gives rise to hyperfine splitting, which can provide evidence for covalency in bonding.

Vibrational spectroscopy

19. For a vibration to be IR active there must be a change in dipole moment.
20. A change in polarizability is necessary for a vibration to be Raman active.
21. The rule of mutual exclusion – if a molecule has a centre of symmetry then there cannot be any vibrational modes which are both IR and Raman active.
22. The frequency of vibration depends on the strength of the bond and the masses of the vibrating nuclei – the stronger the bond and the lower the masses, the higher the frequency.

X-ray diffraction

23. X-rays are scattered by electrons in a crystal.
24. The most important use of X-ray diffraction is in the determination of molecular structure – bond lengths, bond angles, etc.

Neutron diffraction

25. Neutrons are scattered by nuclei in a crystal. The neutron scattering ability of nuclei is not dependent upon the size of the nuclei.
26. Molecular structure determination by neutron diffraction is possible. Light atoms, in the presence of heavy ones, can be located.

Mass spectroscopy

27. The molecular ion peak in a mass spectrum gives the relative molecular mass of a molecule.
28. Fragmentation and isotopic patterns can provide information about molecular structure.

Index